Three
Generations,
No Imbeciles

Three Generations, No Imbeciles

Eugenics, the
Supreme Court,
and *Buck v. Bell*

Paul A.
Lombardo

THE

JOHNS HOPKINS

UNIVERSITY PRESS

BALTIMORE

© 2008 The Johns Hopkins University Press
All rights reserved. Published 2008
Printed in the United States of America on acid-free paper

Johns Hopkins Paperback edition, 2010

9 8 7 6 5 4 3 2 1

The Johns Hopkins University Press
2715 North Charles Street
Baltimore, Maryland 21218-4363
www.press.jhu.edu

The Library of Congress has catalogued the hardcover edition of this book as follows:
Lombardo, Paul A.
    Three generations, no imbeciles : eugenics, the Supreme Court, and *Buck v. Bell* /
Paul A. Lombardo.
        p. ; cm.
    Includes bibliographical references and index.
    ISBN-13: 978-0-8018-9010-9 (hardcover : alk. paper)
    ISBN-10: 0-8018-9010-1 (hardcover : alk. paper)
    1. Buck, Carrie, 1906–1983—Trials, litigation, etc. 2. Sterilization (Birth control)—
Law and legislation—Virginia. 3. Insanity (Law)—Virginia. I. Title. II. Title: Eugen-
ics, the Supreme Court, and *Buck v. Bell.*
    [DNLM: 1. Buck, Carrie, 1906–1983. 2. Sterilization, Involuntary—legislation &
jurisprudence—United States. 3. Eugenics—history—United States. 4. Eugenics
—legislation & jurisprudence—United States. 5. History, 20th Century—United
States. 6. Sterilization, Involuntary—history—United States. WP 33 AA1 L842t 2008]
    KF224.B83L66 2008
    344.7304'8—dc22   2008006546

A catalog record for this book is available from the British Library.

ISBN 13: 978-0-8018-9824-2
ISBN 10: 0-8018-9824-2

*Special discounts are available for bulk purchases of this book. For more information, please
contact Special Sales at 410-516-6936 or specialsales@press.jhu.edu.*

The Johns Hopkins University Press uses environmentally friendly book materials,
including recycled text paper that is composed of at least 30 percent post-consumer
waste, whenever possible. All of our book papers are acid-free, and our jackets and
covers are printed on paper with recycled content.

For AJL and DAL

CONTENTS

INTRODUCTION

"I Wanted Babies Bad"

Woman Told of Her Sterilization

—*Charlottesville (VA) Daily Progress,* February 24, 1980.

That headline interrupted my breakfast one day in 1980, and it continues to echo more than twenty-five years later. At the time I had only a passing acquaintance with the U.S. eugenics movement, but I would soon learn
that its legal high point was the 1927 United States Supreme Court case of
Buck v. Bell. Carrie Buck was sterilized following the Court's validation of
a Virginia law mandating the surgery for people who had been declared
"socially inadequate." The article in my morning newspaper mentioned the
case and explained the lawsuit that had been filed to overturn it. Carrie
Buck's sister Doris was quoted in the article; she joined the new lawsuit because she had been sterilized also.

Thinking this story might provide the topic for a seminar paper, I finished my coffee and hurried to the university's library. Finding the case was
easy enough; reading it took even less time, because what is now widely
considered one of the most infamous decisions in Supreme Court history
is also one of its shortest. With scandalously little justification, and in an
opinion of less than three pages, the Court approved the power of a state to
erase the parental hopes of its "unfit" citizens.

Despite its brevity, the opinion was too powerful to be ignored. Oliver
Wendell Holmes Jr., then the most celebrated judge in America, wrote it. He
was famous for pithy legal maxims, and the *Buck* case provided an occasion
to live up to that reputation. "It is better for all the world," said Holmes, "if
instead of waiting to execute degenerate offspring for crime, or to let them
starve for their imbecility, society can prevent those who are manifestly
unfit from continuing their kind." Carrie, her mother, and Carrie's "ille-

gitimate" infant fit Holmes' category of the "manifestly unfit," all touched by mental and moral defect. He did not hesitate to declare his conviction: "Three generations of imbeciles are enough."

A famous judge, a Supreme Court decision, an over-the-top sound-bite in the opinion—surely this was worth writing about. I leapt into the research, which revealed that Carrie Buck, a poor girl, was separated from her family at an early age and was living quite literally on the wrong side of the railroad tracks. Not long after she turned sixteen, she got pregnant; when the pregnancy became obvious, she was sent away to a state institution. Like other girls sent to the Virginia Colony for Epileptics and Feebleminded, Carrie was thought to be mentally defective and was considered a "moral degenerate."

Virginia enacted a eugenic sterilization law in 1924, just in time for Carrie's arrival at the Colony. The law was based on the theory that defects— criminality, poverty, illegitimacy, and the like—were passed down from people like Carrie to their children. The Colony was founded as a place to shut such people away to live celibate, childless lives. Carrie's mother was already there, accused of being a prostitute and pauper. Carrie's baby, barely six months old, also qualified for a spot at the Colony. Experts had examined the infant, saying she was "below average" and "not quite right," setting the stage for Holmes to condemn all three generations of the family. The *Buck* case confirmed the theory of hereditary defect, providing legal approval for operating on more than sixty thousand Americans in over thirty states and setting a precedent for more than half a million other surgeries around the world.

I completed the paper, but I couldn't let go of the story. Too many questions remained. Was this case an honest legal test of eugenical theory, or was it merely a charade put on by self-proclaimed "Progressive" social engineers? Was Carrie Buck's problem too little mental power or too much sex? How could seven other Supreme Court Justices, including William Howard Taft and Louis Brandeis, join Holmes in his caustic decision without one word written in dissent?

For the next two years I mined the papers of the lawyer who orchestrated the *Buck* case and completed a thesis that explained his role in writing the sterilization law and being its advocate in court. Then I went on to law school myself, but I soon stole time from my studies to find Carrie Buck and talk with her in the last weeks of her life. A tiny obituary in the same local paper that had first introduced us drew me to her funeral, where the handful of mourners made no mention of her past shame.

By then, even more driven to understand how this case had come about, I found Carrie's grade school report cards and her daughter's honor roll record, proving the Holmes opinion false. Years later, I found the photos of the infamous "three generations" buried in the archives of a former eugenics expert; that collection showed how he had manufactured evidence to fit the state's case against Carrie Buck. Other research revealed handwritten minutes of an official meeting between Carrie's appointed lawyer and her adversaries. Those records confirmed that the case was not just a tragedy. It was a legal sham.

The evidence presented at Carrie Buck's trial was transparently weak; with even a modest amount of effort, a competent lawyer could have challenged both the scientific premises of the Virginia sterilization law and the description of Carrie Buck's family offered by the state. But that defense never materialized, in large part because Carrie's lawyer had no intention of defending her. He offered no rebuttal to the state's arguments for surgery; he called no witnesses to counter the experts who had condemned the Buck family; he never explained that Carrie had not become a mother by choice, but that she had been raped.

That lawyer, Irving Whitehead, was a founding member of the Virginia Colony's board of directors and a major supporter of the sterilization campaign of Dr. Albert Priddy, the Colony's superintendent. Following his failure at trial, Whitehead secretly met with Priddy and the Board and voiced satisfaction that the case was proceeding as planned. He had betrayed his client, defrauded the court, and set in motion a series of events that history has uniformly condemned.

Yet the Supreme Court was unaware of these details and, in a nearly unanimous vote, approved the state sterilization law. In retrospect, it would be reasonable to conclude that the result in *Buck* was inevitable. In fact, the very opposite is true. Carrie Buck was the victim of an elaborate campaign to win judicial approval for eugenic sterilization laws, but the campaign need not have succeeded. For almost twenty years before the trial, scientists and policymakers across the country had debated the tenets of eugenics—the "science of good breeding." Several state laws allowing sterilization had been struck down in the courts, and others that remained in force were considered obsolete. By 1924 the brightest minds in the new field of genetics were turning their backs on the simple-minded conclusions of those who championed eugenic sterilization, and their criticisms were far from secret. It was a commonplace observation that the national movement was, if not dead yet, very weak.

Prominent eugenical activist Harry Laughlin had written a Model Sterilization Law he hoped would be adopted by states to replace their earlier legislation. Laughlin believed it was well designed to survive legal attacks. With funding from a Chicago judge, he published a book laying out his plan for a national network of eugenic bureaucracies that would sterilize millions. Virginia's Dr. Albert Priddy had already begun to operate, but a malpractice lawsuit put a temporary end to his freelance surgeries. He read Laughlin's book and borrowed the framework of Laughlin's bill, lobbying behind the scenes for a statute that would give him legal immunity for future operations. Priddy's early campaign for "tort reform" and the claim that sterilization would lead to tax cuts provided immediate motives to revive the calls for a Virginia sterilization law.

When Carrie Buck's case reached the highest court in the land, Supreme Court Chief Justice (and former president) William Howard Taft, himself an active member of the national eugenics movement, was in charge. He encouraged Justice Holmes to write an opinion that concentrated on the Buck family's inherited defects. Holmes learned the attitudes so forcefully exhibited in the *Buck* opinion at the knee of his father, the famous physician and poet. He took up Taft's challenge, crafting an opinion even his colleagues found "brutal."

The *Buck* decision was applauded in major newspapers throughout the United States; in its wake, opposition to sterilization seemed to melt away. In the next ten years, more than a dozen states would follow Virginia's lead; by the mid-1930s, the majority of states had laws allowing operations designed to eradicate the unfit.

The reach of the case was international. Canada, Switzerland, Japan, and each of the Scandinavian countries passed sterilization laws. More than four hundred thousand German sterilizations performed under Hitler's 1933 law "For the Prevention of Hereditarily Diseased Offspring" formed the backdrop for the scores of operations done in death camps under the rubric of "medical experimentation"—procedures later condemned at the Nuremberg trials as crimes against humanity. Lawyers for the Nazi doctors read from the *Buck* opinion, claiming the U.S. precedent in their clients' defense.

But neither scientists nor the public connected U.S. laws to German atrocities. Fifty years after *Buck,* more than a dozen compulsory sterilization laws were still in force, and surgeries were documented in institutions as late as 1979. Far from being a legal dead letter, *Buck* has never been overturned.

After twenty years of speaking and writing about the *Buck* case, I continue to see shock, surprise, and moral outrage in the faces of people when they first hear Carrie Buck's saga. Adults are stunned to think it could have happened in the United States; students are angered, believing that it has been hidden from them. *Buck* is regularly mentioned in books on law, science, and medical history but has never been adequately explained in a well-documented book of its own. This book tells that story, a notorious—and still open—chapter of U.S. history.

The *Buck* case represents one of the low points in Supreme Court history —on a par with *Plessy v. Ferguson,* which announced the now-discredited legal doctrine of "separate but equal," and the *Korematsu* case, which permitted the internment of Japanese citizens during World War II. *Buck* earns a place in the legal hall of shame not only because Holmes' opinion was unnecessarily callous but also because it was based on deceit and betrayal.

Advocates for eugenic sterilization endured many setbacks, but in the end a small group of vocal and determined zealots took advantage of the seductively hopeful message of eugenics that was well known to a full generation of Americans. Those who carried the banner for eugenic sterilization promised a future in which crime, disease, and poverty would decrease, lofty social values would flourish, and taxes would all but disappear. Their agenda appealed to reformers of every political stripe: from suffragists to "social purity" moralists, from temperance workers and fire-and-brimstone preachers to natural resource "conservationists"—indeed, anyone who lived with fears of a country in decline, facing a death spiral of impending degeneration.

The proponents of sterilization recited a variety of motives in favor of sterilization but two arguments were a constant theme. One had to do with sex. Sterilization would reduce sexual misconduct, it was said, by eliminating the kind of people who engaged in masturbation, fornication, incest, homosexuality, and prostitution. Sterilization satisfied political yearnings during the Progressive era because the eradication of moral defects appealed not only to those who embraced conventional religious standards but also to those for whom modern science had taken over the role of religion. Both religious and secular moralists could support the promise of a social fabric cleansed of problematic behavior by applied medical hygiene.

The second argument, by far the most prevalent, was clothed in the language of economics. Sexual surgery was a means to reducing social costs and the community burden posed by the unfit progeny of unfit citizens. Eugenic sterilization was embraced as an efficient remedy for the ills of society, a marriage of science and law that would beget progress as its offspring. Constantly repeating these themes, the supporters of sterilization saw its career morph from a near-defunct social experiment to a triumph in the country's highest court to a permanent stain on the national record. The regular invocation of sex, morals, and money as the justification for sterilization policies allowed laws to remain in effect long after heredity had disappeared as a reasonable basis to justify sterilization. When calls for sterilization reoccur today, those same motives play a key part in the debate.

Recently, attention to the history of eugenics in the United States has led to legislative resolutions rejecting past law and repudiating the Holmes opinion. Governors have apologized for previous sterilization policies, and legislators have called for reparations to the surviving victims of eugenics. At every point where eugenic history is reexamined, *Buck* provides an egregious example of governmental malfeasance under color of law.

The recent triumphs in genetic science announced alongside revelations and recriminations about eugenics—and happening alongside ongoing challenges to reproductive rights—suggest that we still have much to learn about *Buck v. Bell.* This book is the starting point in our lesson.

Three
Generations,
No Imbeciles

The Expert Witness

Aubrey Strode was desperate. After twenty-five years as a lawyer, he knew that there was no substitute in court for strong evidence. For months he thought he had the perfect case to prove the power of heredity: three generations from the same family, all feebleminded. Even the world's most prominent authority on sterilization had never seen one like it, with a grandmother, a mother, and a daughter all touched with defect. But with only six weeks until the trial, serious questions appeared. Then, with less than twelve days before Strode was due in court, his most important piece of evidence seemed to have melted away.

The Buck family papers weren't very helpful. Emma Buck was a widow, and neither her dead husband nor her parents had left any real records. It was not clear who was the father of her daughter Carrie, the girl the State of Virginia wanted to sterilize under the new eugenics law. With so many broken branches in the Buck family tree, it would be hard to prove that anything was hereditary. Worse than that, the witness who was supposed to evaluate Carrie's baby was getting cold feet. Now she couldn't remember saying that anything was wrong with the child, and no one else was available to complete the crucial evaluation.[1] Without it, convincing a judge to order the Buck girl to be sterilized might be difficult. To top it all off, Strode's client, Dr. Albert Priddy, was dying of cancer.

On November 6, 1924, Strode received a letter from Arthur Estabrook that changed everything.[2] Estabrook had a strong reputation as an expert in eugenics. His Johns Hopkins University doctorate was followed by training in the first class of field workers at the Eugenics Record Office in Cold Spring Harbor, New York. Nearly fifteen years of experience had made him a recognized authority among eugenicists. He was not a shy man, and when he heard that Aubrey Strode would direct the trial of Virginia's new eugenic sterilization law, he boldly offered his assistance. The suit could finally

Telegram alerting Arthur Estabrook of request to appear as an expert witness at the Buck trial, 1924. Courtesy Arthur Estabrook Papers, University at Albany Libraries' Special Collections.

settle the legality of a state plan to operate on "socially inadequate" people to prevent their irresponsible breeding.

Estabrook's whole career in eugenics had prepared him for this case. He understood the issues it would explore, he knew the people who would be involved, and he was familiar with the town where it would take place. For several years he did field studies on "defectives" in Amherst County, Virginia, the venue for the trial. He had worked with the townfolk there and had corresponded regularly with his former student, Louisa Hubbard Strode, the lawyer's wife. The lawsuit would highlight a family of "degenerates"—precisely the type of people on which his research and extensive writing focused. Estabrook was convinced that he was the right man for the job, and he must have been disappointed when Strode rejected his initial offer of assistance.[3] But in a few days, his disappointment turned to elation when his wife's telegram reached him in rural Kentucky.

Superintendent Priddy Lynchburg Virginia wants expert witness sterilization test trial Amherst Virginia November eighteenth Can you attend wire Laughlin

Jessie Estabrook[4]

"Superintendent Priddy" was Dr. Albert Priddy, who directed the Virginia Colony for Epileptics and Feeble-minded near Lynchburg. More than anyone, he was responsible for getting the sterilization law introduced in the Virginia Legislature, and he would be a key participant in the lawsuit. "Laughlin" was Estabrook's colleague Harry H. Laughlin, of the Eugenics Record Office in New York. The Virginia law that would be tested in this case mirrored key features of Laughlin's Model Sterilization Act, and in planning the litigation, Strode relied heavily on Laughlin's book *Eugenical Sterilization in the United States.*

Estabrook dropped his fieldwork in Kentucky and saddled his horse for the day-long ride to the rail line. Completing the trip to the Virginia Colony by train, he examined Carrie Buck, the girl who would be the focus of the case, and observed her mother Emma Buck, also a resident of the Colony. He knew that Priddy and Laughlin had already concluded that both were "feebleminded." Estabrook's task was to confirm their conclusions about the inherited traits of promiscuity and mental defect that would justify Carrie's sterilization.[5]

To finish his analysis of the Buck family, Estabrook visited Carrie's hometown of Charlottesville, where he interviewed her sister Doris and other relatives as well as teachers and neighbors who knew the Bucks. His most important observations occurred in the house where Carrie had grown up. Her infant daughter, Vivian—born out of wedlock just seven months earlier—remained there with the woman who also raised Carrie. After he examined Vivian, Estabrook was ready to testify.

The trial in the Amherst County Circuit Court began at ten o'clock on the morning of November 18, 1924. It took just over five hours, including time for lunch. Estabrook approached the witness stand at about one o'clock to report his impressions of the Buck family and to offer his opinion on the State's plan to sterilize Carrie. The evidence Estabrook provided would be crucial to the outcome of the case. None of the nine people who had testified before him, nor Dr. Priddy, whose testimony concluded the trial, had met and evaluated all three generations of the Buck family. Only Estabrook could offer a firsthand, expert perspective on hereditary defect as it appeared in a woman, her daughter, and her granddaughter.

Strode introduced Estabrook to the court and took him through the formalities required of an expert witness, establishing his competence as a scientist who could assist the judge in understanding the complexities of heredity and eugenics that were key to this case. At age thirty-nine, Estabrook was still young enough to exude vitality but old enough to appear distinguished and be taken seriously. He was a resident of New York and a member of the scientific staff of the Carnegie Institution of Washington working at Cold Spring Harbor, Long Island. He did research "along scientific lines" in the department of genetics there, particularly projects "leading to formulation of the laws of heredity." For fourteen years, he told the court, he had studied groups of "degenerates" and "mental defectives," and his most noteworthy studies had been described in two books.[6]

Estabrook's first book, *The Nam Family*, described a "highly inbred rural community" in New York, one marked "by alcoholism and lack of ambition."[7] His second book recalled the many social woes of an infamous New York pauper family, the Jukes of New York's Hudson Valley. Since Richard Dugdale's 1875 study, specialists and the public alike knew the Jukes as the model of all "degenerate" families. They descended from a woman known popularly as "Margaret, the Mother of Criminals," whose decadent ways had given rise to six generations of trouble. The inherited deficiencies of this family generated "convicts, paupers, criminals, beggars, and vagrants," who by the 1870s had cost the state hundreds of thousands of dollars.[8] After revisiting the "tramps," "wantons," and "ne'er-do-wells" of that clan, Estabrook brought Dugdale's research up to date in *The Jukes in 1915*. His book concluded that personal characteristics like the Jukes' sexual "licentiousness" were caused by a hereditary factor and that they had cost the state of New York more than two million dollars.[9]

The Jukes personified certain "definite laws of heredity" proving that feeblemindedness was the basis of bad conduct, the cause of criminality and poverty. The "strain of feeblemindedness," said Estabrook, is a "trait that is present in the germ plasm." He even sketched a chart to show how it was passed down within a family. We know the "actual elements in the germ cell that carry heredity," he asserted.

By the time court adjourned for lunch, Estabrook had explained the theory of heredity, the operation of those "little chemical bodies" that made up the cells, and the power of the "germ plasm" to define a family's future. He drew a diagram for the court, showing how "normals" who mated with the "feebleminded" multiply into "defective stock" in later years.

When court reconvened, Strode's first question to Estabrook focused specifically on the Buck family. He asked Estabrook: "Have you personally made any investigation of Carrie Buck and her ancestry?"

Estabrook's response detailed his examination of Carrie Buck and her mother Emma at the Colony, including his review of case histories prepared by Dr. Priddy. He explained that he had traveled to Charlottesville to gather information on Carrie's sister and half-brother as well as several other relatives from Emma's family. Like the Jukes, he said, the "defective strain" giving rise to feeblemindedness and sexual promiscuity was carried on the mother's side. After administering a mental test to Carrie's sister Doris, Estabrook declared that she was feebleminded too. Though he did not meet her brother, reports by others concerning his behavior in school led to the same conclusion. Estabrook also assumed that Emma's father, though long dead, was "of a defective makeup."

Estabrook had not administered an IQ test himself, relying instead on the one done at the Colony. So he hesitated when asked by Strode whether he had given Carrie Buck herself "any mental tests."

> *Estabrook:* Yes sir. I talked to Carrie sufficiently so that with the record of the mental examination—yes I did. I gave a sufficient examination so that I consider her feebleminded.

Asked how he defined the term "feebleminded," Estabrook included anyone "who is so weak mentally that he or she is unable to maintain himself or herself in the ordinary community at large."

The "socially inadequate" were also marked for sterilization in the Virginia law. By Estabrook's definition, that group included anyone "who by reason of any sort of defect or condition is unable to maintain themselves according to the accepted rules of society."

Strode then moved on to Carrie's baby, Vivian, whom Estabrook had visited immediately before the trial and photographed for his souvenir album.

> *Strode:* Were you able to form any judgment about that child? . . .

> *Estabrook:* I gave her the regular mental test for a child of the age of six months, and judging from her reactions to the tests I gave her, I decided she was below the average for a child of eight months of age.

Estabrook's direct testimony concluded with a few lines about the scandalous Kallikak family—the other tribe of social outcasts often referenced by eugenicists. That group's "defective side" displayed the same negative traits as the Jukes, he said. Estabrook's invocation of the Jukes and the Kallikaks consciously placed the Buck family alongside the most notorious examples of hereditary defect.

Strode ended his questioning with a gesture designed to fortify Estabrook's credibility for the court. *The Jukes in 1915* was an oversized and impressively bound blue volume packed with statistical tables and spilling over with pedigree charts of the legendary Jukes clan. Displaying it to the court, Strode asked Estabrook to identify it as his own work, dramatizing the author's expertise.

> *Strode:* I hold in my hand a monograph entitled *The Jukes in 1915;* I wish you would say whether or not this book sets forth your investigations of that family?
>
> *Estabrook:* It does.

Like the other groups Estabrook had studied, the Jukes were a special case in sexual misbehavior. His book declared that promiscuity was inherited, just like its opposite, chastity. The argument advanced at Carrie Buck's trial echoed the story of the Jukes and similar problem families. Loose women were weak in mind, weak in will, and all too likely to bear similarly afflicted progeny who would fill the jails, asylums, and poorhouses for generations to come. When his book had appeared almost ten years earlier, Estabrook had recommended remedies to the hereditary recapitulation of vice and poverty in such families: segregation in institutions and surgical sterilization, both to prevent reproduction. Even though sterilization of the feebleminded would require them to give up less freedom than they would if kept in permanent custody, Estabrook believed that "public sentiment" still stood in the way of the surgical option.[10]

By the time of Carrie Buck's trial, some believed that "public sentiment" had shifted. By the time the case was finally decided by the United States Supreme Court two-and-a-half years later, few could question that shift. The decision, announced by celebrated senior Justice Oliver Wendell Holmes Jr., owed its most powerful line to the assertions contained in Estabrook's expert testimony about Emma, Carrie, and Vivian Buck. A feebleminded mother and her daughter, followed by a "below average" granddaughter, said Estabrook, justifies eugenic sterilization. To which Holmes responded even more emphatically: "Three generations of imbeciles are enough."[11]

1

Problem Families

Francis Galton (1822–1911) coined the term *eugenics.* The English gentleman scholar's elaborate definition encompassed "all influences that tend in however remote a degree to give to the more suitable races or strains of blood a better chance of prevailing speedily over the less suitable."[1] The term would eventually become popular as a label for the field concerned with "better breeding" or, as its Greek roots suggested, being "well-born."

Biographers have often dwelt on Galton's "enviable pedigree," which included his half-cousin Charles Darwin. The "scientific imagination" of Galton's family seems to explain more fully why he was among the first to make a serious statistical study of inherited family traits.[2] He thought of eugenics as a "virile creed full of hopefulness" that should appeal to our "noblest feelings."[3] Galton became identified with ideas of "hereditary merit," the cluster of desirable traits he wished to replicate through "positive eugenics."[4]

But in addition to exhorting the "better classes" to mate and breed liberally, Galton also suggested the notion, without pressing the point too forcefully, that setting limits on the fertility of the less fortunate might be an appropriate role for government. Just such a scheme of negative eugenics was portrayed in his novel *Kantsaywhere,* set in a Utopia where hereditary worth was measured by tests that determined a citizen's place. Only the genetically gifted won diplomas from the Eugenic College in that fictional realm. Genetic failures were shunted off to labor colonies, where enforced celibacy was the rule; childbirth for the "unfit" was a crime.[5]

Kantsaywhere was not the only place where Galton hinted at his enthusiasm for reproductive interventions. While voicing opposition to compelled marriages, Galton agreed with "stern compulsion" to prevent unions of those with "lunacy, feeble-mindedness, habitual criminality and pauperism." He counted measures that would "check the birth rate of the Unfit" among the first objects of his new science of eugenics.[6]

Even before Galton developed a name for his "science of good breeding," a rich mythology had developed in America about the power of heredity to generate laziness, lawlessness, a weakness for liquor, and an appetite for unbridled sex. Anxiety about those who failed in the contest of life, relying on charity and inflating the taxes of everyone else, was widespread. When an author proposed that the "dangerous classes" could blame their ancestors for lives of crime and poverty or that where there was mental defect, "there must have been sin," he was doing nothing more than echoing the well-received theory of "degeneracy" that had been used to explain social degradation since the seventeenth century.[7] Degeneracy theory gave a human face to the biblical curse condemning children to inherit the sins of their fathers.[8] The curse carried particular power when those sins were demonstrated in dissolute living leading to legal transgressions, disease, and "pauperism," as poverty was typically known.

By the nineteenth century, Charles Darwin's evolutionary theory enhanced fears of degeneration. Many worried that repeated generations of debased living could reverse evolutionary processes, yielding a corruption of blood, a polluted "germ plasm" that would be passed down. Degeneracy was an idea that seemed to blame simultaneously heredity *and* environment for personal as well as social ills; it could lead to the reappearance of bad health, bad behavior, and bad environments for years to come.[9]

The well-informed readers of *Scientific American* knew all about degeneracy. They could hardly have been surprised in early 1875 to read that statistical reports were being compiled about the kinfolk of a vagrant girl known simply as "Margaret." Margaret's descendants were clustered in a part of New York's Hudson River Valley where crime ran rampant. Constant residents of local jails and almshouses, they represented a living model of the "hereditary disposition to crime."[10]

Commentary in the popular journal despaired at the "pestilent brood of human vipers" that descended from children of "Maggie." These "sturdy sinners" made up a "hereditary caste" that multiplied "like vermin" and justified drastic remedies to dam up the flood of ill-spent passion that "makes for unrighteousness." The magazine's prescription to prevent more "lusty savages" from filling the prisons: interrupt the parade of vice by "the knife remedy."[11]

If readers of *Scientific American* found this florid account of interest, so too did members of the medical fraternity who read Margaret's story in the

American Medical Weekly or the *Boston Medical and Surgical Journal.* According to those periodicals, her nine hundred descendants included two hundred confirmed criminals as well as numerous "idiots, imbeciles, drunkards, prostitutes and paupers."[12]

We know that one prominent doctor, also famous as a poet and essayist, read the medical journals carefully. Dr. Oliver Wendell Holmes used revelations from the medical press on "Margaret, the Mother of Criminals" as part of an extended essay on the causes of crime, written for the *Atlantic Monthly.* Dr. Holmes agreed entirely that "deep-rooted moral defects" appeared in the "descendants of moral monsters" in the same way that less problematic traits did. Holmes lectured his readers on the workings of heredity, arguing that an evil disposition was inherited just as surely as were physical afflictions—like hemophilia or premature white hair. The absence of positive traits in one's family was just as tragic, since a person born without "moral sense" could no more develop it than someone born deaf could become a musician. "Moral idiocy," declared Dr. Holmes, "is the greatest calamity a man can inherit," and vice is "more contagious than disease."[13]

Dr. Elisha Harris, an expert in vital statistics, sanitation, and public health, launched the study that so excited Holmes and other opinion leaders. His research in county jails was continued by Richard L. Dugdale, who eventually identified forty-two families whose bloodlines could be traced back to the calamitous Margaret. Dugdale published his findings in 1877 as *The Jukes: A Study of Crime, Pauperism, Disease, and Heredity.*

Dugdale created the pseudonym "Jukes" as a label for a clan "so despised by the reputable community that their family name had come to be used generically as a term of reproach." The Jukes' world was mired in crime and poverty, shot through with the habit of illicit sex. Dugdale's litany of evils listed "crime, pauperism, fornication, prostitution, bastardy, exhaustion, intemperance, disease and extinction" as other common features of their lives. Dugdale renamed "Margaret," the mother of the clan, as "Ada Juke." He highlighted Ada's legacy with a chart that showed nearly a quarter of her descendants beginning life as illegitimate children. Other charts filled pages of *The Jukes,* displaying the family genealogy along with compilations of criminal histories and medical pathology. To summarize the social effect of this ragtag army of vice, Dugdale created a ledger listing costs to the communities where the Jukes lived, costs that could be traced to their degraded lives. The bill for jails, almshouses, stolen or destroyed property, and medical or legal expenses paid by the state added up to more than $1.3 million over seventy-five years, "without reckoning the cash paid for whiskey" or other hidden expenses.[14]

Dugdale's solution for all but the "insane and the idiot" was to change the environment. He believed that modern sanitation would clean up slums and infant education would deter vicious habits. In time, these methods could check the downward spiral of degeneracy in families like the Jukes. Eradicate poverty, said Dugdale, and you would eradicate the breeding ground for crime. On balance, his prescription was hopeful. But writers like Holmes read Dugdale not as an environmentalist but as a determined hereditarian.[15] Even Galton, the father of eugenics, was certain that Dugdale was on the right track in showing how "criminal nature tends to be inherited."[16]

How heredity could lead to a "criminalistic" class descending from the likes of Margaret was far from clear. But John Harvey Kellogg was certain it had to do with sex. The family name of this young doctor would later become linked to All-American breakfast cereals, but he earned early personal fame as a stalwart of the eugenics movement and founder of the Race Betterment Foundation in Battle Creek, Michigan. Kellogg's commentary on *The Jukes* declared that "the throngs of deaf, blind, crippled, idiotic unfortunates who were 'born so,' together with a still larger class of dwarfed, diseased, and constitutionally weak individuals, are the lamentable results of the violation of some sexual law on the part of their progenitors." Following the degeneracy theory of his day, Kellogg asserted that children "begotten in lust" were destined for abnormality.[17]

Kellogg's observations were consistent with other studies of problem families that appeared, mimicking Dugdale's *Jukes*. The Reverend Oscar McCulloch told of his experience with an Indiana tribe that had inherited "wandering blood" from a "half-breed" mother. He christened them "the Ishmaels," noting that their familial illegitimacy stretched back to the fifth generation. Some were prostitutes, others merely beggars, but in all cases McCulloch located the cause of their degradation in an inherited tendency to "parasitism." The Ishmael stock was perpetuated, he said, by the corrupting charity of a "benevolent public," which rewarded the line of social bottom-feeders for indiscriminate breeding.[18]

McCulloch's "Indiana gypsies" were joined in the roll call of degenerates by other families publicized by reformers and criminologists. There was the mixed-breed "tramp family" called the "Smokey Pilgrims" of Kansas and the "Hill Folk" of Massachusetts, a clan descended from a "shiftless basket maker" marked by alcoholism and "reproduction of the grossly defective."[19] Each of these stories consciously harkened back to *The Jukes*, reminding the reader of the cost of poverty and disease, drunkenness and unclean living.

Removing problem families—especially problem women—from the community posed a challenge. Special schools were favored by professionals to sequester those women whose "weak will power and deficient judgment" made them "easily influenced for evil." Lifelong custody could be justified to some "as a matter of mere economy" to save the expense attached to a life of crime. For children identified early and "predestined" to become social pariahs, a kind of "moral quarantine" would protect the community. "Permanent and watchful guardianship" during the child-bearing years was particularly necessary for females, whose tendency to become "irresponsible sources of corruption and debauchery" was especially perilous. The promise that custodial institutions would prevent the birth of defective children was considered economically prudent. Every "hundred dollars invested now saves a thousand in the next generation."[20]

At its first meeting of the new century, the American Academy of Medicine listed among its goals a "scientific process of selection" to address the growth of the "unfortunate classes" that resulted from the "evils of indiscriminate charity." Marriage regulation, immigration restriction, and "asexualization" were listed among the possible remedies, with speculation that the latter solution would likely soon be made legal.[21] A consensus had also begun to develop among the public caretakers—prison officials, poorhouse directors, and the like—who provided shelter for problem families. Physicians also joined the chorus. "Degenerates," whose growing numbers were seen as "a menace to the well-being of the state," should be segregated. This proposal included "the shiftless poor" who would be sent to "colonies" where they could live a "normal, homely, country life" in a "world apart, a world of industry, a celibate world."[22] By 1900, the colony movement that had begun in the northeastern United States began to resonate with reformers in Virginia, who took up the challenge to create a specially designed colony where the state's degenerates could be isolated.

The Virginia Colony

Virginia had long been concerned with the character of its citizens. By the nineteenth century, Virginians had come to appreciate the importance of the "pedigrees of men," just as they understood the value of thoroughbred stock. The Virginia attitude was summed up in the nineteenth century by historian John Fiske. Said Fiske, no one could reasonably expect the law to raise "a society made up of ruffians and boors" to the level of a society populated by "well-bred merchants and yeoman, parsons and lawyers. One might as well expect to see a dray horse win the Derby."[23] Eugenicists agreed; Charles Davenport described the history of Virginia in his own text on eugenics. Some convicts had been among the early Virginia settlers, but "a better blood soon crowded into Virginia to redeem the colony," wrote Davenport. Royalist refugees arriving after the execution of King Charles I in 1649 "enriched a germ plasm which easily developed such traits as good manners, high culture, and the ability to lead in all social affairs, traits combined in remarkable degree in the 'first families of Virginia.'"[24]

Of course, the decrepit and the disabled could hardly fit in among such high-born citizens, and before the American Revolution, Virginia was in the vanguard of states providing government-supported institutional care for the mentally ill. The first hospital in North America built specifically to house the insane opened in the colonial town of Williamsburg in 1773 as the Publick Hospital for Persons of Insane and Disordered Mind. Until 1824, it remained the only institution of its type in America supported entirely by public funds. In 1828, with Virginia's population expanding beyond the Shenandoah Valley, a second hospital for the insane was opened in the town of Staunton. A third facility, specifically for "Negroes," opened in 1868 in Petersburg, and Virginia's fourth mental hospital was founded in the southwest town of Marion in 1887.[25]

As the twentieth century approached, those Virginia mental institutions were severely overcrowded. Physicians, clergymen, and reform-minded politicians began to agree that warehousing people together who suffered from a wide variety of social, psychological, and medical maladies could no longer be tolerated. Deviants like the criminally insane could not humanely be kept in the same asylums alongside those suffering from mental retardation or physical disability. By the same logic, prisons and almshouses could no longer be used to house epileptics along with criminals, the poor, or those diseased with tuberculosis.

As early as 1893, one state hospital director proposed a study to investigate the care of epileptics in Virginia and to review the newly developed colonies of other states.[26] Efforts toward legislation that would fund a colony for epileptics continued for a decade. Hospital superintendents exhorted Virginians to consider the life of an epileptic. The disease was "well-nigh incurable," and the epileptic's condition tended to lead to "dementia, imbecility, insanity, physical and moral degeneracy." The affliction was "hereditary to a marked degree" and involved some malady "being transmitted from parent to offspring with distressing certainty." It was hard to imagine "a more deplorable condition" to endure.[27]

The economic rationale for building efficient state "colonies" to segregate problem citizens was linked with dramatic and pitiful descriptions of the suffering epileptic. A colony seemed the only proper setting for the "humane, rational and scientific treatment" of such people.[28] A financial rationale provided the most salient motive for building institutions. It gave those with less-developed instincts toward philanthropy a more solid fiscal motive for reform. Colonies would alleviate crowding in existing institutions, offer a less confining environment for patients, and allow them to work to subsidize the cost of their care. Politicians supported the colony movement, particularly when new facilities could be built within the districts they served. New institutions were magnets for state revenue, jobs, and commercial development.

Aubrey Strode was one politician who understood this formula for local prosperity. He studied liberal arts and law at the University of Virginia, then opened a law practice in Amherst, Virginia. In 1905, as a new member of the Virginia Senate, he found a use for arguments that favored building a colony in his legislative district.[29] The son of the first president of Clemson University, Strode was a social progressive and a political pragmatist. His motives for supporting a new colony may also have been personal. Both of his parents had died in state institutions only a few years earlier, minds enfeebled with the infirmities of the aged.

Strode's bill to establish a colony for three hundred epileptic patients near Lynchburg, Virginia, became law in 1906.[30] Strode secured an appropriation to begin construction on the first colony buildings where "care, treatment, and employment" of inmates could occur.[31] Physician and former state legislator A. S. Priddy was selected as first superintendent for the new colony.

Albert Priddy was born in 1865, less than fifty miles from the site of Lee's surrender to Grant at Appomattox, just eight months after the conflict locally recalled as the "War Between the States" ended there. After completing his medical education at the tender age of twenty, Priddy spent fifteen years building a prosperous medical practice, during which he earned a local reputation as a bold and innovative surgeon. His understanding of mental illness grew as a physician in the Virginia Asylum at Marion, and his political skills were honed during two terms as delegate to the General Assembly. He learned about the courts as an expert witness, testifying regularly in insanity cases.[32] Priddy became the superintendent of the new state Colony on April 15, 1910.

Priddy described his epileptic patients as "the most pitiful, helpless and troublesome of human beings," tracing their increase to "bad heredity" following the intermarriage of the "insane, mentally defective and epileptic." In 1911, Priddy made his first official call for a sterilization law applying to all prisons and charitable institutions.[33] This announcement coincided with the arrival of patients at the Virginia Colony's new buildings, which were spread over three hundred fifty acres on the bluffs overlooking the James River.

The first inmate at the Virginia State Epileptic Colony was a thirty-seven-year-old man who arrived on May 16, 1911. His home for the previous thirteen years had been a Virginia mental hospital. When he was not in the hospital, he was sometimes held in jail, even though he had committed no crimes. Jail was the only place that seemed safe for him and separated him from the community that feared his affliction. At the Colony, even though he was a willing worker, he suffered from "frequent convulsions" and was "very much demented." His seizures could not be brought under control, and he eventually died and was buried at the Colony.[34]

The history of Inmate no. 1 was very similar to the ten men who joined him as the first inhabitants of the new Virginia Colony. Their ages ranged between twenty-seven and forty-one, and their epilepsy was often traced to specific traumas of youth. One man "fell from a high table backward" at age five, and "insanity" set in after a "lick on the head." He was often "befogged and full of hallucinations." Like the others, he spent time in jails and mental hospitals, and his family seemed relieved to find a place where he could receive specialized care. So desperate were families to find a safe refuge that in one case, a boy and his father walked twenty-five miles to the train line that would take him to the Colony.[35] One at a time, these men died, of hemorrhages after repeated seizures, of pneumonia or other infectious diseases, of accidents, of general exhaustion.

Approximately one hundred fifty men were admitted during the Colony's first year, all with the diagnosis of epilepsy, most having been transferred from other institutions. Within a year, the Colony charter would be amended to allow the admission of women. The condition that qualified them for a place at the Colony was "feeblemindedness," which, like epilepsy, was a source of great concern. Priddy called the feebleminded a "blight on mankind," and by 1914 more new buildings were constructed at the Colony to accommodate them. Priddy warned that the rapid growth of the feebleminded would only continue "unless some radical measures are adopted." In time, "non-producing and shiftless persons, living on public and private charity," would generate "a burden too heavy . . . to bear."[36]

Feeblemindedness

The plight of the "feebleminded" ranked alongside epilepsy as a topic of special interest to professionals working in America's institutions. According to Massachusetts physician Walter Fernald, those defined as feebleminded endured all manner of "congenital defect" ranging from "the simply backward boy or girl but little below the normal standard of intelligence to the profound idiot, a helpless, speechless, disgusting burden, with every degree of deficiency between these extremes." Fernald noted the distinction between idiocy and imbecility (imbeciles had slightly higher intellectual capacity) and concluded that the term *feebleminded* was "a less harsh expression, and satisfactorily covers the whole ground."[37] Worries about hereditary feeblemindedness fed into concerns about sexual misconduct.

One nineteenth-century trend known as the "Purity Crusade" focused on promoting chastity and ending the double standard that condemned a woman's sexual activities while ignoring the similar habits of men. Purity crusaders also worked to eradicate prostitution, often melodramatically described as "white slavery," an insidious force that lured vulnerable women into the urban sex trade. Fears of deteriorating sexual morality and shifting social expectations for women accompanying America's increasing urban-ization and mobile immigrant populations were only exacerbated by the new emphasis on the danger of feebleminded women.[38] The Purity Crusade's anti-prostitution movement coincided with the increasing eugenic fixation on feeblemindedness, and the claim was commonly accepted that most prostitutes were in fact mentally defective.

More than a dozen studies of prostitutes by leading researchers supposedly demonstrated that up to 98 percent of all streetwalkers were of "subnormal" mentality. While these figures were ultimately challenged and discredited, the image of the "moral degenerate," a woman defective in mind as well as morals, remained a powerful rallying point for various kinds of reformers who would ultimately endorse the twin policies of segregation and sterilization.[39] No IQ tests were done on the customers of prostitutes, so at least to some reformers, "moral degeneracy" remained a peculiarly feminine trait.

The Purity Crusade of the nineteenth century fed into the social hygiene movement of the twentieth. While the purity crusaders openly highlighted the moral values of chastity and sexual restraint and concentrated on sexual education as a tool to save women from sexual ruin, the social hygiene movement was led primarily by medical professionals concerned over the health effects of sexually transmitted diseases to women and their children. When the American Social Hygiene Association was formed in 1913, it took as its purpose the promotion of both "public health and morality."[40]

Social hygienists focused their efforts on making the Wassermann test for syphilis widely available and passing marriage laws requiring a laboratory test and a "clean bill of health" as a condition for receiving a marriage license, along with related public health measures. Both the moral tone of the purity crusaders and the public health message of the social hygienists echoed among eugenicists, who saw eugenic reforms like segregation and sterilization as remedies that addressed both morals and medicine. Removing purportedly feebleminded—and certainly "feebly inhibited"—women from the streets would make the world safer for those to whom the eugenic ideal of a pure life came naturally. Conditions like syphilitic insanity and congenital blindness in infants added to the ranks of the "defective," and sterility among married women amplified concerns about falling birth rates among the favored classes—so-called race suicide. Locked away in institutions, unclean women would be prevented from passing their diseases on to men, and the supposedly hereditary propensity to bad behavior embedded in their germ plasm would not be transmitted to a future generation. A eugenic program could address goals of both the Purity Crusade and the social hygiene movement.[41]

While not discarding the rhetoric of sexual moralism they adopted from the purity crusaders, many eugenicists began to repeat arguments from social hygienists that emphasized scientifically based expertise as a pragmatic foundation for legal reform. This shift was entirely consistent with the even larger social movement that was a key underpinning of eugenic activities: Progressivism.

Progressivism had many faces. Not so much a unified movement as a set of ideas pointing the way to reform, it arose in the last years of the nineteenth century and for nearly thirty years had an impact on U.S. political and social institutions. One trend that emerged under the banner of Progressivism was the desire to apply principles of efficiency to the management of government and to delegate the control of social welfare programs to a professionally trained class of experts.[42]

The emergence of feeblemindedness as a topic of public concern signaled a changing role for physicians, educators, and social workers who had ministered to the "less fortunate classes." The philanthropic motive dropped in priority, giving way to the need to protect society from "the menace of the feebleminded."[43] This shift was simply a "matter of self preservation" needed to protect the country "from the encroachments of imbecility, of crime, and of all the fateful heredities of a highly nervous age."[44] Feeblemindedness also opened clear avenues of activity for a professional class of reformers that could guide government policy in a Progressive direction.

In Virginia, the shift took the form of a governmentally funded system of social welfare to connect the existing patchwork of private and public institutions that provided "care, custody, or training of the defective, dependent, delinquent or criminal classes."[45] A new agency, the Virginia Board of Charities and Corrections, would study the problem groups and collect statistics that lawmakers could use in crafting public policy.

The Reverend Joseph Mastin was the board's first executive. A Methodist minister and superintendent of a Richmond orphanage, Mastin surveyed over four hundred doctors, asking them to report the number of "epileptics, idiots, feeble-minded, and cripple children . . . in their care." Another questionnaire collected data from teachers, asking how many children were "unfit for education in the public schools." A "Diagnosis for Mental Deficiency" was included with the questionnaire to aid teachers in locating "backward, possibly feeble-minded children." A wastebin full of factors were to be considered in that "diagnosis," including "blinking, twitching of the mouth, squinting, nervous movements . . . spasms, fits, hysterical crying and laughing . . . cold and clammy hands . . . drooling . . . carelessness, indolence, inattention . . . excessive exaggerations, falsehood, pilfering, and poor moral sense." Brandishing his surveys, Mastin insisted that the data proved the need for an institution to care for the feebleminded. Because of their reproductive recklessness, he claimed, feebleminded women represented a potential burden to the community. Recalling the mythical Jukes family, Mastin reminded his audience of the costs of hereditary defect.[46]

Lawmakers responded to Mastin's work, providing funding for additional investigations and further reports.[47] The Virginia Medical Society supported Mastin's proposal, calling for an institution where the feebleminded could be "segregated, cared for, studied, and treated."[48] A 1912 law authorized bringing feebleminded women to the Colony two years later. The new law gave implicit recognition to theories that stressed the hereditary nature of feeblemindedness and the need to segregate affected females from the general population. Echoing the reproductive interventions implicit in the laws of other states, it specifically directed the admission of "women of child-bearing age, from twelve to forty-five years of age" as the first patients.[49]

As Reverend Mastin's campaign matured, he repeated warnings about the dangers of feebleminded adults being allowed to remain at large. They led to more crime and insanity and would strengthen the "grip of the social evil," as prostitution was known. Known "facts" about feeblemindedness demonstrated that it was hereditary and incurable; it afflicted up to half of "lawbreakers," and many prostitutes were feebleminded. They gave birth to more of the epileptic, the insane, and others with "all forms of neurotic degeneracy."[50]

Mastin hired a field worker to trace the family histories of streetwalking women, all the while arguing for abolition of the "red-light districts" in Virginia cities where legal prostitution flourished. He demanded laws to forbid marriage and prevent childbirth among the "feeble-minded" and urged the use of the Binet-Simon intelligence test to identify them.[51]

The legislature requested a study that would explore the scientific means to achieve Mastin's goals.[52] Mastin's report, *Mental Defectives in Virginia*, restated the conventional wisdom: mental defect was hereditary; charity only encouraged people to multiply irresponsibly; excessive tax money was spent on social welfare—and the amount was growing. The report also included a detailed a study of all Richmond, Virginia, schoolchildren undertaken by a special investigator trained to use the new intelligence tests by psychologist Henry Goddard.[53] The *Eugenical News* listed Mastin's *Mental Defectives in Virginia* among "Eugenical Publications" and praised it as "one of the best" of such state studies.[54]

Mastin's report caught the attention of Aubrey Strode, the Colony's voice in the Virginia General Assembly and Priddy's personal lawyer. Strode's support to establish the Colony had been critical, and he was honored in 1914 with the opening of a Colony building bearing his name.[55] In the wake of *Mental Defectives in Virginia* and with explicit directions from Priddy, Strode introduced legislation that would establish a legal definition for feeblemindedness and expand the procedures for committing feebleminded patients.[56] Other Strode initiatives during the 1916 legislative session expanded the powers of the State Board of Charities and Corrections to register and track all feebleminded persons in the state.[57]

Most importantly, the new law expanded Priddy's medical authority to treat inmates. With his board's approval, Priddy could now impose whatever "moral, medical, and surgical treatment" he felt was appropriate for his patients.[58] However benevolent this language may have seemed to the average citizen—or to legislators, for that matter—it was borrowed from already existing state laws in Indiana, California, and New Jersey and was meant to cover surgical intervention at a doctor's discretion. Soon, Priddy would interpret it with great liberality, as an authorization to sterilize patients.

2

Sex and Surgery

The idea that surgery could be used to eradicate sexual deviance and the mental illness it was thought to cause was not uncommon among doctors in the nineteenth century, though it was controversial. For example, J. H. Kellogg of Battle Creek, Michigan, believed that "imbecility and idiocy" resulted from "that solitary vice," masturbation. He described one frustrated father who took matters into his own hands and castrated his offending son for such behavior. "The remedy was efficient," noted Dr. Kellogg, drily adding, "though scarcely justifiable."[1]

Experimental surgery on both women and men had also been performed in a number of institutions, but the medical and legal communities often opposed it. The outcry that arose following an 1893 report by the Pennsylvania State Hospital for the Insane was typical. Dr. Joseph Price operated on four women there, removing their ovaries in the hopes of restoring their sanity. Fifty other patients were scheduled for the operation. The legal advisor to the Pennsylvania Board of Charities Lunacy Committee declared that surgeons were at risk for criminal prosecution and condemned the proposed operations on poor, mentally ill women as "illegal and unjustifiable." The *Journal of the American Medical Association* concurred, noting that most medical superintendents in mental hospital settings regarded with disfavor "the castration of a woman as a cure for mental disorders."[2]

Not all physicians agreed with this opinion. At least a few believed that if "surgeons of acknowledged skill and integrity" advised the operation, a court order should be sought to perform it. This option seemed particularly expedient in cases where the insanity was of "a markedly erotic type," such that both the woman and the public would be safeguarded.[3]

Regardless of the patient's gender, to some doctors surgery seemed the proper prescription for unseemly sexuality. Several began to operate for "therapeutic" reasons long before surgery was legally authorized. In 1890, Kansas physician Dr. F. Hoyt Pilcher gained attention for castrating fifty-eight children—forty-four boys and fourteen girls—in the Winfield, Kansas, Institution for Feebleminded Children.[4] Pilcher's surgeries were also an attempt to deal with masturbation, thought to be a contributor to mental deficiency.[5] An asylum physician in Virginia also reported that masturbation had been connected to "mental disturbance." Doctors in his state had attempted "heroic measures" such as castration as a "cure" for the practice without producing promising results.[6]

Some doctors operated as therapy for epilepsy. Dr. Everett Flood reported twenty-six cases of "asexualization" in his Massachusetts asylum for epileptics, with circumcision of some men being provided as additional "treatment." Flood claimed that the surgery decreased the number of epileptic seizures, eradicated the sexual appetite, and helped one "solitary" inmate to develop "a more social disposition." It was also said to cure kleptomania.[7]

Pennsylvania's Dr. Isaac Kerlin operated not only to curb an "epileptic tendency" in the patient but also to remove her "inordinate desires which [were] . . . an offense to the community." He challenged states to take the lead in legalizing surgery "for the relief and cure of radical depravity."[8] Kerlin's assistant Martin Barr surveyed sixty-one institutions in an attempt to discern whether a consensus existed on what surgical technique should be used to stop procreation and to whom it might be applied. The majority of those who replied favored castration, although most respondents were understandably coy when asked whether they had actually performed the surgery themselves. Fully three-fourths gave their "unqualified approval" to the proposal that surgery should be legally authorized.[9]

In some cases, sexual surgery was clearly proposed to prevent reproduction. An 1897 Michigan bill called for "asexualization" of three-time felons, rapists, and male or female inmates of the Michigan Home for the Feeble Minded and Epileptic. The frank and simple object of the proposed operation was to insure "that such persons shall cease to reproduce their kind." The proposed law was sent to asylum and prison superintendents in the United States and Canada as well as three hundred physicians. Almost two hundred responded, all but eight favorably. The bill was passed by the Michigan House of Representatives but fell six votes short of passage by the Michigan Senate. A similar bill was introduced in Kansas, but it also failed.[10] Clearly, at the end of the nineteenth century castration was practiced on an experimental basis in many institutions, and its use was under "earnest discussion" for a variety of ailments related to mental and physical disease.[11] The legal community was also aware that castrations were taking place in almshouses and elsewhere "in a quiet way" to rid the world of feebleminded children who constituted the "progeny of worthless stock."[12]

The radical nature of the operation had definite drawbacks. In addition to the legal problems it might engender, it could cause pronounced psychological reactions in patients. Philadelphia surgeon J. Ewing Mears related the story of an English patient overtaken with melancholy and "brooding over the loss of his testes" who killed the surgeon who operated on him. At a meeting of the American Surgical Association at the New York Hospital, doctors exhibited patients carrying "celluloid testes." The prosthetic gonads were apparently traded for the originals in an attempt to avoid a reaction that might "dislocate the mental equilibrium."[13]

Undeterred by the potential "dislocation" disgruntled patients might cause physicians, Pennsylvania considered a "Bill for the Prevention of Idiocy" in 1901 that specifically focused on "mental defectives" in institutions.[14] The bill declared that "heredity plays a most important part in the transmission of idiocy and imbecility," and it authorized surgeons "to perform such operation for the prevention of procreation as shall be decided safest and most effective." The bill passed both chambers of the Assembly but was returned by Governor William Stone "for some trifling technicality." It failed to become law.[15]

The legislation offered in 1901 was reintroduced in 1905 and endorsed by both sides of the Pennsylvania legislature "without meeting any very determined opposition," although fifty legislators abstained.[16] The Philadelphia press noted that Governor Samuel Pennypacker would be given an opportunity to consider the precedent that the law might set. One editorial writer suggested that the idea could easily be extended to "incorrigible criminals, habitual drunkards and other persons filled with tainted and debased blood."[17]

Pennypacker, a former judge known for his pointed wit, rejected the "Act for the Prevention of Idiocy," as it was now called. If legislation alone would prevent idiocy, he argued, all civilized countries would have long since changed their laws. Pennypacker was wary of giving the medical profession carte blanche in deciding what methods would be "safest and most effective" in preventing procreation. Plainly "to cut the heads off the inmates" would fit the vague language of the Act. The law allowed unethical experimentation that was flawed, he said, because "idiocy . . . is due to causes many of which are entirely beyond our knowledge."[18]

Pennypacker should probably be celebrated for this early criticism of faulty eugenics legislation, but he is probably remembered more for a related anecdote that demonstrates his adeptness at handling the press. At the end of his term, he attended a "roast" put on by the reporters who had chronicled his administration. His introduction was met with the barrage of catcalls and boos common to such occasions, to which Pennypacker is reported to have calmly replied, "Gentlemen, gentlemen! You forget you owe me a vote of thanks. Didn't I veto the bill for the castration of idiots?"[19]

Public sentiment seemed to support society's right to check the proliferation of criminals and others of supposedly "defective" heredity, but many people remained squeamish about operations that mutilated these most personal human organs. A new technology delivered the solution. Surgeons like J. Ewing Mears explained that it was possible to render a man infertile by tying off the spermatic cord without removing any organs. This process would relieve those men who suffered from chronic "over-stimulation" because they had "over-indulged the sexual appetite."[20]

In 1899, Dr. Albert Ochsner suggested a different operation to achieve infertility. In a discussion focused on treatments for prostate problems, Ochsner described how he surgically removed a portion of the cord technically known as the vas deferens, thereby permanently removing a route for the sperm to travel. Ochsner's surgery may have been the first reported vasectomy, but he specifically prescribed it as a means of preventing the procreation of convicted criminals. Since castration ignited "the strongest possible opposition," he recommended that vasectomy be considered not only for criminals but also for "chronic inebriates, imbeciles, perverts and paupers."[21]

Others endorsed vasectomy as an antidote to the absurd "sentimental objections" raised against castration. Criminals and "mental incompetents" were "simply excrementitious matter" fit only for elimination lest they contaminate "the body social."[22] Vasectomy could purge society of "hordes of social parasites" and would avoid the furor over punitive laws.[23]

The prospect of wholesale sterilization of "sexual perverts and other dangerous defectives" was not universally acclaimed, even among surgical enthusiasts. Some considered anything less than full castration a "half measure" that would free those in institutions to wreak havoc once released from custody. Opponents of sterilization by vasectomy also feared the potential for spreading venereal diseases among people for whom pregnancy was no longer a worry and for providing an incentive to "all forms of consenting vice."[24] But the supporters of vasectomy discarded these concerns in favor of an operation they judged "less serious than the extraction of a tooth."[25] This new, less drastic operation would replace castration as the favored medical intervention to make a man sterile. It had the effect of neutralizing some of sterilization's more vociferous critics.

State Laws

The person most famous for popularizing vasectomy was Harry C. Sharp of the Indiana Reformatory. Sharp's surgeries began in 1899, and his patients were prisoners suffering from the dread condition of "Onanism" or habitual masturbation. His experience operating on several hundred prisoners set the stage for Indiana's 1907 eugenical sterilization law, the first such enactment that survived both public outcry and legislative scrutiny.[26]

Sharp had the fervor of an evangelist and a rhetorical flare equal to that of his Indiana predecessor, hereditarian Reverend Oscar McCulloch. Certain that the criminal class had inherited deficient faculties of self-restraint, Sharp believed that regardless of the virtue in their hearts, some criminals simply could not contain their passions. Their propensity to crime and other forms of immorality was inborn since, as Scripture taught, "nature punishes those who violate her laws," and the "punishment is visited unto the third and fourth generation."

Sharp operated without anesthesia. In 1909 he claimed that after some 456 vasectomies, he had observed no "unfavorable symptoms." For women whose weak character made them unable "to resist the importunities of unprincipled men," he recommended a similarly innocuous operation.[27]

Dr. J. N. Hurty, secretary of the Indiana Board of Health and later president of the American Public Health Association, supported Sharp in his crusade. Hurty's interest in eugenics increased during his twenty-six years as state health officer, during which he observed "the stupidity of mankind, [and] the worthlessness and filthiness of certain classes of people."[28]

The 1907 Indiana law inspired by Sharp and supported by Hurty shared several features of the Pennsylvania bill that failed in 1901 and 1905. It began with a declaration that was nearly identical to the preamble of the ill-fated Pennsylvania legislation: "Heredity plays a most important part in the transmission of crime, idiocy and imbecility." The law also gave surgeons wide discretion "to perform such operation for the prevention of procreation as shall be decided safest and most effective"—another obvious borrowing from Pennsylvania. In fact, most of the language of the Indiana law was taken directly from the bill that Pennypacker had vetoed two years earlier. Indiana Governor J. Frank Hanly did not share Pennypacker's concerns.

When Sharp spoke to his fellow prison doctors, he trumpeted his "brilliant success" as a surgeon and praised Governor Hanly's sense of "civic righteousness."[29] At later gatherings he tutored physicians in surgical technique, emphasizing the safety of the procedures he used on girls as young as eleven.[30] Sharp's comments met with a mixed response; some doctors objected strenuously even to the discussion of coercive surgery.

Nevertheless, some members of the medical press applauded Sharp, dismissing the "sentimental mollycoddles" who naively hoped that hereditary criminals could be reformed.[31] Lawyers also praised the Indiana law and reported the "wonderful unanimity" in favor of sterilizing criminals to prevent the replication of their social disease.[32]

Other states followed Indiana's lead. Washington passed an unusually brief single-paragraph statute in 1909 focusing on rapists, child molesters, and "habitual criminals." Those convicted of the named offenses would be subject to an unspecified "operation" designed for the "prevention of procreation."[33]

The Washington law was tested in the case of Ralph Feilen, a serial child molester. Following Feilen's conviction, a court assessed a twenty-year sentence, adding vasectomy as an enhancement. The lawyer who appealed the charge called the law one of the "freak enactments" that had been rushed through the legislature by a group of youthful prosecuting attorneys. If the purpose of such laws was eugenics, he argued, it would be better to castrate financiers who robbed large numbers of unsuspecting citizens with fraudulent stock promotions.[34] The Washington Supreme Court disagreed, upholding the sentence and dismissing the idea that vasectomy was a "cruel" punishment forbidden by the state constitution. The opinion quoted from popular literature on eugenics, echoing the assessment of vasectomy as an "office operation" that was favored by a "wonderful unanimity" of the medical press.[35]

California adopted what was perhaps the most expansive legislation of all state acts permitting sterilization, covering people in prisons, state hospitals, and the Home for Feebleminded Children. It allowed "asexualization" surgery whenever institutional superintendents thought it beneficial for the "physical, mental or moral condition" of the inmate.[36] Connecticut passed its sterilization law the same year, allowing physicians in the state prisons and state hospitals for the insane to choose men for vasectomy or women whose ovaries should be removed.[37]

New Jersey's law was passed in 1911 and was immediately signed by Governor Woodrow Wilson, a eugenics enthusiast.[38] The law shared many features of the Pennsylvania proposals and the Indiana statute already in place. It required Wilson to appoint a "Board of Examiners" consisting of a surgeon and a neurologist who would meet periodically to determine for whom "procreation was inadvisable" among the "feebleminded (including idiots, imbeciles and morons), epileptics, rapists, certain criminals and other defectives" in New Jersey institutions. The examiners could sanction any operation it found "most effective."[39] Opinion leaders like New Jersey inventor and scientific luminary Thomas Edison applauded the mandate for sexual surgery that would lessen the number of criminals.[40]

Within a month Wilson appointed the required board of examiners. It quickly set the law into motion, choosing Alice Smith, a resident of the New Jersey State Village for Epileptics, as the first sterilization case. Smith lived at the home for nine years but had suffered no seizures in the five years preceding her legal encounter. Disregarding this evidence, the board agreed unanimously that procreation was inadvisable for Smith and voted to perform a salpingectomy, removal of a portion of the Fallopian tubes, on her.

When performed by an experienced surgeon, vasectomy was a localized operation of relative simplicity. Patients generally suffered little discomfort, and postoperative complications were unlikely. The sterilization of women was more problematic. The operation of salpingectomy was an advance over earlier surgical techniques for women because it did not involve removal of the ovaries (as in an oophorectomy) or the uterus (hysterectomy) and was therefore not a castrating operation. But it still involved an abdominal incision and the exposure of internal organs; the potential for complications after surgery was significant.

The attorney appointed to represent Smith immediately appealed the decision to the New Jersey Supreme Court, which promptly struck down the state law. The court's opinion described the proposed operation in detail. A salpingectomy, it said, "must be performed in both sides of the body, and hence is in effect two operations, both requiring deepseated surgery, under . . . prolonged anesthesia, and hence involving all of the dangers to life incident thereto."[41] There was a danger from infection, inflammation, and surgical shock. The operation was performed under involuntary anesthesia, performed "by force" in order to "completely destroy all liberty of will or action." It threatened "possibly the life, and certainly the liberty" of Alice Smith.

Having described the case in dramatic terms, the Court went on to survey the other legal issues it raised. This was the first time a court had had to address the question of whether a person who had committed no crime could be subjected to an involuntary sterilizing operation.[42] Endorsing such a conclusion opened a wide vista of possibilities. Many other diseases besides epilepsy could provide a motive for surgery in order to purify coming generations. Indeed, in principle, the range of options was hardly limited to disease. The hungry could be sterilized, to prevent future hunger; whole races could be sterilized to do away with their discomforting presence. In fact, a California physician in search of a solution to the "Negro Problem" had proposed such a solution the year before passage of the New Jersey law.[43]

The New Jersey court sidestepped the question of whether *all* sterilizing operations were unconstitutional. It decided instead that since the law at hand only applied to people in institutions and not those living elsewhere, it represented a scheme that was "singularly inept" at accomplishing its supposed goal. Many more people with epilepsy lived outside state facilities than in them, and only those with no other means of support were taken in by charitable institutions. The few poor epileptics living at state expense were, by institutional confinement, already prevented from having children, while those outside were "more exposed to the temptation and opportunity of procreation." A sterilization plan that applied only to the confined poor, concluded the Court, must violate the Equal Protection clause of the U.S. Constitution. The only other reason the Court could identify as justification for the law was that it was intended to save the state money by sterilizing all of the inmates and then turning them out to the streets, allowing the institutions to be closed. The "palpable inhumanity and immorality" of such a plan, according to the Court, was impossible to impute to the legislature; New Jersey's law was void.[44] Although New Jersey's attorney general immediately claimed that the case would be taken to "higher Courts," this never happened, and several later attempts to revive a sterilization law in New Jersey failed.[45]

One state resorted to popular referendum to defeat advocates of sterilization. Antivaccination advocate Lora C. Little of Portland, Oregon, successfully rallied popular support—via the Anti-Sterilization League—against eugenic surgery. Her work helped to defeat a 1913 Oregon sterilization law, which was voted down in a ballot measure despite the support of the press and many politicians. Portland lawyer C. E. S. Wood, her colleague in the effort, condemned the sterilization law, saying that "it accomplishes nothing, may be an engine of tyranny and oppression and is ROT."[46]

Other state laws were also challenged in the Courts, with varying results. Nevada copied the Washington State law in 1912 with one amendment. It allowed an operation to be added to the sentence of anyone found guilty of child sexual abuse or rape but limited the surgery by excluding castration.[47] Criminal defendant Perry Mickle had the bad judgment to admit not only that he had raped a woman following a drinking binge but also that he was subject to "epileptic fits." Apparently in an attempt to spare the man from the consequences of a lie he may have concocted in a bid to win the court's sympathy, the sentencing judge in *Mickle v. Hendrichs* gave the defendant an opportunity to retract his statement about epilepsy. However, Mickle insisted, "hope to God to die," that he could prove that he had epilepsy. Thereupon the judge sentenced him to prison for life and added that "to protect society" from the man's offspring, the sentence would include what the court reporter referred to as "Baseptomy."[48] But a federal court reversed this sentence, saying that the Nevada law violated the prohibition of "cruel and unusual punishment" in the state constitution. The court declared that the possibility of reformation of the criminal was just as important to society as "the eugenic possibilities of vasectomy."[49]

The first sterilization case to reach the U.S. Supreme Court reviewed an Iowa law "for the Unsexing of Criminals, Idiots, etc." enacted in 1911 to allow vasectomy or tubal ligation of convicts and other people in institutions.[50] It was revised in 1913 to call for an examination of the records and "family history" of those whose procreation was considered "improper or inadvisable." The new law included a sweeping description of people at risk, encompassing those likely to produce children "with a tendency to disease, deformity, crime, insanity, feeble-mindedness, idiocy, imbecility, epilepsy or alcoholism" as well as the syphilitic and "moral or sexual perverts."[51]

Rudolph Davis had been convicted of one felony for breaking and entering in another state prior to the enactment of the Iowa statute. He committed a second felony in Iowa after the law's passage. The Iowa attorney general challenged the constitutional propriety of applying the law to Davis. The federal court hearing *Davis v. Berry* judged even a vasectomy "cruel and unusual" because its purpose and the inevitably associated "shame, and humiliation and degradation and mental torture" were commensurate with castration. The court concluded that the law was also unconstitutional because it did not provide for a proper hearing to protect the prisoner's due process rights. Prison officials were ordered not to perform the operation.[52]

Barely two weeks after the court's decision, the Iowa legislature repealed the law Davis had challenged and enacted a third sterilization law significantly different from the first two.[53] These developments seemed to make *Davis v. Berry* moot, yet the appeal of the federal court's ruling was allowed to proceed to the Supreme Court. The debate in the high court stretched for a full year through briefings that challenged the rights of federal trial courts to set aside legislative enactments of the states.[54] The arguments provided an education in the history and politics of sterilization laws for the Supreme Court Justices and demonstrated yet another example of the fervor with which state legislative prerogatives were defended. But because the action of the Iowa legislature had removed any threat to sterilize Davis, the Court directed that the suit be dismissed. The decision in the case was announced in a two-paragraph opinion focusing on the procedural defects in Iowa's case, written by then seventy-five-year-old Justice Oliver Wendell Holmes Jr.[55]

3

The Pedigree Factory

In 1865, Gregor Mendel (1822–1884) published the paper on the inherited characteristics of sweet peas that laid the foundation for studying genetics. During his lifetime, however, the German monk's work went almost entirely unnoticed. Scientists working independently in England, Holland, Germany, and Austria rediscovered Mendel's seminal paper in 1900.[1] It provided a biologically based scientific model, complete with "laws" of heredity.

Those laws became common parlance to people interested in heredity. According to Mendel, people inherited certain characteristics or "traits" from their parents in a predictable pattern. Those traits were passed down as independent units via some physical "determiner" or "factor." The factors, later named "genes," were the essence of heredity and in the aggregate made up what came to be called the "germ plasm"—the physical link connecting the generations. Though Mendel's work involved tracking features of pea plants such as height or the color of their flowers, people who studied humans initially focused on eye color, hair color, and other easily observable traits to analyze the workings of heredity. They talked of the one-to-one correspondence between the determiner (or gene) and the "unit character" (or trait).[2]

Mendel's theory of inheritance, Francis Galton's family study methods, and the general passion to eradicate social problems came together in an American institution dedicated to the study of eugenics. In 1910, biologist Charles Benedict Davenport established the Eugenics Record Office (ERO) in Cold Spring Harbor, New York.

Davenport was a credentialed member of America's scientific elite. He earned a Ph.D. at Harvard in 1892, then taught there and at the University of Chicago. He was named director of the Station for the Experimental Study of Evolution, funded by the Carnegie Institute of Washington, in

1904. Over his career he held memberships in the National Academy of Sciences and the National Research Council.[3]

Davenport began his study of heredity as a disciple and correspondent of Francis Galton in the late 1890s. In 1904 Davenport wrote that "unit characters" were the "essence of individuals."[4] Less than a year before the formal founding of the ERO, Davenport gave a lecture at Yale University that summarized his position on the aims and the format of his brand of eugenics. He proposed a system that would survey family traits. Such a plan would "identify those lines which supply our families of great men." But studying the great families was only one goal of eugenics; Davenport also urged tracing the origins of "our 300,000 insane and feebleminded, our 160,000 blind or deaf, the 2,000,000 that are annually cared for by our hospitals and Homes, our 80,000 prisoners and thousands of criminals that are not in prison, and our 100,000 paupers in almshouses and out."[5]

According to Davenport, some 3 to 4 percent of the American population came from these groups, and they represented "a fearful drag on our civilization." Should we merely stand by to watch philanthropists shower their beneficence on "the delinquent, defective and dependent classes" or even raise the taxes of ordinary Americans to that end? Better to follow the lead of science and work to "dry up the springs that feed the torrent of defective and degenerate protoplasm."

Davenport speculated that research in institutional records and the archives of schools and insurance companies would pave the way for eugenic legislation that would prevent "idiots, low imbeciles, incurable and dangerous criminals" from having children. Preventive methods could include institutional segregation and surgical sterilization. The options for self-protection extended, according to Davenport, even to executing criminals, in order to "annihilate the hideous serpent of hopelessly vicious protoplasm." Davenport predicted that preventive medicine, guided by eugenic principles, would replace palliative philanthropy.[6]

Eugenic agenda in hand, Davenport successfully approached Mary Harriman, then the recent widow of railroad magnate E. H. Harriman, for funding. Between 1910 and 1918, the Harriman contribution to endow and maintain the Eugenics Record Office reached nearly $650,000. Additional money from the Carnegie Foundation, John D. Rockefeller, and other philanthropists provided substantial support for the research institution that for nearly thirty years would be a focal point for the eugenics movement in the United States.[7]

The first board of scientific directors of the ERO included an extraordinary array of scientific talent, starting with Alexander Graham Bell as the chairman. Bell was famous among the general public for inventing the telephone, but the scientific world also knew him as a student of heredity.[8] He was particularly interested in the genetic basis for familial deafness and specialized in studying the hereditary features of sheep.

Joining Bell as vice chairman of the board was William H. Welch. The first dean of the School of Medicine at Johns Hopkins University, Welch was a giant in the field of public health. At the time of his appointment to the Eugenics Record Office board, he was president of the American Medical Association. He had already served as president of the American Association for the Advancement of Sciences and would later also lead the National Academy of Sciences.

The other members of the board were Welch's Hopkins colleague anatomist Lewellys Barker, Yale economist Irving Fisher, fruit fly geneticist and later Nobel Prize winner Thomas Hunt Morgan, and Harvard Medical School psychiatrist E. E. Southard. These luminaries were joined by many others who supported the infant science of eugenics. For example, Davenport's position on eugenics was echoed by former president Teddy Roosevelt, who wrote in a letter that "society has no business to permit degenerates to reproduce their kind."[9] Publicly, Roosevelt was no less bold, saying "I wish very much that the wrong people could be prevented entirely from breeding."[10]

As the founder of the ERO, Davenport was its staff director, but the man he named as its superintendent and entrusted with its daily operation was Missouri Normal School teacher Harry Hamilton Laughlin. Laughlin's advocacy in favor of eugenical sterilization laws would become legendary, and his expertise at drawing pedigrees would be recognized internationally. Born to a Missouri preacher and a mother whose moral passions focused on the prohibition of alcohol, Harry Laughlin began his career as a schoolteacher and student of biology. Davenport invited him to move east after observing his work during summer study sessions at the Brooklyn Institute of Arts and Sciences Biological Station at Cold Spring Harbor. Laughlin took over the management of the ERO, conducting research, shepherding publications, coordinating staff, and teaching field workers.

Harry Hamilton Laughlin and Charles Benedict Davenport at the Eugenics
Record Office, 1913. Courtesy Truman State University.

Initial plans for the ERO included the collection of thousands of family records. Some submitted "Records of Family Traits" to the ERO; other
data was gathered by field workers acting with the cooperation of asylums,
hospitals, and other institutions. Once assembled, family information was
displayed in genealogical charts using symbols to denote troubling characteristics that might reappear within related groups, such as criminality,
sexual immorality, feeblemindedness, insanity, and infectious diseases like

syphilis or tuberculosis. A special fireproof vault was built as part of the ERO building to store the catalogued pedigrees.[11] Davenport would eventually claim that the eugenicists' collection of "human records to be used in the study of heredity in man . . . led the way to the great development of human genetics."[12]

The office also had an educational orientation. Resident experts trained visiting field workers to collect data on both good and bad hereditary characteristics. The ERO published research results in the form of bulletins about eugenics. The bulletins would be of value, according to Davenport, "to show the heads of institutions, boards of control, private investigators, genealogists, and legislative authorities" what "modern methods" of eugenical investigation could accomplish. By studying problem families in their home environments, field workers could assemble convincing data as the basis for "wise legislation" that would prevent the reproduction of "defectives."[13] The Office could also offer advice to individuals concerning the "eugenical fitness" of their proposed marriages.

In many ways, the Eugenics Record Office offered a perfect setting to put Progressive theory to work. It would engage in constant studies, employing the most up-to-date statistical techniques to make sense of human behavior and the rules of heredity. It would provide a welcome home to experts who could assimilate the accumulated knowledge of psychology, biology, and the new fields of genetics and social work and become a "eugenist," as specialists in heredity were named. It would train a cadre of new experts, whose practical knowledge could be put to work pursuing further research and educating others in the need for eugenic reform—the most efficient means of solving social problems. Above all, it would enlist the forces of science to collect important data that would form a basis for policy recommendations, to advise government officials and lawmakers, to lead the way to reform of laws and regulations.

The ERO became a practical study center encouraging collaboration among experts in the field of eugenics. Cooperative work between the ERO and the Training School for the Feebleminded at Vineland, New Jersey, began as soon as the ERO opened. Vineland was also the first field station to which workers were sent on training excursions.[14] The ERO's first bulletin was Vineland psychologist Henry Goddard's pamphlet on the "Heredity of Feeble-Mindedness."[15] It contained dozens of pedigree charts composed of material collected from the families of residents of the Vineland School.

The second bulletin, "The Study of Human Heredity," was a joint effort of the ERO and Vineland. It showed the field worker how to collect and analyze family data and ingratiate herself (the vast majority of field workers in training were women) with research subjects in order to gain their trust. Specific instructions on drawing pedigree charts—a method for graphic representation of family traits—were also included, as were model charts and a key to symbols signifying the various pathologies and behaviors field workers were trained to catalogue.[16]

Later ERO manuals such as "The Trait Book" contained a taxonomy of human characteristics ranging from "physical beauty" to "criminality," each listed with a code number. Using the proper code, a field worker could describe, chart, and label any condition found in research subjects. "The Trait Book," like other early Davenport work, emphasized identifying "unit characters" or "indivisible" characteristics passed down as discrete hereditary units.[17] "The Family History Book" was designed to assist studies that would parallel the work of Dugdale on the Jukes and McCullough on the Ishmaelites.[18] This type of fieldwork would generate "pedigrees of sufficient accuracy and detail to justify prediction as to the hereditary potentialities of selected strains and individuals."[19]

The pedigree charts of "defective" families that were such a key feature of ERO training materials consisted of detailed family genealogies. The charts grew out of methodical investigation of individuals with suspect traits and were drawn to demonstrate graphically the way traits reappeared in families. The pedigrees dramatized what were thought to be the biological bases of social problems. They were often accompanied by information designed to alert readers to the costs of deviant behavior—primarily in the form of higher taxes to support public institutions. They sometimes contained a calculation of public funds for welfare support, criminal activity, and institutional care traceable to a specific family, just like the claims made about the Jukes.

Harry Laughlin's pedigree work became part of the curriculum at the ERO. He and other faculty there used the bulletins as guidebooks in the basic training program of field workers who traveled to Cold Spring Harbor for summer seminars. Each student learned how to assemble a pedigree chart; students were also instructed in the use of accepted symbols indicating conditions such as epilepsy, feeblemindedness, insanity, alcoholism, syphilis, sexually immorality, and an impulse to "nomadism" or aimless wandering, all thought to be hereditarily determined.[20] As part of the ERO educational outreach efforts, instructions for compiling pedigrees of normal families were also made available to high school teachers and college instructors to aid in courses in biology, sociology, and psychology.[21]

Revisiting the Jukes

One man trained at Cold Spring Harbor would become the most prominent of all field workers and a master of pedigree studies.[22] Arthur Howard Estabrook, a native of Leicester, Massachusetts, earned his bachelor's degree at Clark University in 1905, took a master's degree there in 1906, and then received his Ph.D. in 1910 from Johns Hopkins.[23] Estabrook learned eugenic research techniques in the first class of field workers the summer before the ERO opened and was soon hired to begin his career as a staff member there.

Estabrook's first major fieldwork resulted in *The Nam Family* (1912), a book he coauthored with Charles Davenport. It described a "highly inbred rural community" in New York State characterized by "alcoholism and lack of ambition," a degenerate family much like Dugdale's Jukes. The book included a massive circular pedigree with four hundred fifty family members with detailed descriptions of people in the Nam line. Since sexual misbehavior appeared so regularly as a trait of the Nams, Estabrook devised special notations for his pedigree charts to show multiple sexual partners (Sx_1), licentiousness (Sx_2), prostitution (Sx_3), and erotomania (Sx_4) in addition to indicators of illegitimacy. Liberal use of the abbreviation for "shiftlessness" was also necessary to properly characterize this low-achievement clan. The book contained financial comparisons with Dugdale's Jukes, with a specific tally for productivity lost to drinking. "Nam Hollow" was described as an epicenter of degeneracy, a "social pest spot" whose conditions spread like a virus. Nurses and social workers could add a "veneer of good manners" to the inhabitants, but Estabrook looked to institutional confinement of the Nam youth and prevention of childbirth as the best solutions to future decay. "Asexualization" was mentioned as an equally efficient but not likely remedy, since it was out of favor with the public.[24]

Another early research project undertaken by the ERO was an attempt to locate the original notes of Richard L. Dugdale, author of *The Jukes*, so that they could "take up the work where Dugdale left off."[25] When the papers on the Jukes were discovered in a prison basement, updating their story became Estabrook's next major project.[26] Since "Jukes" was a pseudonym, Estabrook used Dugdale's coded text containing the actual family names to track down the descendants of the family that had formed the basis for the earlier work.

As his study progressed, Estabrook wrote an article about one family he surveyed. "A Two-Family Apartment" described the life of a woman who lived in a shack she shared with a litter of pigs. The two families, pigs and humans, were said to be "rivals" in the amount of filth they generated, and the woman was described as "an illegitimate child of a feeble-minded mother, herself now the mother of feeble-minded children." A picture of the woman, holding one of her infants while looking at a group of pigs, bore the caption "After their own kind." Estabrook concluded the article with the dramatic revelation that the woman belonged to the infamous Jukes family.[27] By then, Estabrook told his colleague Harry Laughlin that he was completely convinced of the "inheritance of immorality."[28]

Estabrook presented preliminary results of his work to a meeting of the Eugenics Research Association, and the completed study finally appeared as *The Jukes in 1915*.[29] The volume was a bold restatement of the classic study of poverty and disease. Estabrook confirmed some of Dugdale's environmentalist conclusions, determining, for example, that removal from the rural confines of the original family could have beneficial effects on some of the Jukes descendants. Others, however, succumbed to nature and followed the path of "criminality, harlotry and pauperism" determined by their heredity. Like the Nams, the Jukes were a special case in sexual misbehavior. One chart noted the concurrent Nam family traits of sexual licentiousness, alcoholism, feeblemindedness, and general criminality. The heredity of contrasting characteristics such as chastity, declared Estabrook, "seems assured in the main." By then the remedies to this recapitulation of vice and poverty were obvious: segregation or sterilization. But even though sterilization would be less restrictive of a person's liberty than custodial segregation, Estabrook again concluded that "public sentiment" still stood in the way.[30]

Estabrook's books, like those of Davenport, furnished substantial scholarly references and models for field workers learning to chart problematic pedigrees at the ERO. Along with books like *The Jukes*, they became required reading not only for those who ventured to Cold Spring Harbor for instruction but also for the millions who would study eugenics in American schools and colleges. Another volume that was required reading in the eugenics curriculum was *The Kallikak Family* by ERO collaborator Henry Goddard. It eventually joined Dugdale's *Jukes* as the second premier example of family studies of the "cacogenic," or "ill-born."[31]

Just two years before Arthur Estabrook entered Clark University, Henry H. Goddard received his Ph.D. there in 1899 as a student of the pioneering psychologist G. Stanley Hall. While activities at the ERO revived interest in *The Jukes*, Goddard's work at the Vineland School led to the next chapter in the mythology of defective families.

The Kallikaks

The keystone of Goddard's book was Deborah Kallikak, a girl whose family story was gathered through extensive fieldwork and interviews by one of Goddard's assistants. Born in an almshouse, the comely Deborah was a resident of the Vineland Training School, where fourteen years of records, complete with illustrative photos, showed her progress. While tracking the path of her family line, Goddard happened upon the two branches descended from Martin Kallikak Sr., and the story that would make him famous. Deborah was a feebleminded girl of the "high grade, or moron" type. In 1909 Goddard had coined the term *moron*, from the Greek word for "fool," to address the vagueness of the commonly used term *feebleminded*.[32]

At his Research Laboratory of the Vineland Training School for Feeble-Minded Girls and Boys, Goddard was the first American to appreciate the potential uses of the intelligence tests developed by French psychologist Alfred Binet.[33] Goddard became famous for developing a measure for "feeblemindedness" and applying the tests to what would become the most popular successor to the line of families that began with Dugdale's *Jukes*.

Combining two Greek words—*kalos*, meaning beauty or goodness, and *kakos*, meaning bad or evil—Goddard named his New Jersey tribe the Kallikaks. He considered their story superior to previous family studies because it was a "natural experiment" in heredity. One line of the family displayed positive inherited features, while the other exemplified negative traits. Together they represented an object lesson in the workings of Mendelian theory and the social danger posed by "defectives," community discards who lived on the margins of society.[34]

Both sides of the family were sired by one Martin Kallikak, a Revolutionary War soldier of "good English blood." During his travels with the militia, Martin stepped away from the "paths of rectitude" to be entertained by a feebleminded girl often described as a "tavern wench." The result of this dalliance was Martin Jr., later known as "Old Horror" for the swamp of decrepitude he left to posterity. Described by one associate as "simple . . . not quite right," Martin Jr. was well remembered as a victim of strong liquor and his own weak will.[35] Goddard traced the story of Martin Jr.'s similarly deviant descendants, completing the degenerate pedigree.

Goddard also sketched the family history of Martin Sr.'s other children, born of "a woman of his own quality," his eventual "lawful wife," a good Quaker woman.[36] True to hereditary form, those offspring were moral exemplars and models of social success.

Goddard recounted the contrasting lineages like a biblical genealogy. Son of a feebleminded harlot, Martin Jr. begat ten children possessed of various defects; they in turn begat several generations of feebleminded relations. By the time of Goddard's investigation, four hundred eighty descendants were numbered among Martin's issue, and only forty-five of them were "normal."[37] The rest were overcome with alcoholism, epilepsy, criminality, or prostitution; many were "peculiar," that is, of questionable cognitive ability or "feebleminded"; the women often "sexually immoral."

By contrast, the happy results of Martin Kallikak Sr.'s married life constituted a parade of successes. According to Goddard's research, "doctors, lawyers, judges, educators, traders, landholders, in short, respectable citizens" made up the vast majority of Martin's legitimate heirs.[38] There was no feeblemindedness among them, no illegitimacy, no immorality, no criminality. Goddard's moral was clear: good blood begat good people; feeblemindedness and sexual license begat nothing but trouble. The psychologist illustrated his book with pedigree charts, capturing the flow of generations graphically, labeling each member "N" for normal or "F" for feebleminded.[39]

"There are Kallikak families," Goddard wrote, "all about us." If all slums were demolished, he continued, slums would reappear, "because these mentally defective people" create the slums, and they know no other life. In *The Kallikak Family*, Goddard shifted attention from the "low-grade idiot, the loathsome unfortunate" who filled institutions to the high-grade moron—the social class that posed the most significant threat. In contrast to those who would execute idiots, Goddard recommended an operation to sterilize morons that would render them harmless to the community. While a clear understanding of how mental characteristics were passed down through heredity remained elusive, Goddard pointed to new insights derived from Mendel's work as the path for future research. Though sterilization was not likely to provide a "final solution" to the problem of hereditary feeblemindedness, it could yield a "makeshift" approach, necessary because "conditions have become so intolerable."[40]

Some reviewers noted an absence of "strict scientific evidence" for some of Goddard's assertions and waited for the "more technical volume" that he promised for a professional audience. For now, the lay reader for whom Goddard's parable of degeneracy was written would be well served by his straightforward narrative.[41] It was repeated in some of the most popular schoolbooks in America; for generations of readers—the book went through twelve editions—*The Kallikak Family* offered solid examples to justify familiar Old Testament wisdom about unclean living and inheritance.[42] As the Good Book said: "I, the Lord your God, am a jealous God, visiting the iniquity of the fathers on the children, and on the third and the fourth generations."[43]

Scholars who awaited the detailed evidence from Goddard soon encountered a second book of more than six hundred pages, a seemingly irrefutable mass of data: In *Feeble-Mindedness: Its Causes and Consequences*, Goddard defined feeblemindedness as "a state of mental defect existing from birth or from an early age and due to incomplete or abnormal development in consequence of which, the person affected is incapable of performing his duties as a member of society in the position of life to which he was born."[44]

People with nonspecific social problems, the "ne'er do wells" who made up the "shiftless, incompetent, unsatisfactory and undesirable members of the community," were also suspected of being feebleminded. The new Binet intelligence tests, later to be popularized by Goddard himself, helped identify exactly who those people were. Goddard explored the effects of underlying mental defect on illegitimacy, sexual immorality, and drunkenness. After an extensive discussion of cases, he presented the qualified conclusion that feeblemindedness, like intelligence generally, was a unit character, inherited in a predictable Mendelian fashion.[45]

Goddard offered several solutions to address his finding that hereditary feeblemindedness was at the root of most social problems. He proposed a tiered scheme. For the lowest grade of intelligence approximating a child from birth to age two, the *idiot,* and those of intelligence comparable to a child from age three to seven, the *imbecile,* Goddard recommended custodial care in a setting of controlled environment segregated from society. For those with intelligence comparable to a child from age eight to ten, the *moron,* Goddard had a different solution. They could be put to work in a simple occupation, such as supervised farming, housework, or manufacturing—the setting created in a farm colony. Of course, these settings were out of the way, and they kept morons separated from normal people. But even if many could be put in colonies, those who could not would become parents. To eliminate that possibility, Goddard cautiously pointed to the need for a "very general practice of sterilization" that would free society of the need to supervise the remaining feebleminded and set them free to marry without fear of their progeny.[46]

Kindly reviewers found a "pleasing absence of the rant" in Goddard's treatment and approved of the "sober, dignified" manner with which his data was presented.[47] Less friendly voices critiqued the "method of pedigree investigation"—pioneered for field study by Goddard and prescribed by the ERO—and dismissed the "unit character" thesis for the heredity of feeblemindedness.[48] But Goddard's colleague Charles Davenport found it to be generally "a useful piece of work" that all students of social problems would value.[49] He included *The Kallikak Family* and *Feeble-Mindedness* in the library of the ERO as important examples of the science of eugenics and the techniques of pedigree study. Eugenic advocates in other states also looked to them as references that held the keys to identifying defective citizens, a precondition for initiating social policy.

4

Studying Sterilization

Even before the Eugenics Record Office began its work, Charles Davenport was involved with the American Breeders Association Committee on Eugenics, and he used that forum to advocate further research to advance the new field.[1] He hoped to deter the potential "volunteer army of Utopians, freelovers, muddy thinkers" and the like who had already tried to claim the "banner of Eugenics." Insisting on "hard headed, critical and practical study," he urged readers to contribute pedigree charts showing unusual familial conditions and inherited diseases.[2]

ABA President Bleecker Van Wagenen chaired the Committee on Eugenics. He met Davenport during philanthropic work at the New Jersey Village of Epileptics. The committee soon expanded to include eugenic activists H. H. Goddard and Harry Laughlin. Davenport told his colleagues that six hundred collected pedigrees provided plenty of evidence that feebleminded "defectives" should "be restrained from passing on their condition."[3] Laughlin later made the implication of Davenport's statement concrete, proposing the appointment of a committee to study the results of the "experiment in sterilization" made by the few states who had initiated legislation in that area.[4] Van Wagenen's group began compiling data on sterilization in 1911. Within three years the committee would issue the most thorough study of sterilization ever done in America, and it would form the basis for an extensive treatise on that subject by Harry Laughlin. The report would also provide the road map for states that wished to establish the constitutionality of eugenical sterilization.

Van Wagenen summarized the preliminary sterilization study in a paper presented to a meeting of the First International Congress of Eugenics in London in 1912. The group he chaired included three physicians and over a dozen notable consultants such as geneticist Raymond Pearl and anatomist Lewellys Barker, both of Johns Hopkins University; Nobel laureate surgeon Alexis Carrel of the Rockefeller Institute; Yale economist Irving Fisher; and Henry Goddard.[5]

At the meeting, Van Wagenen declared that people of "defective inheritance" should be "eliminated from the human stock." Included among the "socially unfit" were the feebleminded, paupers, criminals, epileptics, the insane, the congenitally weak, people predisposed to specific diseases, the deformed, the blind, and the deaf. U.S. Census data from previous decades demonstrated that the number of people in institutions—such as prisons, hospitals, and asylums—totaled over 630,000 and was growing as a percentage of the population. Another three million people of "inferior blood" were not yet in institutions, and seven million others—10 percent of the total population—were carriers of hereditary maladies. All told, this mass of problematic heredity was "totally unfitted to become parents of useful citizens."[6]

The remedies Van Wagenen proposed to prevent proliferation of such hereditary "defectives" ranged from the laissez-faire approach of doing nothing to segregation for life to prevent childbirth. Other options included restrictive marriage laws, euthanasia, environmental betterment, and eugenics education. Sterilization, the committee's focal point, was another possible remedy. Van Wagenen gave a short history of state sterilization laws accompanied by a brief legal commentary on the constitutional feasibility of a new law. Several case studies filled out the report, complete with pedigree charts of families related to criminals and defectives who had been sterilized in Indiana and New Jersey.[7]

Van Wagenen admitted that public support for sterilization was still undeveloped and that legal change to date had been the result of a small and "energetic group of enthusiasts" whose efforts had amounted to a "hobby."[8] He concluded by predicting that the sterilization laws in the United States would have to face legal challenges in the courts before many more compulsory operations were performed. Van Wagenen suggested that much more research was needed to determine the physical and psychological results that followed an operation. He believed that existing sterilization laws had been adopted prematurely and argued for further research, but no immediate action.[9] The fuzziness of eugenic pronouncements led commentators who heard about Van Wagenen's 1912 London presentation to remark on the number of "bright people" who "vied with each other to show the world how little is known" about sterilization.[10]

Charles Davenport also attended the London meeting, and he presented a summary of state laws that restricted marriage on a variety of grounds, including mental or physical disability. Critical of much of the legislation, Davenport declared that the "only way to prevent the reproduction of the feeble-minded is to sterilize or segregate them." Law, he said, must "take lessons from biology."[11]

In what would become one of the most public battles over the value of the new eugenic theories, David Heron of London's Galton Laboratory assailed the methods of eugenicists in the United States. He dismissed papers on mental defect given at the International Congress, condemning their "careless presentation of data, inaccurate methods of analysis, irresponsible expression of conclusions and rapid change of opinion." Heron's assault focused on Charles Davenport's misuse of Mendelian theory, the "unsatisfactory manner" in which data at the ERO had been collected, and the "slipshod fashion" in which it was presented.[12]

The *New York Times* printed Davenport's detailed defense with a headline announcing an "English Attack on *Our* Eugenics."[13] The debate shifted to the pages of *Science,* which quoted an indignant Davenport condemning the "stupid, captious and misleading" comments and "delusions" of a European counterpart who dared to question the scientific bona fides of the U.S. movement as it gathered public attention and approval.[14] Heron was particularly pointed in highlighting Davenport's seeming hypocrisy in supporting and then opposing sterilization.[15] Changing course, Davenport insisted that his position on legal change was conservative and admitted that "we know very little of the laws governing heredity and feeblemindedness."[16] Yet Davenport had endorsed the idea of a medical certificate as a marriage requirement and called for a eugenics board in each state, official physicians to issue marriage licenses, and eugenic field workers to investigate the family background of all license applicants. In Davenport's scheme, citizens who had children after a marriage license was denied would be penalized with sterilization.[17] His positions on sterilization changed often; within a year he was again declaring that the institution at Cold Spring Harbor would "push to the limit" its program in favor of sterilization, devoting "gigantic efforts" toward the enactment of sterilization laws nationwide.[18]

With Van Wagenen and Davenport in London, Harry Laughlin kept busy back at Cold Spring Harbor training a new cohort of field workers to do family studies.[19] His attention to sterilization was much more than a hobby. He believed that the data collected by field workers would enable states to decrease the number of "social misfits" rather than build more institutions for them.[20] But before that data could be used, Laughlin exhorted members of the public to "get right on Eugenics."[21]

Public education in eugenics had many supporters, but the law was the avenue of choice for the social changes the eugenics group desired, and marriage laws were not a new idea. Early proposals to regulate marriage on "eugenic" grounds combined attacks on crime, venereal disease, and problematic heredity. Connecticut passed a law in 1895 prohibiting marriage of any "epileptic, or imbecile, or feeble-minded" people, focusing specifically on unions where the woman was under forty-five years of age, which ruled out most chances of childbirth.[22] Public health advocates presented a bill to Congress in 1897 that would enable the collection of statistics about marriages that could "produce physically and mentally defective offspring; and any information leading to race improvement through better marriage selection."[23] For the first forty years of the twentieth century, the highest officials in the U.S. Public Health Service were among the most enthusiastic supporters of the claim that every social ill—crime, poverty, syphilis, mental disorder—could be cured by "genetic" interventions.[24] They argued that eugenics was "fact, not fad."[25] "Medical supervision" of marriage was commonly suggested to check the "indiscriminate marriage of those who have hereditary illness."[26] In 1915 the movement to regulate marriage received a boost when U.S. Public Health Service Surgeon General Rupert Blue awarded a eugenic certificate for marriage to a Washington, D.C., couple.[27]

But despite the insistence of public health physicians, it was often difficult to separate the medical significance of eugenics as a science that focused on hereditary traits and the various popular understandings of eugenics as a general term employed to signal many different concerns about good health. In an era that witnessed major efforts to clean up squalid cities and eradicate the "social evil," as prostitution was known, eugenics was also often linked to diseases that could be passed from parent to child via congenital infection—including some sexually transmitted diseases. Although what was then known as "venereal disease" was not inherited in a genetic sense, it certainly had an impact on the health of children and was clearly connected to conditions parents carried. To combat that scourge, "society women" who had "entered the fight against vice" pushed for the establishment of the National Society for the Promotion of Practical Eugenics; their leaders included women like Mrs. Woodrow Wilson and Mrs. William Jennings Bryan.[28]

To many, *eugenics* became a catchall term that could be used to justify new kinds of laws; as a result, laws requiring testing for syphilis or other conditions that affected fertility were routinely labeled "eugenic marriage laws." They were written in the same spirit as enactments against spitting in public, which were designed to combat infectious diseases like tuberculosis. By 1911, one hundred and fifty cities and at least three states had passed anti-spitting ordinances.[29] "Eugenic" laws were enacted as the natural successors to the earlier efforts at sanitation.

Since these laws were so clearly linked to sexual morality, they drew support from organized religion. In 1912, Walter Taylor Sumner, dean of the Protestant Episcopal Cathedral of Sts. Peter and Paul in Chicago announced that he would perform no marriages there until the parties to the union obtained a "clean bill of health." The document must certify that they are "normally physically and mentally, and have neither an incurable nor communicable disease." Sumner's supporters applauded the decision, which would allow "the Church do what it can to protect the innocent and clean from the guilty and diseased."[30]

Rev. John Haynes Holmes of New York's Church of the Messiah endorsed the measure to remove "from the stream of life the poisonous elements of physical and psychological decay." Echoing public health arguments, he declared that governmental power was adequate to the task, just as it was to quarantine a "victim of smallpox or yellow fever."[31] "Mawkish sentiment must unquestionably yield to the high issues involved in eugenics," chimed in Rev. George C. Peck, a Methodist pastor.[32] Though some questioned the Sumner plan, before long clergymen began to band together to push the "eugenic certificate" idea.[33] Heeding the advice of religious advisors, some couples rose to meet the challenge.[34]

Not every man of the cloth embraced eugenics. An Episcopal pastor in Charlottesville, Virginia, ventured up to the gates of Jefferson's home at Monticello, where in a grand gesture of protest he burned his robe and his prayer book. He denounced his church for turning its back on the poor and "wasting time on eugenics."[35] His coreligionists dismissed the comments, defending the value of "the lessons of eugenics," qualifying the critique as "so foolish and wild as not to deserve attention."[36] But the clerical plan to license marriages was soon overshadowed by schemes for more direct interventions to interrupt the parental hopes of society's least fit. Chicago doctor G. Frank Lydston succinctly summarized one strain of public opinion when he said that the "Court of Appeals to which adverse certificates of matrimonial qualifications should be referred is the surgeon's knife."[37]

The work of the Eugenics Record Office seemed organized to marshal support for sterilization. Among its goals was study of the "best methods of restricting the strains that produce the defective and delinquent classes of the community."[38] Pennsylvania asked marriage license applicants about their mental, physical and financial conditions.[39] Soon, the Wisconsin legislature linked its passage of marriage laws to a sterilization law.[40] As Americans began to echo the language of eugenic improvement, Charles Davenport applauded the "vigor of the popular reaction to the eugenical ideal."[41]

The First National Conference on Race Betterment was organized and hosted by Dr. J. H. Kellogg, at Battle Creek, Michigan. That 1914 meeting provided Harry Laughlin with an audience eager for more details on the ABA sterilization study. The paper Laughlin presented included a plan to eliminate "the great mass of defectiveness . . . menacing our national efficiency and happiness." His calculations assumed that the lowest 10 percent of "human stock" was so poorly prepared for civilization that its survival represented "a social menace." His charts showed the extraordinary growth in institutions for the "socially inadequate" as compared to the general population. By Laughlin's calculations, a systematic program to purify the human "breeding stock" would require fifteen million sterilizations over approximately sixty-five years.[42]

Laughlin's statistics were illustrated with pedigrees showing the source of various types of "defectives." In one chart, hereditary epilepsy plagued a "Poorhouse Type" clan; a second showed alcoholism, criminality, and feeblemindedness typical of a "Hovel Type" family.[43] The "Poorhouse" pedigree illustrated the problematic "increase of defective children by defective women." Using a crude but homely analogy, Laughlin argued for selective sterilization of women:

> As a rule the tax on the female dog is two or three times greater than that on a male dog. Such difference in taxation is not made because of a difference in individual menace, but rather because of a more *direct* responsibility for repro-

duction. The females of such homeless strains are not protected, and consequently they increase very rapidly. Consorting freely with equally worthless mates, their progeny are often excessive in numbers, and of a worthless, mongrel sort. The castration of one-half of the mongrel male dogs would not effect a substantial reduction in the number of mongrel pups born.

The unprotected females of the socially unfit classes bear, in human society, a place comparable to that of the females of mongrel strains of domestic animals.[44]

Laughlin's projected sterilization program would start in institutions, run for about ten years, and then be extended to the population at large. A conservative phase-in of the law would allow each state to compile an inventory of "cacogenic," or ill-born, "human stock" and train eugenical experts, paving the way for enhanced public sentiment in support of eugenics. Laughlin anticipated that surgery limited to inmates of institutions might raise legal issues and could violate the constitutional assurance of equal protection. Should the U.S. Supreme Court conclude that a sterilization program confined to those living in institutions was discriminatory—this had, after all, been the conclusion of the New Jersey Supreme Court—the program would have to be widened to the general population earlier. Doubting that such a result was likely, Laughlin hoped instead to launch a more cautious program based on thorough eugenical analysis of families and careful identification of those who made up the "worthless one tenth" of the people.[45]

News coverage of the Battle Creek conference was extensive. New Yorkers saw Laughlin's prediction that sterilization was "humane" and would eventually be widespread, and Chicagoans read about his talk supporting sterilization alongside Dr. Kellogg's advocacy of "pedigree marriages" and a "registry of human thoroughbreds." Readers in Cincinnati learned about eugenics in reportage on hookworms and the perfectly proportioned woman. Detroit residents saw Laughlin's proposal for "preventing the reproduction of degenerates" as part of the concern for a bleak future of famine and bankruptcy, even a "babyless age." Some papers also gave space to detractors of sterilization and covered the controversial side of eugenics. They found Laughlin's predictions an unlikely eventuality, writing that the public would oppose any comprehensive sterilization plan.[46]

Back at Cold Spring Harbor, Charles Davenport protested the public ridicule the sterilization proposal had engendered. When the *New York*

American highlighted Laughlin's plan to sterilize some fifteen million people, Davenport claimed that he was misquoted in the article and declared that not only had the ERO never launched a sterilization campaign, it was in fact opposed to such programs. He demanded an apology for those, like himself, who had been "dragged into association" with something he "had never heard of."[47]

Davenport's protest was disingenuous at best. As early as 1910 he had defended sterilization, writing that "where the life of the state is threatened extreme measures may and must be taken." He called for legislation to authorize segregation or sterilization of "idiots, low imbeciles, incurable and dangerous criminals." At one point Davenport declared that he actually preferred castration to sterilization by vasectomy, because it ruled out sexual relations altogether.[48]

Davenport's posturing must have appeared particularly hollow to those who were familiar with *Eugenics Record Office Bulletin No. 10,* published in February 1914. The bulletin, which carried the long-winded title "Report of the Committee to Study and to Report on the Best Practical Means of Cutting Off the Defective Germ-Plasm in the American Population," listed Laughlin as its author. It provided extraordinary detail on the program Laughlin had outlined for his Battle Creek audience only a month earlier and incorporated all the pedigrees and statistical charts Van Wagenen had used for his London presentation in 1912.

By far the largest of all bulletins published by the ERO, Laughlin's Report was 215 pages long. Its first part described "The Scope of the Committee's Work," which covered such fields as medicine, physiology, biology, and anthropology but also included a "woman's viewpoint." The sterilization committee even staged what may have been the first bioethics consultation, asking experts to review the "ethical, moral and ontological aspects of sterilization" as well as the limitations on marriage adopted by the clergy for eugenic purposes.

The substantive portion of the report began with an analysis of the "phenomenon of heredity" and its role in increasing the numbers of "socially inadequate" people in America. Thirty pages were devoted to classification of the "cacogenic" class and enumerating the ill-bred varieties of the human population. As of the 1910 census, more than 840,000 people were living in U.S. institutions for the "anti-social and the unfortunate classes." Many more had left those same institutions and were living with similar people who needed institutional care but had never been taken into custody. Other normal parents were also part of the community, but even they had the potential to "produce defective offspring." Taken together, said Laughlin, this "great mass of humanity is not only a social menace to the present generation, but it harbors the potential parenthood of the social misfits of our future generations." The defective traits common to these "socially inadequate" groups were inborn and must be cut off. "This is the natural outcome of an awakened social conscience; it is in keeping not only with humanitarianism, but with law and order, and national efficiency."[49]

Several remedies were available to address the growing menace of the socially inadequate. Restrictive marriage laws and positive eugenical education were deemed useful but ultimately ineffective in achieving long-term changes in population quality. Euthanasia and polygamy—though effective—would exact "too dear a moral price." The nation's "blood will deteriorate" under a laissez-faire policy in a society whose charitable institutions subsidized the lives of the deficient. The committee's solution was to link long-term institutional segregation to sterilization, using the surgical option only in cases where, through appropriate legal procedures, a person could be declared socially inadequate and taken into state custody.[50]

The committee's conclusions stressed a broad program of education, segregation, marriage prohibition, and selective sterilization that would "largely but not entirely eliminate from the race the source of supply of the great anti-social human varieties" within two generations.[51] First, there should be early identification and commitment of the socially inadequate in order to segregate them and prevent their reproduction. All institutionalized persons supported by public funds were to be examined and their family backgrounds investigated. Those "found to be potential parents with undesirable hereditary potentialities and not likely to be governed by the highest moral purpose" would be sterilized.

Laughlin calculated that his institutional sterilization program would eventually require approximately 150 operations per year for every 100,000 people in the general population.

Laughlin realized that support—both legal and financial—would depend on popular consensus on eugenic goals. Unless eugenics could become part of the "American civic religion" and the citizenry develop a "quickened eugenics conscience," the program would fail.[52] But he made it clear that those who objected to this radical solution must face the economic consequences: paying the cost of institutions to segregate inmates during their entire reproductive lives.

The Model Law

The second volume of the committee's report was a plan for the "ultimate extinction of the anti-social strains" in the population. At 150 pages, it was the longer portion of the work of Van Wagenen's committee. Inadvertently undercutting Davenport's repeated attempts to distance himself from sterilization laws, Laughlin thanked the ERO director in the report for his "frequent consultation and valuable cooperation." "Legal Legislative and Administrative Aspects of Sterilization" was entirely devoted to sterilization law, "the most promising agency for reducing the supply of defectives."[53]

Laughlin began with a description of existing sterilization laws and analyzed their legislative histories. He listed bills for sterilization that had been rejected by legislatures and others that had been vetoed by governors. He summarized the potentially problematic features of all existing laws and concluded with his own solution to these problems: A Model Sterilization Law. The objective of that law was spelled out in its preamble: "An Act to prevent the procreation of feebleminded, insane, epileptic, inebriate, criminalistic and other degenerate persons by authorizing and providing by due process of law for the sterilization of persons with inferior hereditary potentialities, maintained wholly or in part by public expense."[54]

The text of the Model Law proposed the creation, in every state, of a eugenics commission to examine both institutionalized people and the population at large. Such a commission would determine, through its eugenic expertise, which people "because of the inheritance of inferior or anti-social traits, would probably become a social menace, or a ward of the state." It would then be within the power of the commission to order the sterilization of any patients in the problem group. Existing expertise in eugenics, said Laughlin, was "so advanced" that the "hereditary potentialities" of any person could be predicted. Eugenical field study would allow the state to meet its legal burden of proving the "potential parenthood of defectives" before surgery could occur.[55]

Not everyone shared Laughlin's confidence in the progress of eugenics. Even Laughlin's mentor, Princeton biologist Edward Grant Conklin, urged caution. Though he did not desire to disparage eugenics, "that infant industry," Conklin was still wary of overemphasizing inheritance.[56] Conklin's concerns were described by an editorial writer at the *Journal of the American Medical Association* as an aversion to "the extreme modern doctrine of eugenics," with its "dreary contentment with an alleged inevitable inheritance."[57]

Some scientists who believed in a more optimistic brand of eugenics raised perplexing arguments against sterilization. Committed Darwinians who believed that the most fit members of a species would survive could not at the same time hold that the least fit would eventually pose a threat to that survival. Others realized that even if all recognized "defectives" were sterilized, the result would probably be insignificant. Using the mathematical formulation known as the Hardy-Weinberg principle, geneticists quickly determined that carriers of latent defective traits that had not manifested in the current generation would eventually surface in future progeny. Even a massive program like the one Laughlin's report proposed would have relatively insignificant impact on the incidence of undesirable traits.[58]

A study at the University of Washington found that much of the legislation "intended to prevent the breeding of a degenerate race" was based on biological facts that were obsolete and "worse than useless."[59] The State Charities Commission of Illinois condemned the popularization of eugenics and pushed for "clearly demonstrated facts" and "quickened public conscience" rather than an "attempt to impose by law upon the public poorly understood or wholly misunderstood principles of human conduct and life."[60] A review of the new Pennsylvania marriage law found that only one marriage license had been refused in the first eight months the law had been in effect. Apparently, applicants simply lied when asked if they were possessed of any of the listed infirmities, such as financial "ability to support a family." The legislation was criticized as ineffective, impractical—"a failure."[61]

The debate over the potential legal applications of eugenics intensified with the publication of Davenport's book *The Feebly Inhibited.* Comprised of studies done by eugenic field workers, the book linked social problems to inherited temperament and supplied a companion term for *feeblemindedness.* Said Davenport, "The chief problem in administering society is that of disordered conduct, conduct is controlled by the emotions, and the quality of the emotions is strongly tinged by the hereditary constitution." Human reactions, "whether violent or repressed," were determined by heredity. The book was full of the usual pedigrees showing families with hereditary "peculiarity," epilepsy, or a "shiftless" instability.[62] Davenport continued to assert that moral traits were hereditary, along with physical and mental traits. And despite denying that he was an advocate for sterilization, he pushed for state eugenic boards to control reproduction and sterilize those who had had children in defiance of regulators.[63]

Reviewers of *The Feebly Inhibited* found flaws with Davenport's methods of collecting and interpreting data and noted that the book had omitted important information necessary to understand the working of "complex human traits." Observation might lead a researcher to believe in inherited temperament, but Davenport's work failed to prove "the mechanisms whereby this inheritance takes place."[64] Samuel Kohs, a psychologist at the Illinois House of Corrections, sounded a warning to "some of our over-confident friends who are advocating a eugenic program based on slippery and insecure foundations."[65]

Legal Critique

Debate over the growing literature on sterilization was common in the legal world, and critics of the procedure raised pointed objections. Many felt that any measures to allow state-compelled sterilization were likely to be arbitrary because of the complexity of heredity and the novelty of theories describing its workings.[66] Choosing clear diagnostic criteria for selecting surgical cases was also difficult, and critics feared they could not "easily be formulated in the words of a statute."[67] And because Laughlin lumped the "criminalistic" into the list of "socially inadequates" liable for sterilization, attention was often focused on how the law would be applied to people convicted of crimes.[68]

As Laughlin was preparing his Model Law for publication, one of the most extensive criticisms of the sterilization laws already in force was published by New York attorney Charles A. Boston, who would later become president of the American Bar Association. Boston took particular offense at the law enacted by the "legislative wiseacres of Indiana" allowing doctors to operate "after forming a $3.00 opinion." Boston fixed his aim on the "probably ignorant, if not to say malicious, wardens and superintendents" who would carry out the law. He found the legislation "premature" and the theory upon which it was based "possibly unsound." It was of "doubtful utility," because if used conservatively, it would miss cases and sterilize too few; if used liberally, it would penalize too many unnecessarily. In hard times, less powerful people would suffer the "penalty for bad economic conditions." Boston called the New Jersey law a "flirtation with danger." His summary reminded the reader that "prisons and insane asylums have been the most shameful institutions of so-called Christian civilization."[69]

Though some commentators were sure that proposals for sterilization had passed the "academic stage" and could no longer be considered the radical idiosyncrasy of "some 'crank' warden, doctor or alienist," other lawyers seconded the Boston protest.[70]

Two physicians reviewed a thousand cases, seeking evidence of "direct inheritance of criminalistic traits." They launched a very penetrating medical critique that focused on the ERO and the pedigree study method. They could find no case where "anti-social" tendencies appeared in succeeding generations without concurrent "striking environmental faults" or other physical and mental problems that might play a part in aberrant behavior. They dismissed the use of family charts in the absence of thorough developmental histories and called the idea of hereditarily transmitted criminal traits an "unsubstantiated metaphysical hypothesis."[71]

All this controversy over sterilization pointed out the need for a systematic, serious study. A committee of the American Institute of Criminal Law and Criminology (AICLC) was formed to analyze proposals for the sterilization of criminals. The committee included a probation officer, a judge, a social worker, a theologian, and a member of the clergy. Representatives of the pro-sterilization forces included the American Breeders Association's Bleecker Van Wagenen, Indiana sterilization pioneer Dr. Harry Sharp, and the ERO's Harry Laughlin. The committee's first report provided readers with a primer on the substance of the sterilization debate, with comments by supporters as well as critics of the procedure. It also included the text of Laughlin's Model Law.[72]

There was unanimous agreement that while anecdotes were plentiful, no rigorous research had been done to determine the medical advantages or disadvantages of sterilizing operations to those who might endure them. Data on the value of surgery to those who faced it was crucial, because if there was no therapeutic benefit, it would be hard to claim that sterilization was anything but punitive. This was a serious problem for sterilization advocates; it had already resulted in the invalidation of some state laws aimed at criminals as examples of unconstitutional "cruel and unusual punishment."

An AICLC report was published in the *American Bar Association Journal* in early 1916. It again decried the absence of "any careful study of individual cases" that would verify claims about the therapeutic value of surgery. The legal situation was unsatisfactory to both sides of the debate. Fifteen states had adopted sterilization laws, but ten were not in use, because they had either been invalidated in court or simply fallen out of favor with the doctors who could have used them. In light of these findings, the committee recommended more research by proper experts and called for a moratorium on laws authorizing sterilization of criminals unless they were supported by scientific evidence.[73]

A stinging dissent from Dr. William Belfield of the Chicago Medical Society claimed that the "dominant sentiment" of students of heredity was at variance with the report. He argued that a plan to "check the ominous flooding of the nation with irresponsibles" that had in only six years been adopted in twelve states and endorsed by two state supreme courts was worthy of "attentive consideration and virile treatment" by the AICLC.[74]

The committee's final report declared that consensus could not be reached; no further work would be fruitful. Pointing to the "general reluctance" of officials in California and Indiana to apply the criminal sterilization laws, some members stated flatly that experts do not believe in "inherited criminality as a trait."[75] Harry Laughlin took the opportunity to repeat the position of the Eugenics Record Office: While "criminality as a unit trait is not inherited . . . certain factors that go to make anti-social individuals are inherited." Among those factors were "wanderlust, specific types of feeble-mindedness, lack of sex control and the lack of other moral inhibitions." Dr. Belfield dissented again. Likening the group to a "hung jury," he proposed that they be replaced by "men whose views on other topics do not incapacitate them for the study of a problem in public welfare."[76]

By then the debate on sterilization had moved to other quarters. One medical journal editor was effusive about the new popularity of the field of eugenics and the value of eugenic surgery, noting that "eugenics, the youngest of the sciences, has become the giant of them all. Today the practical wisdom of the stock-breeder is sung by many tongues. . . . Eugenic books are crowding out libraries; eugenic plays are filling our theatres." Those who favored sterilization included prominent physician G. Frank Lydston, popular author Jack London, and Anthony Comstock, the self-proclaimed "terror to evildoers" who headed the Society for the Suppression of Vice. The governors of Utah, Texas, Washington, and Wisconsin agreed. In contrast, Governor Edward Dunne of Illinois said, "I am emphatically opposed to sterilization of criminals and defectives by statute. I regard it as barbarous and inhuman, and attendant with the dangers of rank injustice."[77]

Disagreements over sterilization also led to rifts within the eugenics movement. The radical nature of Laughlin's proposals led Alexander Graham Bell—one of the founding scientific directors of the ERO—to complain that too much effort had been spent on the negative side of eugenics. The public impression of the field was so colored that the very word *eugenics* suggested to most people "hereditary diseases and objectionable abnormalities." Many also associated it with "an attempt to interfere, by compulsory means, with the marriages of the defective and undesirable," said Bell.[78] Disgusted with propaganda excesses, Bell would eventually leave the eugenics movement and disassociate himself from the ERO.

Geneticists also began to question eugenic policy. William Bateson, who had been among the first to rediscover Mendel's work, condemned the "violent methods" adopted in some states to check reproduction, declaring that no fact "yet ascertained by genetic science justifies such a course."[79] Bateson's comments reverberated as others too began to see "legislative tyranny and its handmaiden, brutality," at work "under the guise of applied eugenics."[80] He warned against a population of "one uniform Puritan gray," should the eugenicists get their way.[81]

Thomas Hunt Morgan, whose work on the genetics of the fruit fly would earn him the first Nobel Prize for work in genetics, disagreed as early as 1915 with the "reckless statements and unreliability" of reports made by the Van Wagenen committee. Morgan criticized both the pedigree study method and the eugenicists' conclusions on the origins of mental differences. Morgan's dissatisfaction with the shoddy science of many eugenicists led him, like Bell, to an early break with the movement.[82]

Neither public censure nor private controversy deterred Laughlin, and though the press regularly printed critiques of so-called eugenic marriage laws, disagreements among scientists remained muffled, generally hidden in the pages of obscure journals.[83] The public heard other voices, such as baseball-star-turned-evangelist Billy Sunday regaling his flock with life lessons gleaned from eugenic texts. In one sermon heard by thirty-five thousand men, Sunday described the scandalous lives of the hapless Jukes and the power of "one God-forsaken, vicious, corrupt man and woman to breed and propagate and damn the world by their offspring."[84] The *New York Times* commented that Sunday spent so much time on the influences of heredity that talk of science "almost overshadowed the denunciations of sin."[85] Laughlin's message also resonated in states like Virginia, where sterilization had been a subject of legislative debate for some time.

The Mallory Case

Linking public health concerns to eugenics was commonplace in Virginia, where lawmakers were quick to grasp the power of arguments using the public's health as a justification for restrictive legislation. Early attention to the dangers of transmitting diseases prompted attempts to regulate all manner of intimacy. A 1902 bill considered by the Virginia General Assembly would have ruled out kissing except between those who could prove by a doctor's testimony that they were free of disease.[1] Another measure threatened to prohibit the kissing of Bibles during the administration of oaths.[2] As in most states, fear of tuberculosis led to anti-spitting laws.[3]

In Virginia, laws prescribing surgical sterilization had initially failed. In 1907, a doctor called for state-mandated surgery to address fears of interracial crime in the Old Dominion. He advocated castration for black men who assaulted white women, adding as part of his prescription the requirement to "cut off both ears close to the head."[4]

Race was not the issue for prison surgeon Charles Carrington, when, mixing punishment with therapeutics, he began operating on inmates in 1902. He waited until 1908 to reveal his practice publicly to assembled prison doctors and then, pointing to inheritance as the key to crime, urged legalization of "the one sure cure."[5] He exhorted Virginia doctors to prevent the breeding of rapists, murderers, burglars, arsonists, and trainwreckers. Carrington recounted his own experience with "curing" with sterilization the "wildest, fiercest, most consistent masturbator" he had ever seen. Pointing to the experience in other states, he predicted that a sterilization law "is bound to come in Virginia."[6] While Carrington described sterilization as "further punishment," others were quick to note that the prison doctor's proposal was not punitive but eugenic—"to stop the breed."[7] Not everyone concurred with this proposal; the president of the American Bar Association called it "offensive because of its barbarism."[8]

With the support of his state's medical journal, Carrington campaigned for a law in 1910 modeled after Indiana's, with a similarly broad mandate for surgery.[9] Ignoring criticisms of the Indiana law, the Carrington proposal would have allowed any state institution to perform surgery on "a confirmed criminal, a rapist, an idiot or imbecile" for the prevention of procreation and to choose the procedure that was "safest and most effective."[10] The bill was passed by Virginia's Senate but failed in the other legislative chamber, the House of Delegates. Observers blamed "blind sentiment" and "strong prejudice" for the legislative defeat. The bill apparently seemed too radical to the "lay mind" of legislators, who were unable to distinguish the eugenic effects of coercive surgery from punitive motives.[11]

The crusade for sterilization gained traction in Virginia with doctors repeating the national cry for laws to combat the hereditary diseases of alcoholism, syphilis, feeblemindedness, and immorality.[12] Some insisted that "feeble-minded women are notoriously immoral" and filled the almshouses like a "horde of parasites." They echoed the call for sterilization to prevent state bankruptcy.[13] The drumbeat increased with other doctors insisting that the maxim "blood will tell" was understood nowhere better than Virginia. A proposal was made for establishment of a "eugenics bureau" to regulate marriages as part of the state health department.[14] Carrington, awed at "what a fearful thing heredity is," gave talks on "Eugenic Marriages."[15] In 1913 Virginia clergy endorsed a plan requiring a health certificate of bridegrooms as a condition of marriage.[16]

The following year, Virginia Governor William Hodges Mann highlighted his own support of a "eugenic marriage act" in the annual message to the General Assembly.[17] Like the clergy, he limited the scope of the law to potential grooms as a means of preventing syphilis, the "terrible disease which stamps its horrid curse upon the innocent offspring unto the third generation."[18]

The inaugural volume of the *University of Virginia Law Review* echoed the chorus of professionals, with one lawyer asking wearily how the "blessings of liberty, or full domestic tranquility" could be enjoyed if persons "civilly unfit" were permitted to "procreate their species and scatter their kind" among normal citizens. His solution: the legislature should enact scientifically based sterilization laws. Repeating the well-worn story of the Jukes family, the author concluded that he must "proclaim to the world the story of sterilization."[19]

Albert Priddy was well aware of the changing sentiment toward sterilization since he had taken over the Virginia Colony for Epileptics and Feeble-minded. His first official call for sterilization had coincided with Carrington's initiative. He asked lawmakers to consider "the application of legalized eugenics" in 1911. They should restrict marriages of the insane, the feebleminded, and confirmed alcoholics and, following the lead of states like Indiana, review the "practicability of a law permitting the sterilization of inmates" in asylums and prisons. Two years later his advocacy seemed tempered by the critics of sterilization, and he admitted that the process could "never become a general procedure on account of the many objections" that had been raised against it. His reservations about an extensive program, however, did not prevent him from later endorsing sterilization on feebleminded women who could earn their living outside of an institution and thus save the state tax money.[20]

Priddy was in touch with experts on mental deficiency from all over the country. Guests from New York toured the Colony grounds and studied the methods there. Priddy told them about the woman he had hired to compile family histories of Colony inmates and "investigate groups of prostitutes."[21] He also traveled to California and Indiana to stay abreast of legal developments in his field and to consult with colleagues.[22] As a result of his expansive research, by 1916 Virginia's statutes shared features of California's law, which allowed surgery designed to benefit the "physical, mental or moral" condition of inmates.[23] Virginia law also gave institutional physicians wide discretion. It reflected the medical prerogatives suggested first in Pennsylvania and successfully built into the early laws of Indiana and New Jersey, where doctors had been allowed to perform any operations that were "safe and effective." Virginia's 1916 legislation governing Colony procedures was consistent with earlier Virginia legal provisions admitting "women of child-bearing age" to the Colony to prevent them from reproducing. Before patients were discharged, a simple operation would insure that society was "protected" from their potentially degenerate progeny.

Priddy had experience operating on patients scheduled for parole to local families, and sterility was sometimes noted as an incidental result of surgeries he had performed to relieve "chronic pelvic disorder."[24] Though Virginia's 1916 enactment did not mention sterilization, Priddy believed that it allowed him to perform operations at will. He petitioned the Colony's Board

of Directors to operate as soon as the law went into effect, and they authorized salpingectomy for eight women Priddy had chosen.[25] Priddy eventually confirmed that he interpreted "a logical and plain construction" of one
provision of the new law to justify his "therapeutic" prerogative. By the end
of the year, Priddy had sterilized "twenty young women of the moron type."
He explained that in most cases some "pelvic disease" was found that made
an operation for "removal of the tubes necessary" to relieve pain.[26]

The records Priddy left behind reveal the goals and scope of his sterilization program. Women committed to the Colony and destined for sterilization were invariably described as "immoral." Evidence for that trait might
include symptoms as vague as "fondness for men," time spent in a "sporting
house," or a reputation for "promiscuity." Some were accused of a tendency
to be "over-sexed" or "man-crazy." Formal diagnoses sometimes indicated
"nymphomania" or "sexual degeneracy" or positive tests for sexually transmitted diseases. Most were also diagnosed as feebleminded. They were often homeless women who fled their families following physical or sexual
abuse.[27] The parents of inmates scheduled for sterilization frequently had
trouble with alcohol, and they were characterized as "insane," "defective,"
"weak minded," "incorrigible," "wayward," "illegitimate," "untruthful," or
criminal. All of these traits were thought to be hereditary. It was not unusual for young women to arrive at the Colony in the throes of pregnancy;
some blamed their condition on older adults, in one case even implicating
the family physician.[28] The babies born at the Colony were adopted by the
mothers' relatives or by Priddy's friends; as a last resort they were sent to
foundling homes.

One sixteen-year-old girl known for her "general backwardness" came
to Priddy from an orphan home. She was considered "entirely unsafe" because of her desire to be with boys. She had the troubling habit of writing
notes and would regularly "steal opportunities of talking to the little boys"
in the orphanage.[29] Declaring that the legislature opened the Department
for the Feebleminded for "just such cases" as this, Priddy chose her for
sterilization.[30]

Priddy also decided to operate on some married women. One suffered
from "wanderlust," was given to telling "coarse stories," and showed signs
of "homosexuality."[31] Another arrived at the Colony in search of a cure for
her seizures. She asked the doctor to release her into her husband's care, but

Priddy refused, stating that "a crop of defectives" would result if they lived together. The husband eventually agreed to her sterilization, on the condition that she not be told of the operation's purpose.[32]

Evidence of "hereditary" problems for Colony inmates was often weak, and hiding the true nature of the surgery from them was common. One woman, though from a "fearfully degenerate" family, was the mother of a child who was found to be "unusually bright."[33] After her release, she wrote Priddy for permission to marry; she was not aware that she had been sterilized.[34] Another patient married following her discharge from the Colony and consulted a doctor when she was unable to have children. Though he described her as "mentally bright," Priddy had not made it clear to her that she was now sterile.[35]

A diagnosis of "feeblemindedness" was an optional qualification for sterilization. In one case, Priddy did not consider a woman feebleminded "in any way," but her admission papers suggested that her mother may have been "immoral," and letters the girl had written to men at the age of thirteen scandalized a probation officer. This was all the reason Priddy needed to operate. He wrote to a relative desiring her discharge: "For the protection of defective girls who leave the institution and to prevent harmful heredity we have sterilized a great many of them." He later gave the family a choice whether the operation would occur but was careful to relay his own convictions clearly: "I am so fixed in my belief as to heredity that I believe waywardness is inherited just as much as mental defectiveness which is revealed by mental examination, and with the sad history of her mother's career which I have had, I do not think [she] should be allowed to run the risk either legitimately or illegitimately of bearing offspring."[36] On another occasion, Priddy clearly stated that a questionable IQ test and a "record of immorality" was sufficient to declare a woman feebleminded within the meaning of the law.[37]

Priddy left no doubt about his goals at the Colony. He regularly described patients such as the "good looking girl . . . of the higher defective grade known as a moron." Such girls, he said, were "sexually very passionate" and weak "when exposed to temptation." These characteristics explained why so many of this type were prostitutes, said Priddy, adding that "at least one half of all children born of such women are feebleminded, epileptic, insane or criminals." The Virginia Colony for Epileptics and Feeble-minded was founded for the care of these women until their ability to have children had passed. Priddy had "a duty to society" to prevent such a girl from being set free "to breed defectives."[38]

The double standard could not be clearer: the sexual activities of women were a cause for commitment, while similar behavior in men prompted no legal response whatsoever. In one case, a girl reported to Priddy by a police officer was quickly committed to the Colony because she posed a "danger to society" for transmitting syphilis. The girl was apparently considered more of a danger than the disease, since the officer's son and other boys who had contracted syphilis from her were allowed to remain in the community.[39]

This gender discrepancy notwithstanding, Priddy's male patients were not immune from the threat of surgery. Included among the earliest surgical cases were "four males showing vicious and dangerous tendencies."[40] The first boy sterilized at the Colony was admitted in July 1913, suffering from epilepsy thought to be caused by a fall from a tree.[41] Though he was judged to be of average intelligence, in 1916 the Colony Board still voted to sterilize him.[42] Another boy had various injuries to his head before the age of ten and would endure several operations at the Colony. Before his vasectomy, he had his tonsils and adenoids removed and his appendix taken out; he was also circumcised (apparently as a treatment for epilepsy), though "with only temporary benefit." He eventually escaped, but when Priddy learned that he had joined the Army, his records were marked "discharged—improved."[43]

Perhaps the most dramatic case involved a twenty-four-year-old man who had been jailed for theft, was of limited intelligence, and had other behavioral problems. Like most of Priddy's patients, his major difficulty involved "moral delinquency" and "excessive sexual indulgence." He escaped from the Colony several times during his first year there. On one such "elopement," he was accused of attempting to assault a twelve-year-old girl. Following that episode, Priddy reminded the man's father—a Baptist minister—that the law allowed him to sterilize patients if they were "dangerous to the public." He had already received the approval of the Board to sterilize this inmate and another one "by the complete method" because their habits of escape made them "especially dangerous to the safety of women if allowed to go at large." The father of one of the men knew Priddy, and he consented to the "double castration" for his son.[44]

On one occasion, the Board voted to sterilize a boy who was only nine years old. Priddy initially rejected him for admission after reviewing his commitment papers, judging it "quite clear" that he was an "idiot" and not suitable for the Colony since he could not count to ten. Nevertheless, he was admitted and his mental "deficiencies" were attributed to his mother, "a drunken prostitute . . . probably illegitimate." Three years later, he was reevaluated as a "very bad boy . . . but does not seem feebleminded." After his discharge, his foster parents judged him a "very bright boy" and sent him to a private academy, where he "gets a star every day." At age fourteen, he scored 96 percent in his schoolwork. Priddy admitted that despite being committed as feebleminded, he was "practically normal."[45]

By spring of 1917, the Colony Board had approved at least fifty sterilizations for Colony inmates. A few, however, avoided surgery. One boy had suffered from spinal meningitis, was blind in one eye, had a cleft palate, and was generally described as "physically defective." His mother declared, "He is not crazy," and threatened to sue the committing official, Sarah Roller of the Richmond Juvenile Court. Priddy grudgingly agreed that the boy was neither "feebleminded" nor "defective." Although his mother was a "pauper," the boy had tested one year ahead of his natural age and was "precocious." He was discharged after twenty months, his potential for parenthood intact.[46]

Two other Colony inmates who had been chosen for sterilization also came to Priddy via Sarah Roller. They did not escape surgery, and their family's legal objections would bring great public embarrassment to Priddy for stepping too hastily through the "therapeutic" loophole he invoked to justify his operations.

Mallory v. Priddy

A network of informants sent patients to Priddy at the Colony. The group included law enforcement officials, court employees, and bureaucrats who monitored aberrant behavior and administered state laws in Virginia's communities. No referral source was more helpful to Priddy than Sarah Roller, a probation officer at the Richmond Juvenile and Domestic Relations Court. After the Colony expanded to accommodate the "feebleminded," Roller wrote to him regularly, alerting him to problem girls who might become future Colony inmates. She often appeared as the petitioner on legal papers when young people were committed to the Colony.

Priddy contacted Roller just after he engineered the 1916 change in Virginia law that authorized him to provide expanded "moral, medical and surgical treatment" to Colony patients. He noted that of the 185 patients admitted to the Colony during the previous year, fifty-seven had come from the city of Richmond. Priddy gloated at his success in helping Roller "clear up the rubbish" in Virginia's capital. Bragging aside, the main purpose of Priddy's letter was to alert Roller that he was ready to relieve her of the "burden of the Mallory crowd."[47]

Sarah Roller was familiar with the Mallory family's thick file at the State Board of Charities and Corrections. Willie Mallory, the forty-two-year-old matriarch of the clan, had survived twelve pregnancies. Nine of her children lived through infancy, but two were stillborn and one died of diphtheria at age three. At age twenty-seven, Willie spent five months in a mental hospital suffering from "acute melancholia" and was "suicidal" following childbirth. Doctors blamed her condition on "ill health and overwork."[48] Her husband George earned his living as a sawyer, cutting wood at a mill. At least once his drinking caused a serious problem, and Willie called in the police to settle the matter.[49]

In 1916 the Mallory family fell on hard times. Their house and all their belongings were destroyed in a fire, and George relied on charity for a month to rent a place for his family to live.[50] For a while he worked as a day laborer, eventually securing a regular job at a sawmill in a nearby town. He traveled home every other week to visit Willie and eight of their children who remained in Richmond.

George was away on September 27, 1916, when a strange man came to the door, asking about renting a room. He was never identified, though later accounts suggested that he was working with the police in an attempt to entrap the Mallorys on a morals charge. Two family friends—the fiancée of Irene Mallory and his brother—were visiting that evening.[51] When Mrs. Mallory took the stranger upstairs to show him a room, two city detectives came into the home and arrested her, taking the two visitors and the eight children into custody as well. Mrs. Mallory was charged with running a brothel. Irene and the two family friends were charged with "disorderly conduct" and fined five dollars each in a Richmond court the next day. The testimony of the policemen was later undercut by the charge that one had made an "indecent proposal" to attractive nineteen-year-old Irene, allegedly offering to drop the charges if she would comply with his wishes.[52]

The next day, Sarah Roller and two doctors appeared before a justice of the peace to convene a "Commission of Feeblemindedness" to determine if the Mallorys were proper candidates for treatment at the Colony. Later, Mrs. Mallory described their examination and the "mental test" she was given.

Lawyer: What Happened at the Detention home?

Willie Mallory: Two or three people examined my mind. All the little children were standing around me. One boy was sitting in my lap and the others clung around me. Little Bertha had her arms around my neck; the rest were as near as they could get, crying.

Lawyer: What else?

Mallory: Then a doctor examined my mind and asked if I could tell whether salt was in the bread or not, and did I know how to tie my shoes. There was a picture hanging on the wall of a dog. He asked me if it was a dog or a lady. He asked me all sorts of foolish questions which would take too long for me to tell you.

Lawyer: What happened?

Mallory: Then, the doctor took his pencil and scratched his head and said "I can't get that woman in," and Mrs. Roller said to them, "Put on there, 'unable to control her nerves,' and we can get her in for that." That is about all.[53]

The suggestion that Willie Mallory could be committed to a mental institution for being "unable to control her nerves" reflected Roller's experience in sending people to state hospitals under the previous state Lunacy Act, which allowed such "medical" judgments to be used as the basis of a commitment order.[54] In contrast, the new state law required that a petition must be filed and a written examination report submitted to the court before a commitment order would be valid. Ignoring these new legal provisions, Roller's "Commission" declared the Mallory trio feebleminded and committed them to the Colony.[55]

The Juvenile Court found that the younger Mallory children had been "exposed to vicious and immoral influences."[56] It turned them over to the Children's Home Society, a private agency that operated with the approval of the State Board of Charities and Corrections; it specialized in placing orphaned, neglected, or abandoned children in foster homes.[57]

Willie Mallory and her two daughters became inmates 691, 692, and 693, arriving at the Colony midway through 1916—the first year of Priddy's sterilization policy. According to her commitment papers, Willie's career included begging, public charity, and "streetwalking." Her father had been married four times, and her mother, five of her children, and several other family members were labeled "feebleminded." "Heredity" was listed as the cause of the Mallory "mental deficiency."[58]

Jessie Mallory was fifteen years old that fall. Five feet three inches tall, she weighed only eighty-seven pounds, and her clothes were covered with vermin. She was anemic and infected with gonorrhea; she also had a heart murmur. The doctor who examined her summarized her general appearance as "unhealthy." The intelligence test she took at the Colony showed a mental age of ten years, two months. Her father was described as an "alcoholic."[59] Thirteen-year-old Nannie was the third Mallory sent to the Colony. No record accompanied her; her main qualification for Colony residence seemed to be the Mallory name.

It was weeks later before George Mallory learned exactly what had happened to his family. When he finally contacted Priddy, he reported that he had "rented a good horse" and was ready to come to bring them home. But Priddy refused to release the Mallorys, saying that they had been found "detective" and had been arrested "as menaces to public morals."[60] In time, the Mallory family closed ranks, and Priddy was besieged by grown children and other members of the extended Mallory family demanding release of their relatives.

Though she would later deny that she had ever consented to surgery, some Colony medical records indicated that Jessie had "repeatedly insisted on having an abdominal operation performed" to alleviate her pain. When Priddy operated, he claimed that she was suffering with chronic salpingitis—an inflammation of the Fallopian tubes caused by an infection such as gonorrhea. Weakening Priddy's assertion was the section of the medical form reserved for a diagnosis. On that line, infirmary nurses had simply written "sterilization."[61]

After the operation, Jessie was sent to live with a woman in West Virginia but soon violated the terms of her parole by returning to Richmond. Priddy told Sarah Roller to "round her up and arrest her" for return to the Colony, but she eluded capture.[62] Even though Priddy told Jessie's fiancée of her operation, within the next year they were married.[63]

Priddy told other family members that the law authorized him to operate on women of childbearing age "to prevent the transmission of mental diseases to their offspring"; he planned to sterilize Willie as well.[64] In an attempt to avoid this fate, on April 12, 1917, Willie attempted to escape from the Colony. She made it only as far as nearby Lynchburg, where she was captured and returned to Priddy. She would later claim that he locked her in a "dungeon" and threatened to keep her there for a month as punishment.[65]

The news of impending surgery reached Willie's daughter Irene, who wrote Priddy, pleading with him not to operate. Unwilling to allow Willie Mallory to return to the neighborhood where she was known for "poverty and immorality," Priddy refused to release her. A few days later, Priddy removed Willie's Fallopian tubes and one of her ovaries. Two weeks after the operation, she was released to the care of her son-in-law. In the usual case, Priddy demanded no more than ten dollars for the cost of transporting a patient to a new home. For the Mallorys, he required a hundred-dollar bond on the condition that Willie not be allowed to contact her children or live with her husband.[66]

George Mallory did not abandon his efforts to earn young Nannie Mallory's freedom, but Priddy's responses grew even more harsh. He said that when the girl was sent to the Colony, Mallory "lost all control of her."[67] As the anniversary of the Mallory arrest approached, George met with a lawyer. By mid-October of 1917, Priddy had heard from friends in Richmond that he was being sued for five thousand dollars for the harm he had inflicted on Willie Mallory.

Priddy immediately launched a frenzied attempt to secure the proper paperwork that would justify holding the Mallory family. The absence of a valid court order of commitment for the Mallorys was a major concern. Priddy suggested that the Juvenile Court judge was also liable since he had cooperated in the commitments even though his court had no jurisdiction. Priddy claimed he had only received the Mallory women at the judge's direction. Saying that these "troublesome cases" were the cause of much difficulty, Priddy made it clear that he expected the judge "to protect me in this matter" and obtain the necessary legal documents.[68]

Next, Priddy wrote to his friend Sarah Roller at the Juvenile Court. Marking the letter "Personal and Confidential," he told Roller that he was not worried about the lawsuit but had no desire to attend court in Richmond. Admitting that his failure to wait for the proper court "to technically commit them" was "rather irregular," he nevertheless claimed to have acted "strictly within the law" and claimed to be a proper custodian for the Mallorys, just as if they had been sent to a jail under court order. He demanded

Roller's support, leaving no doubt that he expected to be insulated from legal consequences. "I shall certainly expect you and Judge Ricks to stand by me, in this and every other case I take in on your account," he wrote.[69]

Under the new law, Priddy had an explicit obligation to examine papers of commitment and return any that needed correction.[70] But skirting the provisions of the commitment law had become a habit with Priddy, who in another case suggested that he had detained a woman simply "as a matter of humanity." The *Mallory* litigation demonstrated that this brand of kindness could bring legal trouble.[71]

Within days, officials at both the Juvenile Court and the State Board of Charities and Corrections wrote Priddy, assuring him that the matter was merely a technicality related to recent changes in the law that had yet to be implemented by court staff. In any similar cases Priddy should just ask and they would "secure the necessary papers" for people already in custody. Priddy thanked them for "getting the record in the Mallory case settled."[72]

The Commonwealth of Virginia, County of Amherst

To the Sheriff of the ~~City of Richmond~~—Greeting:

We command you to summon A. S. Priddy

to appear at the clerk's office of our Circuit Court of the City of Richmond, at the Court Room of said Court in said City, at rules to be holden for said court on the Third Monday in November 1917 next to answer the action of Willie Mallory of a plea of trespass on the case

Damages Five thousand Dollars. And have then there this writ. Witness, E. M. ROWELLE, Clerk of said court at Richmond, the 18th day of October 1917, and in the 142nd year of the Commonwealth.

E M Rowelle Clk by Garland B Taylor D.C.

Summons ordering Dr. A. S. Priddy's appearance in court for the Willie Mallory lawsuit, 1917.

But before long, Priddy returned home to find a subpoena nailed to his front door.[73] He also received a long and impassioned letter from George Mallory seeking Nannie's release:

Dear sir one more time I am go write to you to ask you about my child I cannot here from her bye no means. I have wrote three or four times cant yet hereing from her at all we have sent her a box and I dont no wheather she received them or not. I want to know when can I get my child home again my family have been broked up on false pertents same as white slavery, Dr what busneiss did you have opreateding on my wife and daughtr with out my consent. I am a hard working man can take care of my family and can prove it and before I am finish you will find out that I am. I heard that some one told you lots of bad new but I have been living with her for twenty three years and cant no body prove nothing againts my wife they cant talk enything but cant prove nothing. My laywers said that you treated them very [?] just to think my wife is 43 years old and to be treated in that way, you ought to be a shamed of your selft of opreateding on her at that age Just stop and think of how she have been treated What cause did you have to operateding her please let me no for there is no law for such treatment I have found that out I am a poor man but was smart anuf to find that out—I had a good home as eny man wanted nine sweet little children—now to think it is all broke up for nothing I want to no what you are go do I earn 75$ a mounth I dont want my child on the state I did not put her on them. if you dont let me have her bye easy terms I will get her by bad she is not feeble minded over there working for the state for nothing now let me no at once I am a humanbeen as well as you are I am tired of bein treated this way for nothing I want my child that is good understanded let me know before farther notice. Now I want to know on return mail what are you go do wheather are go let my child come home let me here from her.

Verly Truly
Mr. George Mallory

My last letter to you for my child with out trouble dont keep my child there I have told you not to opreated on my child if you do it will be more trouble then you [?][74]

Priddy notified the Colony Board of the litigation at the next monthly meeting. The Board determined that since he was acting "as an officer of the institution in accordance with law," the state would defend him. He was directed to turn the matter over to George Caskie, whose law firm represented the Colony.[75]

Furious at these developments and still unaware of the details of the Mallorys' lawsuit, Priddy went on the offensive. Now that he had the proper documents to claim that Willie and Jessie Mallory were "regularly and legally committed feeble minded patients," he asked court officials in Richmond to arrest the women and return them to the Colony.[76] He also responded to George Mallory's letter. The letter was "insulting and threatening in its tone," said Priddy. He had received the Mallory family files from the Juvenile Court and the Board of Charities and Corrections and was in no mood to entertain barely literate threats from the man. He threatened Mallory: "Don't you dare write me another such letter or I will have you arrested in a few hours."[77]

Priddy's lawyer began to prepare his defense, arguing that the Richmond Court had no jurisdiction over Priddy; meanwhile, the Doctor found out that "a sensational sort of peripatetic lawyer" and a colleague of similar reputation had filed the Richmond lawsuit.[78] It claimed that Priddy had deprived Willie of her family and her liberty and forced her to work without compensation. When she tried to free herself, he put her in a "dungeon" with the intention of "terrorizing her and punishing her." The operation, done "by force, and violence" against her will, had the effect of "unsexing" her. The five thousand dollars she demanded would compensate her for damages suffered both through the dread of the operation before the fact and the "great pain" she endured in healing afterward.[79]

Even before Priddy could respond to the damage claims, Virginia's Supreme Court of Appeals ordered the Children's Home Society to bring the five younger Mallory children before the Court in response to a writ of *habeas corpus.* An account in the *Richmond Times Dispatch* highlighted the unusual nature of this order, which represented the first time in memory that a child had been called before the high court. George Mallory was "determined to have his family returned to him," reported the newspaper.[80]

Mallory's lawyers then insisted that Priddy release Nannie. They needed her as a witness in Willie Mallory's case and were also "reliably informed" that Priddy was holding Nannie "for the purpose of an illegal operation" he was planning. Early compliance would save the Virginia Court the trouble of issuing a second writ, added the lawyers.[81]

Priddy's lawyer denied that Nannie was facing surgery, saying that she had "no physical disease" that would require an operation. But Priddy was not ready to free her. Mallory's lawyers went ahead with a second petition to

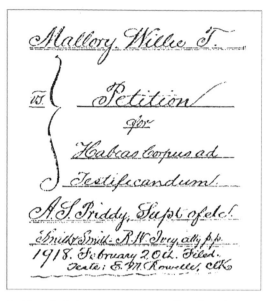

Petition for habeas corpus demanding the release
of Willie Mallory, 1918.

the Court of Appeals, seeking Nannie Mallory's freedom. They described
the Mallory family as "respected citizens . . . but . . . poor, and not over-
plussly [sic] blessed with worldly riches." The "sacred confines" of the home
had been violated, they said, and Willie Mallory—mother of twelve and
grandmother of three—was "abducted and kidnapped" along with her children. Nannie Mallory was illegally confined, working as an unpaid "servant" tending four babies born at the Colony.[82]

In the middle of this flurry of legal papers, George Caskie filed his defense to Mrs. Mallory's lawsuit. In it he reasserted that Willie Mallory was
"feeble-minded," as determined by Sarah Roller and the two doctors con-
vened by the Juvenile Court. That finding gave Priddy the right, he said, to
detain the Mallory women at the Colony. Despite his lengthy correspondence with Sarah Roller about the Mallory family, Priddy endorsed his
lawyer's defense even though it falsely declared that Priddy had no involvement with plans for the arrest of the Mallorys in Richmond. This strategy
was hardly surprising, since a conspiracy to commit a person to the Colony
"unlawfully or maliciously" was a crime and could have made Priddy liable
for fines and jail time.[83]

Priddy denied any ill treatment of Mrs. Mallory but admitted that she had been held in a separate room—not a "dungeon"—as a punishment for violating Colony rules. He also denied that the sterilization was involuntary, claiming that Willie Mallory had requested the operation and that the Board of Directors had approved it. It was not the surgery that destroyed Mrs. Mallory's power to have children, said Priddy; her diseased state had already caused that result. Without the operation, she "would most probably have died."

Though Priddy and his Board had relied on a more expansive interpretation of the 1916 law when the sterilization program was initiated, George Caskie must have found that view of the law questionable. No mention of new statutory authority to sterilize women on eugenic grounds appeared in his response to the *Mallory* suit. Priddy's defense ended with a charge against Mrs. Mallory. She was, the doctor declared, "well-known to the police of the City of Richmond, and the social and charity workers of the State of Virginia, as a deficient, and as a most troublesome and undesirable citizen . . . incapable of leading a clean and proper life."[84]

Encouraged by the Juvenile Court's decision concerning her younger children, Willie Mallory challenged Priddy's claim that she had requested an operation. She reminded him that her daughter Irene had contacted him, saying that her mother was in poor health and too old to endure an operation. Priddy had replied that the operation posed no danger because Willie was in "in perfect health." Mrs. Mallory taunted Priddy, saying that she would have his letter framed as evidence. If she was "in perfect health," she obviously needed no operation; far from being unable to take care of herself, Willie proudly noted that she was working for ten dollars a week. Priddy responded confidently, predicting success in court, saying that he was "gratified to know" that his surgery "did so much good" that Mallory could now "earn an honest living . . . and not live such a life of shame."[85]

Priddy's confidence was less evident in his correspondence with others involved in the Mallory suit. News of his loss in the Court of Appeals moved him to write the Richmond Juvenile Court again, requesting "prompt attention" from the judge and Miss Roller to help him prove his case. He complained to Reverend Joseph Mastin, author of *Mental Defectives in Virginia* and executive of the State Board of Charities and Corrections, that if the *Mallory* suits were successful, "there is no use in the State trying to maintain an institution for defectives of the criminal class." He urged Mastin to contact Mallory's doctor to solicit testimony about the venereal disease Jessie Mallory had contracted. She was, he said to another correspondent, "the worst diseased one we found" among the approximately 125 operations he performed at the Colony.[86]

Priddy also wrote to Irving Whitehead. Of all the supporters of the Colony, Whitehead had the most significant history. He was a member of its first Board of Directors, which had named Priddy superintendent in 1910. In 1912 he traveled with Priddy to the Indiana Colony for Epileptics to review the workings of that state's pioneering sterilization law. He supported sending a Colony employee to Henry Goddard's Vineland, New Jersey, institution to learn the uses of the Binet-Simon test in identifying the feebleminded. As chairman of the Colony Board, Whitehead had endorsed Priddy's expansive interpretation of the 1916 law to authorize sterilizations, and he urged his fellow Board members to approve the first surgeries in the summer of 1916. At his final Board meeting the following March, Whitehead initiated the vote to approve the operations for Willie and Jessie Mallory.[87]

Whitehead left the Colony Board to take a position in Baltimore as lawyer for the Federal Land Bank, but he kept in touch with Priddy by mail and saw him during regular visits to Amherst. As the date for the Mallory trial approached, he wrote Priddy to calm his friend's understandable anxiety about the pending day in court. There was no need for "uneasiness about the matter," he said, because as they and the Board had agreed, the law allowed Priddy to sterilize inmates. Additionally, Whitehead found the Mallory litigation "so palpably 'bold' & without merit" that no matter what a jury might do, the judge would overturn any negative decision.

Whitehead believed that the patients had consented. If they claimed that their lack of mental capacity made consent invalid, Whitehead recommended that Priddy argue that their incompetence also invalidated the right to refuse. Whitehead also believed that the same legal incompetence should cancel their power to sue. Moreover, he said, the patients needed the surgery, which saved their lives. "This is a splendid case to test the law," Whitehead declared. Priddy had had to operate; if he had not, he would have been "guilty of the grossest negligence." Consistent with the position he had taken as a Board member, Whitehead declared that Priddy's "unusual obligation," both as a doctor and as a state official, imposed a "clear duty" to operate on the Mallorys.[88]

Whitehead was a lawyer, but his commitment to the sterilization program blinded him to the dangers of the plan he had endorsed for Priddy. He was dramatically mistaken about both the state of the law and the potential public reaction to eugenical sterilization done against a patient's will. The *Mallory* action was designated as a "trespass"—what today would be considered a case of civil battery. It was well-settled legal precedent in 1916 that for fully capable patients, surgery without consent was unheard of, and operations done in the absence of consent often resulted in a verdict for the patient.[89] For incompetent patients, consent of parents or next of kin was expected, and liability was assumed for operating without it.[90] Damages could even be awarded for surgery on mental patients.[91] No Virginia legal cases had questioned these fundamental premises.

As a member of Priddy's Board, Whitehead had ignored the legal precedents and instead supported a radical interpretation of a vaguely worded statute. While Priddy claimed the right to operate without interference, in other states unauthorized sterilizations were greeted with sensational headlines. When the superintendent of the Montana State Asylum was sued for operating on eleven patients without their consent and in violation of the state eugenics laws, the headline in the *Helena Daily Independent* trumpeted "Butchery of Helpless."[92] But even in the face of litigation, Whitehead did not let his ignorance of the law of medical consent weaken his embrace of a eugenic program.

Other letters of support arrived from Priddy's friends in Richmond. They reported the activities of opposition lawyers and pledged to help him recapture the Mallorys, whom one characterized as "most assuredly, a menace to society." This certainly fit Priddy's assessment of the family, and the doctor promised that his own witnesses would prove that "the Mallory bunch was a set of criminal paupers . . . subsisting on the immorality of their daughters, and by begging."[93]

Leaving nothing to chance, Priddy instructed Sarah Roller to continue surveillance of the Mallory children. He solicited statements from former doctors and nurses at the Colony about Willie Mallory's surgery, coaching them to recall his version of the facts, and providing generous subsidies to pay their expenses related to the suit.[94]

Though Priddy's friends in the Juvenile Court eventually generated a proper commitment order, it was entered nearly a year after the Mallorys were taken to the Colony. George Mallory's lawyers protested that the order was produced only to permit the rearrest of Willie and Jessie so that Priddy could imprison them at the Colony. They condemned the order as "an outrage" issued solely to shield Priddy from legal liability and entered in the records of the Richmond court only after Priddy had received notice that he was being sued.[95]

When Willie Mallory's suit against Priddy finally came to trial, the argument that he had operated out of necessity was apparently effective, as the jury refused to award damages.[96] Medical need rather than eugenics was still the best defense for doctors in 1918.[97] Perhaps the jury harbored fears of the feebleminded. The country had been shocked to learn that testing of Army recruits at the outset of World War I suggested that as many as 47 percent of all recruits were feebleminded, and the news had an impact on popular perceptions of mental defect.[98] Putting eugenic theory and concerns about the menace of feeblemindedness aside, the trial judge did not believe that therapeutic intentions could justify future sterilizations and warned Priddy to stop operating until the existing law was changed.[99] Perhaps the most telling commentary on the impact of the *Mallory* case came from Priddy's friend, Western State Asylum superintendent Joseph DeJarnette. The *Mallory* lawsuit, he said, "frightened all the superintendents in the State and all sterilization was stopped promptly." DeJarnette had his own experimental sterilization program, but after the Mallory litigation he admitted that he and his colleagues were all "afraid to operate."[100]

The Virginia Supreme Court of Appeals eventually voided the detention of Nannie Mallory and freed her because the Juvenile Court had failed to follow the specific procedural provisions of the commitment law. The court had already released the other Mallory children illegally held by the Children's Home Society.[101] The Mallorys were certainly not the most sympathetic plaintiffs, but they were on solid legal ground. The laws violated by the Children's Home Society were reasonably clear, and the failure of the Juvenile Court to follow mandatory procedures for committing Nannie Mallory was obvious. In contrast, it was far from obvious that the new language of Virginia law allowed state hospital doctors to sterilize patients without consent. Had the defense of "medical emergency" not been available to justify Priddy's surgery on Willie Mallory, it is likely that he would have lost that case as well.

The effect of the lawsuit on Priddy's sterilization policy was immediate. Fearful of legal repercussions, as soon as the first *Mallory* case was filed Priddy began taking precautions before operating. In one case he said he would not operate without consent of the family but made surgery a condition of a patient's release. The patient's family eventually capitulated so that Priddy would free her from the Colony, but she left the institution only after they had waived any legal claims against Priddy for his coerced surgery.[102]

After the *Mallory* case, Priddy put the best possible face on the verdict, saying that the jury had "vindicated the management of the institution," but he also admitted that the lawsuit had highlighted "the importance of complying with every technical requirement of law." He continued to recommend patients for sterilization, but in each future instance "pelvic disease" of an undisclosed type was indicated as the basis for the operation, and a written request from the "feebleminded" women for the surgery was required.[103] In a notable change of tone from his early boasts of sterilizing "women of the moron type," Priddy's later reports cautiously noted that sterilizations were carried out only when necessary for the relief of the suffering patient. Without a law that would protect state doctors from being dragged into court, it looked as if a robust plan for eugenical sterilization had no future in Virginia.

6

Laughlin's Book

By 1920, Harry Laughlin had established an international reputation as an expert on eugenic sterilization. Governments as far away as Germany consulted him for information on U.S. sterilization practices.[1] He had even testified before Congress about the importance of eugenics to immigration policy.[2]

In that testimony, Laughlin described the process of family history collection by trained field workers from the ERO who could determine whether a family was "industrious or shiftless." He invoked the names of the Jukes, the Kallikaks, and the Ishmaels. He described a study of the "socially inadequate" he had done for the Bureau of the Census and the publication it yielded: *The Statistical Directory of State Institutions for the Defective, Dependent, and Delinquent Classes.*[3] While that survey was in progress, Laughlin lobbied to change the terms describing the institutional population in the 1920 census.[4] Laughlin wished to rechristen the demographic categories formerly described as the "defective, dependent, and delinquent classes." This "scrap-basket" phrase should be discarded, he said, in favor of the "socially inadequate," a term that would conform to the language crafted by the Van Wagenen sterilization committee. The newer phrase was "shorter and more business-like," according to Laughlin, and properly described the portion of the population who weigh down the productive members of the community. People shouldn't be called "socially inadequate" just because age, youth, or illness might render them temporarily dependent—so long as their families took care of them. Those without family resources who had to rely upon government programs or charity were, of course, exactly the people the term was meant to describe.[5] And many of them were potential candidates for sterilization.

From the time of his arrival at the ERO in 1910, Laughlin had collected every available detail on sterilization law in America. His report for the Van Wagenen committee formed the basis of a state-of-the-art survey of the field, and his Model Law had been distributed to interested legislators and eugenics advocates around the country. By 1920, further research yielded a mound of facts that, from Laughlin's perspective, certainly demanded publication. Along with the history of state sterilization laws, case reports, and judicial opinions, Laughlin had almost a thousand manuscript pages of text. He eventually surveyed more than 160 institutions in the United States seeking exhaustive data on the number and kinds of operations they had performed. Before leaving New York for a summer of research in California, he dedicated his efforts to getting a book into print. But the likely size of the volume and the controversy the subject drew scared potential publishers away.

In the spring of 1920, Laughlin approached the National Committee on Mental Hygiene, asking Thomas Salmon to review his manuscript. Salmon, a former Public Health Service doctor and the medical director of the National Committee, had by then been associated with the eugenics movement for more than a decade, and he praised Laughlin's qualifications to produce the book, saying that the Laughlin material would make an "invaluable source" and reference text for people in social work. But he took pointed exception to including the Model Sterilization Act and the related commentary. As the emphasis on coercive sterilization represented the view of a small minority, he said he would "dissent from nearly every word in it." Salmon rejected the book unless the Act and the "rather strong bias" in favor of sterilization in earlier chapters were removed.[6]

Laughlin also contacted the Rockfeller-funded Bureau of Social Hygiene, prompting the Bureau's executive secretary, Katherine Bement Davis, to write the Macmillan Company on Laughlin's behalf. Macmillan agreed to publish the book with strict conditions. Fifteen pages and two fold-out pedigree charts would have to be cut, and the Bureau was asked to guarantee purchase of the first five hundred copies of the book. After a more thorough review of Laughlin's manuscript, Macmillan demanded a guarantee for a full thousand copies. The cost of subsidizing that printing was estimated at nearly seven thousand dollars, apparently too much for the Bureau of Social Hygiene to bear. As an alternative, Davis suggested that Laughlin shorten his manuscript to one hundred pages and publish it as a pamphlet. But Laughlin was unwilling to shrink his magnum opus so drastically, proposing instead to bring the material completely up to date and seek another publisher.[7]

Laughlin then asked ERO Director Charles Davenport if the Carnegie Institution of Washington would support the volume on sterilization, since it had published several of Davenport's books. Davenport pledged minor staff assistance at the ERO to help Laughlin and suggested that money for publications might be available in conjunction with the International Eugenics Congress planned for 1921, but he assured his colleague that Carnegie would not subsidize any work "that it does not control." He also expressed significant doubts whether Carnegie would want to be identified with the sterilization book, given the preliminary reactions to its content.[8]

Growing frustrated, Laughlin wrote directly to John D. Rockefeller to ask for assistance. Rockefeller had supported activities of the ERO before, and Laughlin had spent a good deal of his time cultivating like-minded philanthropists. His letter was given to Rockefeller's attorney, who saw nothing in the book in which Rockefeller "could properly be interested." The lawyer sent Laughlin's request to Davis for a second opinion. She explained that Laughlin's pleas for a subsidy were not new, noting that none of the people he had contacted cared to take responsibility for the portions of the book containing "direct propaganda favoring sterilization legislation."[9]

Despite hesitating earlier to print parts of his study before the complete volume was released, Laughlin published a thirty-two-page synopsis of his findings in the journal Social Hygiene in October 1920. The somewhat tentative note from the editor of Social Hygiene applauded the forthcoming volume as a "scholarly work on sterilization" that would likely become a "standard authority in its field." But the endorsement contained a significant qualification. Since nearly all the states with sterilization laws extending to "cacogenic persons" rather than criminals had been judged unconstitutional, the editor indicated that the validity of Laughlin's Model Act was at least "open to question."[10]

Laughlin ignored the criticism and forged ahead, completing his final edits and requesting anatomical illustrations from a colleague.[11] He crafted the last revisions hoping that they would provide the ammunition he needed to convince a publisher of the book's value.[12]

In September 1921, Laughlin and his colleagues from the ERO joined Bleecker Van Wagenen to promote sterilization at the Second International Congress of Eugenics in New York. Laughlin gave a reprise of Van Wagenen's 1912 report, summarizing sterilization law to date. He claimed that the "eugenical standard" to determine who would be subject to surgery after medical diagnosis and field investigation was "fairly well established."[13]

Laughlin took the opportunity as head of the Committee on Exhibits at the International Congress to display a poster showing "Eugenical Classification of the Human Stock" with categories of "Socially Inadequate Persons" and a map displaying details of sterilization laws in the United States.[14] Three of Arthur Estabrook's charts detailing studies among "socially inadequate" clans were also included in the exhibit.[15]

By the time the Eugenics Congress was over, it must have been clear to Laughlin that his attempt to publish the book commercially had failed. He was unable to attract support even from his old friends in the eugenics movement. They thought his insistence on a mandatory sterilization law was strident, misguided, and unscientific. But one natural ally with whom he had conferred in New York remained: Judge Harry Olson of Chicago's Municipal Court. Unlike Laughlin's critics, Olson embraced the possibilities of using sterilization as a powerful weapon in the long-term war on crime. In 1922, the year when eugenicists celebrated the centenary of both Mendel and Galton, Olson thought Laughlin's message was exactly what the public needed to hear. He had contacts in the highest echelons of the legal community to ensure that *Eugenical Sterilization* would not be ignored.

Harry Olson

From modest beginnings, Harry Olson had risen to a position of prominence in his home state of Illinois and become a leading national champion of legal reform and advocate of eugenic sterilization. Born in 1867 in Chicago, his father's early death left Olson to make his own way as a teacher and then as a lawyer. His career achievements included cofounding the American Judicature Society, of which he served as chairman from 1913 to 1928. There he worked alongside such giants of the legal world as Roscoe Pound, dean of the Harvard Law School. Another AJS colleague was John Henry Wigmore, renowned scholar of the law of evidence, who praised Olson's domain as "The Most Famous City Court in the World."[16] Olson was a member of the prestigious American Law Institute and a trustee of Northwestern University. Twice he mounted unsuccessful campaigns for mayor of Chicago. His captivation with eugenics would lead to his being named

Harry Olson, chief judge of the Municipal Court of Chicago and advocate of eugenic sterilization, 1929. Courtesy American Philosophical Society.

president of the Eugenics Research Association. Olson would eventually play a key role in publishing and distributing Harry Laughlin's *Eugenical Sterilization in the United States*.[17]

Olson's interest in criminal law began early in his legal career, and he worked as a prosecutor for ten years before becoming a judge. He was an early convert to eugenics, believing that heredity was a key causal factor in crime. The strength of this belief was demonstrated in 1906, when Olson turned down an offer of fifty thousand dollars to defend millionaire Harry Thaw, who had shot famous architect Stanford White. White's affair with Evelyn Nesbit, Thaw's showgirl wife, was the motive for the killing, but Olson wanted to base a defense on Thaw's faulty heredity.[18] The Thaws, not inclined to magnify a purported stain on the family pedigree, refused to endorse Olson's strategy. In the spectacle billed as "the Trial of the Century," Thaw eventually escaped the gallows, arguing—without Olson's assistance—that he was temporarily, not hereditarily, insane.

That same year, Olson was elected chief justice of the newly created Chicago Municipal Court, and he left private legal practice.[19] Several years later, he served on the Chicago Vice Commission with social work pioneer Jane Addams, Sears-Roebuck magnate Julius Rosenwald, and Chicago Commissioner of Medicine W. A. Evans. Alliances among reformers in the name of "social purity" were common, as was overlap among old temperance movement stalwarts and leaders in eugenics; a fundamental part of the eugenics message was social reform.[20] The Vice Commission included eugenics on its list of the medical means for controlling prostitution.[21] Olson's tenure on the commission, with its focus on the public health value of eradicating the sex trade, led to creation of a morals court in 1913 to hear cases involving prostitution and obscenity.[22] Olson's Municipal Court applied ideas of management efficiency and the newest tools of forensic analysis to showcase the "scientific" administration of justice.

In his twenty-four years as chief justice, Olson emphasized the importance of studying heredity as a touchstone for understanding crime. Official publications from his court were filled with such items as an "Illustrative Heredity Chart" with symbols to indicate "alcoholism," "mental defect," and "sexual offender" as hereditarily transmitted traits.[23] In order to collect data for the scientific study of crime, Olson established the court's Psychopathic Laboratory in 1914. As its director he named William J. Hickson, a staunch hereditarian who had worked at Vineland with Laughlin's colleague, Henry Goddard. Olson arranged for Hickson's lengthy reports on criminals examined at the laboratory to appear regularly in legal journals.

Judge Olson's perspective from the bench of a big city court and his familiarity with the literature of criminology led him to opinions that meshed seamlessly with Laughlin's. For example, Olson criticized his own city's Vice Commission for failing to consider the mental status of prostitutes during its investigations. He believed that feeblemindedness and prostitution were linked; he thought the thousands of cases reviewed by the Psychopathic Laboratory proved the point.[24] He also agreed with Laughlin about the dangers of uncontrolled immigration, which, since colonial days, had flooded America with "vagrants, ne'er do wells, women of the streets and boys and girls of the slums and alleys."[25]

To address this blight on the cities, Olson endorsed a preventive policy designed to "preserve the integrity of the race" by segregating "human derelicts" in farm colonies.[26] He celebrated research that revealed the hereditary roots of crime and "the secret of Criminal Defectiveness," a condition responsible for "dependency, unemployability, alcoholism . . . wife desertion . . . mental and physical diseases and accidents."[27] Like Laughlin, Olson believed that future generations could be purged of negative hereditary traits by surgical sterilization. In 1918, Olson appointed Laughlin as a consultant to his laboratory, giving him the title of "Eugenics Associate." In March 1922, Olson announced that Laughlin would be assigned to work in Chicago by his employer, the Carnegie Foundation, with the hope that a Chicago branch of the Eugenics Record Office would be established at a later date. One of the eventual goals for the office, according to Olson, was to consider the sterilization of "morons and dangerous criminals."[28] With Laughlin's help, Olson had already prepared to submit a sterilization law to the Illinois General Assembly in 1923.[29]

The meeting of Laughlin and Olson at the 1921 New York Eugenics Congress provided an occasion to rescue Laughlin's book from oblivion. After several months of editorial work, on January 23, 1923, Olson paid to have the book published as "A Report of the Psychopathic Laboratory of the Municipal Court of Chicago." Five hundred and two pages in all, the book cover featured a photo of a sculpture Olson had commissioned for his Chicago courtroom: "Heredity, Fountain of the Ages." The final page in the book carried a quotation from the father of social Darwinism, Herbert Spencer: "To be a good animal is the first requisite to success in life, and to be a Nation of good animals is the first condition of National prosperity."

The book began with Olson's lavish tribute to Laughlin, whose explanation of heredity and the "science of eugenics" was said to be of "vital importance," as it represented a "signal service" to the nation. The "permanent value" of Laughlin's "scientific research" would be ensured, said Olson. He quoted Irving Fisher, Yale University economist and eugenical expert, who called eugenics "the highest form of patriotism and humanitarianism"; it offered "immediate advantages" to society by reducing the "burden of taxes" spent for people in public institutions and also by increasing "safeguards against crimes" committed against both people and property. To spread this message thoroughly, Olson promised that the book would be printed in sufficient numbers to "reach the leaders of the medical, legal and clerical professions" as well as lawmakers and members of the press.[30]

The bulk of the text, some 320 pages, chronicled the history of the legal movement for sterilization in the United States. Every law that was passed by a state legislature and every court opinion—whether upholding or striking down the law—was printed verbatim. Laughlin filled the book with the language of legal amendments and repeals as well as governors' vetoes. Every state that performed operations submitted notes and comments from field workers, asylum physicians, and administrators explaining how sterilization laws were used, tabulating how many operations had taken place, and describing who was covered by the law. Laughlin analyzed the laws by subject, noting the motives of legislatures, whether they focused on therapy, punishment, or heredity. He then explained in exhaustive detail the various lawsuits brought in challenge of sterilization statutes in the states and the family histories—complete with foldout pedigree charts—of people who were chosen for surgery, a group that included epileptic felon Rudolf Davis of Iowa, "feebleminded" Alice Smith of New Jersey, and several "moral perverts" in Indiana, Washington, Oregon, and Nevada.

The final third of the book began with Olson's own contribution, a legal opinion declaring the Model Law valid under the U.S. Constitution. The judge wasted little time in technical argument, dedicating the majority of his "opinion" to praise of his Psychopathic Laboratory for demonstrating the importance of heredity as the cause of crime and reminding readers of the need for legislation to "curb the menace of inferior stock." Next came formal legal opinions from state attorneys general defending the validity of laws in California and Connecticut. A brief comment from a New York lawyer questioned punitive features of sterilization laws, and Laughlin even found space to include a one-paragraph citation to a lengthy article that attacked the laws.[31]

The next chapter laid out an extensive argument in favor of the state powers to "limit human reproduction in the interests of race betterment." Eugenical sterilization, Laughlin asserted, was like other uses of the state's inherent "police power" to protect the common good. States regularly executed criminals for this reason, and governments used military conscription to raise armies, sending soldiers to their death to preserve a nation's safety. Limits on reproduction in the name of "common benefit" require a smaller sacrifice from citizens drafted from the ranks of "irresponsible members" than did wartime service; surely the "interests of the common welfare" justified such a restriction of liberties. The survival of a democracy depended on finding "new social remedies," so eugenical sterilization, within the boundaries of proper legal process, could replace war, famine, and disease as the means to cull humanity of its "degenerate and handicapped strains."[32]

Sterilization was also a public health measure like compulsory vaccination, designed to prevent the public at large from infection by a diseased individual. Both involved seizing people and subjecting them to intrusive operations. In fact, vaccination was more dangerous than vasectomy, said Laughlin. Moreover, vaccination had been approved as a proper use of state power by the U.S. Supreme Court in the 1905 case of *Jacobson v. Massachusetts*.[33]

Similarly, marriage and the accompanying right to procreate were limited in numerous ways by state legal regulation. The "defective" were prohibited from marrying on public health grounds, as were those with venereal diseases. Laws existed to segregate the "anti-social" in institutions, preventing them from reproducing, just as quarantine prevented the spread of infectious disease. Finally, immigration laws insured the "inborn qualities of future generations." They protected the hereditary legacy of the nation from pollution by mixture with supposedly inferior foreign elements. Laughlin acknowledged that all of these legal provisions limited personal freedoms but were nonetheless valid means of protecting the public. So too was sterilization, he argued, which placed restrictions on procreation of the "socially inadequate" in order to block the transmission of hereditary problems to future generations.[34]

Laughlin included a chapter on "eugenical diagnosis" showing how a trained field worker could collect "pedigree facts" and chart them as an aid to determining who was "cacogenic" and thus hereditarily unfit. When called upon to offer an opinion on the "potential parenthood of socially inadequate offspring," the eugenicist would be an expert witness in court.[35]

The "anatomical and surgical" aspects of sterilization took up a full chapter, complete with graphics to guide the novice surgeon through various operations. Laughlin then summarized research on the physiological and mental effects of surgery on those to whom it would apply.

The book concluded with three chapters that listed all of the arguments against sterilization. The text of the Model Law was printed with extensive annotations on each of its sections, and Laughlin added a set of model legal forms that could be adapted by any state to guide administration of its eugenic bureaucracy.

The reader of Laughlin's book had, in a single volume, a complete reference text on a very esoteric subject. This exhaustively detailed treatise was sometimes repetitive and often tedious, but it marshaled all of the "scientific facts" that could be mustered to bolster a public policy in favor of sterilization. Legislators who embraced that policy found a detailed statute, easily adaptable for use in the state assembly. Welfare administrators and social workers got a manual that would help them identify the "socially inadequate" in their communities. For the doctors who ran asylums and mental hospitals, Laughlin offered medical references and practical directions on surgery, complete with illustrations. And for the use of attorneys called on to defend a sterilization law in court, he catalogued arguments for and against the procedure, taking them from actual briefs and opinions in previously decided cases.

Olson ordered three thousand copies printed. He began to publicize the book extensively, sending out an early, favorable review to 280 different newspapers and fifty-five copies of the book itself to the larger city newspapers.[36] The picture of the statue gracing the book cover ran in the *Chicago Daily News*.[37] The *Chicago Daily Tribune* described Laughlin's "elaborate legal defense" of the Model Sterilization Law, noting his contention that it rested "on the same legal principals as those involved in statutes compelling vaccination."[38]

In his regularly syndicated column, "How to Keep Well," former president of the American Public Health Association and *Chicago Tribune* medical columnist Dr. W. A. Evans praised the Laughlin volume as a "valuable book on cacogenetics," a term he apparently coined for his column.[39] Others saw eugenic sterilization in a simpler, more pragmatic light, and headlines eventually focused on economics: "Breeding Better Folks Held Way to Lower Taxes."[40]

Not every response to Laughlin's work was positive. Olson's captivation with the Laughlin sterilization scheme had already been mocked by some commentators, who asked if the categories of the "socially inadequate" might not be increased by one, making room for those who "strut about parading the quaint illusion that science has picked them out as the prize breeding stock of creation."[41] By the time the book appeared, readers of

Harry Laughlin's book *Eugenical Sterilization in the United States*, 1922.

most legal journals had heard about the Jukes and the Kallikaks and the other studies of "degenerate" families. Some commentators, discussing trends in eugenic thought, considered this information obsolete and of "scarcely more than historical interest." Post–World War I standards of psychological analysis had proven that the proportion of criminals who were feebleminded was not much more than the proportion of feebleminded in the general population.[42]

In his review of Laughlin's book for the *Journal of Heredity,* California eugenics expert Paul Popenoe suggested that eugenic sterilization laws are "gradually losing interest for the public"—and for eugenicists as well.[43] But one legal reviewer applauded, judging the book a "real service . . . rendered to the nation." Though Laughlin's expansive definition of the "socially inadequate" was thought an "unwarranted extension" of state powers, the book as a whole was judged an "invaluable contribution" to the eugenics cause.[44]

Olson knew that the impact of the book depended not on reviewers but on the people who could use it to design and implement eugenic laws. Declaring to Laughlin that "the book will have a very great influence on legislation," Olson sent books to dozens of people on his own mailing list, which comprised a virtual who's who of the U.S. legal world. Friends of the eugenics movement like Secretary of State Charles Evans Hughes received a copy, as did former President and then Chief Justice of the United States Supreme Court William Howard Taft.[45]

Over the years Laughlin had written Taft at least a dozen times. When Taft was secretary of war, Laughlin wrote to make suggestions for a "civic triangle" of architecture for Washington, D.C.; Laughlin repeated his suggestions two years later, without success, to President-Elect Taft. By 1914 Taft had completed his term as president and joined the faculty of the Yale University Law School. Laughlin sent him the sterilization committee report and his Model Law, asking for commentary on sterilization—"this new social agency." Taft demurred. In 1921 Laughlin drafted a "World Constitution" and somewhat grandiosely asked the former president for an introductory note that could be published with the text.[46] Again, Taft declined.

When Laughlin finished *Eugenical Sterilization in the United States,* he had good reason to be doubtful of getting a positive reception from the busy Taft. But Harry Olson wrote on the letterhead of the American Judicature Society, recalling that Taft was "interested in the conservation of human resources" and enclosing Laughlin's book. Taft responded to Olson, promising to "read it with great interest."[47]

Laughlin targeted several hundred other notables to receive a complimentary copy of the book. Among them were educators and champions of the eugenics movement like Stanford University Chancellor David Starr Jordan and Clarence Cook Little, former university president in both

Maine and Michigan. Pioneering conservationist governor Gifford Pinchot of Pennsylvania made Laughlin's distribution list, as did representatives of major philanthropies like the Carnegie Foundation, the Rockefeller Foundation, and the Commonwealth Fund. Princeton biologist, eugenics supporter, and Laughlin mentor Edward Grant Conklin had a skeptical attitude toward Laughlin's expansive scheme, but he remained discreet and polite, responding to his copy of the book with confidence that it would "be of the very greatest service in promoting eugenical education and practice."[48]

Not all recipients of the Laughlin volume were famous; in fact, many were merely foot soldiers in the battle to establish eugenic sterilization as public policy. Asylum and hospital directors around the country found room for the book on their shelves, and Olson distributed the available copies quickly. State public health officers and hospital directors in Virginia requested copies. Few would find the book as useful as Dr. Albert Priddy, to whom Laughlin directed a copy at the Virginia Colony for Epileptics and Feeble-minded.[49]

7

A Virginia Sterilization Law

At the time Harry Laughlin's Model Law was first published in 1914, twelve states had enacted sterilization laws.[1] During that same period, four others passed bills that governors subsequently vetoed, and two existing state laws were invalidated by the courts. Between 1914 and 1922, the year that Laughlin's book appeared, five other states passed laws—but one early enactment was repealed, three more were overturned in court, and two other attempts at new or amended legislation were vetoed.[2]

By the 1920s, this very spotty record of legislative progress left proponents of sterilization far from optimistic about the future of legally mandated surgery. Some 3,200 sterilizations had been performed on inmates of prisons, insane asylums, homes for the epileptic or feebleminded, and similar institutions of social welfare by 1922, but almost 80 percent of those operations took place in California.[3] In other states, legal controversy still existed about the extent of patient/inmate rights and whether surgery could ever be a proper punishment for crime. Other problems included widespread disagreement about the value of sterilization in a campaign for "human betterment"—even among self-proclaimed eugenicists. Though some physicians who directed institutions of social welfare had been among the staunchest lobbyists for the laws, many others disapproved of the procedure on both moral and scientific grounds. The most dramatic case of medical resistance occurred in South Dakota. Despite a state law that authorized surgeries in a state institution, physicians who opposed the procedure there simply refused to operate.[4] In the face of these obstacles, by 1923 the extensive and coordinated program Laughlin had proposed in 1914 seemed unlikely to materialize.

In Virginia, Albert Priddy was blocked by the absence of legal authority that would support a wholesale campaign of surgery. Priddy's legal embarrassment in the *Mallory* cases prompted him to lobby the General Board of State Hospitals. That group—made up of representatives from all Virginia mental asylums—endorsed the campaign to sterilize the "unfit" and informally approached legislators for support. Hesitant to get behind what some saw as a radical measure, legislators joked that the law "might get all of us."[5]

Priddy's eugenics crusade stalled at the prospect of being mocked by lawmakers. But lack of legislative sanction for surgery was only a temporary barrier. He requested and received support from the Colony Board of Directors to visit the famous Chicago Juvenile Court of Judge Harry Olson.[6] Then Priddy turned to Aubrey Strode, the Colony's advocate in the Virginia General Assembly, to remedy his legal vulnerability. Strode had served in the Judge Advocate Corps of the Army during World War I. His Army commission, arranged by his boyhood friend Irving Whitehead, had taken him to France for several months.[7] He returned to Virginia for a final session as state legislator following the Armistice. Though Strode was out of the country during the *Mallory* litigation, upon his return he wasted no time renewing his work for Priddy—drafting laws for the Colony.

In 1920 Strode drafted two bills in an attempt to remedy the vagueness of earlier laws that had been scrutinized in the *Mallory* lawsuit. The first bill specified that lawsuits in which inmates of state mental hospitals challenged their confinement must be filed in the city where the inmate was hospitalized. It contained a provision requiring the state to pay the legal costs of hospital superintendents, like Priddy, who were sued. The second bill declared all persons already committed to state hospitals as feebleminded or epileptic by court order or "commissions of feeblemindedness" to be "lawfully committed patients."[8] In other words, patients currently in state hospitals were, retroactively and *en masse,* assumed to have been properly committed. Both bills became law.

The new laws had the effect of protecting Priddy from other lawsuits like the *Mallory* case. Never again would he suffer the humiliation of being dragged into court in another part of the state, nor would he worry about the expense of funding a legal defense. Strode's work gave Priddy a new layer of insulation against legal action and provided him with more power over those in his care.

Priddy approached Strode again in 1921, asking for advice on how to restart the campaign for a state sterilization law. Strode's research revealed that several state laws applying to inmates of public institutions had been declared unconstitutional. They violated provisions for due process or were

Dr. Albert S. Priddy, original defendant in *Buck v. Priddy*, the case that
became *Buck v. Bell*. Courtesy Central Virginia Training Center.

suspect because they did not apply to residents of asylums and other citizens equally. Some laws did not require notice to patients who might face surgery; others failed to provide them with legal counsel to oppose the operation.[9] Strode also remembered from his earlier service in the General Assembly that the 1910 Carrington bill to sterilize prisoners had failed because of negative public sentiment. These factors led Strode to advise Priddy against introducing a sterilization law. The State Hospital Board subsequently dropped the legislative proposal.

A Virginia Sterilization Law | 93

Undeterred by this setback, Priddy openly campaigned for a sterilization act. He argued that "High Grade Morons of the Anti-Social Class"
were "wholly unfit for exercising the right of motherhood." He contended
that most of his own female patients would have made their way to a "red-
light district," had law not abolished those places. Such women could not
be reformed, he said, "because they are defectives." Priddy declared himself
strongly "opposed to a drastic and far reaching law." But women diagnosed
as "incorrigible and delinquent mental defectives" represented an exception
to this cautious sentiment. Sterilization for them was "the only solution."
Priddy admitted that in the previous seven years (1915–1922), he had operated on as many as one hundred women "for pelvic disease." Those sterilizing operations allowed women to be placed in families where they were
self-supporting and "led happy and useful lives," he claimed. The choice
was clear: without the option of sterilization, the state must plan to pay for
their lifelong detention in institutions, at a price that had "no limit."[10]

The involvement of Strode, Priddy, and Irving Whitehead in Democratic Party politics was sometimes difficult to distinguish from their involvement in the fortunes of the Virginia Colony for Epileptics and Feebleminded. Priddy was a former state legislator who remained active in party
politics. When a special state convention was called to select a Democratic
nominee for an open seat in the U.S. House of Representatives, Priddy encouraged Strode to test the political waters. Strode kept Priddy apprised
of his political strategy. Irving Whitehead was also a Strode confidant and
supporter. Whitehead assisted Strode's campaign, at one point by relaying
Priddy's offer to send two carloads of Colony patients to swell the crowd at
an election rally.[11]

The year 1922 was difficult for Strode. He failed to win the party nomination, and his wife died in an auto accident soon afterward. But eighteen
months later, he put ill fortune behind and married Louisa Hubbard. That
union would provide Strode and Priddy with one more connection to the
national leaders of the eugenics movement.

Louisa Hubbard was one of the first graduates of the University of Pennsylvania School of Social Work, where she was thoroughly trained in the
eugenic perspective toward social problem groups. She learned the relationship between crime and heredity by reading *The Kallikak Family*. She was
skilled in constructing pedigrees, and she knew how to administer IQ tests

to clients. Several of her classes covered topics in eugenics, all emphasizing the importance of segregating the feebleminded and "prevention of procreation." Her course in psychiatry explained that feeblemindedness was 60 to 80 percent hereditary and that the feebleminded were "often morally weak." She explained on a test how Mendel's laws controlled heredity and feeblemindedness, and she echoed clichés about "the danger of inbreeding inferior stock." To the young student, Mendel's insights into genetics formed part of the "scientific basis of social work."[12]

Following graduation, Miss Hubbard joined the staff of the Red Cross and was assigned to survey families of war veterans returning from Europe. She toured the farmlands of rural Virginia on horseback, collecting information and providing a conduit for Red Cross relief and assistance. Always short on resources, she arranged to use Aubrey Strode's office in Lynchburg whenever business took him out of town.

When she left Virginia in 1920 for a job in North Carolina, Hubbard maintained a correspondence with Strode that blossomed into romance after the death of Strode's first wife. By the summer of 1923, they had confirmed their plans for marriage, but a delicate problem surfaced. Hubbard's medical history suggested that she might not be able to bear children. To assuage her fears, Strode arranged to have his friend Priddy conduct a confidential medical examination. The diagnosis revealed that Hubbard's problems were related to ulcers, not childbearing. On hearing the good news that she could safely be a mother, Hubbard praised Priddy. She joked to Strode about her "craziness" with plans for the impending marriage and the possible need of Priddy's psychiatric expertise. In 1923 Louisa Hubbard and Aubrey Strode were married; Priddy also took a bride that year.[13]

Within a few months, the State Hospital Board was again considering a sterilization law "in view of its importance to the institutions." The Board's change of heart, and what in retrospect Strode credited to a major "change of public sentiment" about eugenic sterilization, could more easily be explained by the political interests of E. Lee Trinkle, governor of Virginia and longtime Strode political colleague.[14] By the time Strode wrote the sterilization bill, he had known Trinkle for more than twenty years. Both were Democrats and shared a number of progressive views; both supported reforms to state health and welfare systems. Together they served in the state senate, at various times endorsing the prohibition of liquor and fighting for women's suffrage. Trinkle's experience included a term on the board of directors of the Children's Home Society, the agency that became entangled in the *Mallory* litigation.[15]

Following his election as Virginia governor in 1922, Trinkle made several unprecedented visits to meetings of the General Board of State Hospitals. His chief budget officer accompanied him, and at each meeting he delivered an analysis of the state's fiscal predicament.[16] The need for economic austerity was no news to the institutions' representatives, who had faced budget deficits for several years.[17] At the September 1923 meeting of the hospital board, Trinkle gave a report on the "critical financial condition of the State" and urged that it was "absolutely necessary . . . to use the most rigid economy" in institutional administration. Discussion ensued on the means by which "increased demand for admission to [state] institutions could be prevented." One of the measures proposed and approved by the board and the governor was a legislative initiative "to legalize under proper safeguards, the sterilization of insane, epileptic and feeble-minded persons . . . to relieve the institutions of their crowded conditions" and so that patients "could leave the institutions, become producers and not propagate their kind."[18]

The role of the State Hospital board in Trinkle's eugenical budget scheme was never clearly outlined for the public. On the eve of the 1924 Virginia General Assembly session, however, it was obvious that the state's finances were stretched. Trinkle's budget message blamed poor trade conditions, a business downturn, a coal field strike, and inflated revenue projections for producing the nearly two-million-dollar shortfall at the end of 1923.[19] He explained that the state budget had "jumped by leaps and bounds" under his two most recent predecessors in the governor's office. Expenditures grew, he said, because the state, "in her desire to be a real mother to all who would tend toward helplessness," initiated new programs that had not previously been a state responsibility. To understand this trend better, Trinkle reported that he had personally visited each state institution in order to assess its needs. He found facilities bulging at the seams, with more needy patients arriving daily.

Past budgets showed that in 1916, expenditures directly related to state hospitals caring for "defectives" constituted about 12.5 percent of the total Virginia budget. By 1920, the total cost had increased by nearly 59 percent but still made up only 13.15 percent of the total budget; this was so because the other costs of state government had also increased by more than 50 percent. While the increase in support for "defectives" was real, it was consistent with inflation in other sectors of government. During the same period, the "public works" budget—for example, money spent on state highways—had almost quintupled.

In general, economic conditions in the South in late 1923 were better than any time since 1919. Agriculture was booming, and unemployment had decreased.[20] The 1924 state budget proposed by Trinkle reversed inflationary trends, shrinking by more than 4 percent; the costs attributed to "defectives" also dropped, but only slightly. Thus, in the year that the Virginia sterilization statute was enacted, costs attributable to the "socially inadequate" were decreasing, rather than increasing. This was the economic backdrop when Priddy told Strode to draft a sterilization bill in time for the next session of the General Assembly. Their hope was that increased understanding of "the laws of heredity and eugenics and changing public sentiment" might lead to a law that could survive legal attack.

Strode would later recall that he drafted the bill "after some study."[21] His legal research was made significantly less onerous by a book Priddy had given him. The language he used and the arguments for the hereditary nature of mental and physical defects were taken, in many places verbatim, from Harry Laughlin's Model Law, printed in *Eugenical Sterilization in the United States.*

Of the fifteen state sterilization statutes in force in 1921, only ten were still in active use. New or revised laws were introduced in Illinois, Minnesota, New Hampshire, and Ohio. Each of these attempts failed. Concern over litigation elsewhere had even slowed the pace of operations in California.[22] But Strode avoided strategies that had failed in other states, intentionally excluding most of the inflammatory language that some eugenicists used. The preamble sounded prudent—applying only to people in institutions and only then in "certain cases." The word *eugenics* was nowhere to be found, nor were technical terms like *germ-plasm* or *genetics.* Instead, the law consisted of a simple argument for the need for sterilization and a lengthy recitation of procedural details to insure the rights of patients.[23]

Strode's law began with a short preamble that summarized the rationale for eugenic sterilization. It declared that "both the health of the individual patient and the welfare of society may be promoted . . . by the sterilization of mental defectives." This assertion contained a qualification that was critically absent from sterilization laws that had failed. Not only would "the common good" be served by sterilization, but the health of the patient would also be advanced. What patients would gain was not explained, and since the time of the controversy that had surrounded Laughlin's early efforts, no credible evidence had emerged to support the view that sterilization provided therapeutic benefits.

Strode then incorporated an additional qualification: sterilization would occur only "under careful safeguard and by competent and conscientious authority." The implication was that only physicians with the best interests of patients in mind would operate. Sterilization would be accomplished in males through the operation of vasectomy; in females, salpingectomy. Though these medical terms were not common to the general public, the lay reader was comforted by the explanation that these "operations may be performed without serious pain or substantial danger" to patients. To make the point even more clearly, another section of the law also specifically prohibited castration.

The eugenic argument came next. The state already cared for and supported many "defective" persons in various institutions. If such people were discharged from these facilities, they would "likely become by the propagation of their kind a menace to society." However, if they were sterilized and then released, they could "become self-supporting with benefit both to themselves and society." Two fears were broached in this section: fear of the social menace—crime, disease, and poverty—that resulted from indiscriminate reproduction, and fear of the rising tax burden that would result from maintaining expensive state institutions to care for more defective people. Even though the potential benefit to patients was not spelled out, the implication was clear that life outside of an institution, even after sterilization, was better than life within one.

The final sentence of the preamble explained how defective people posed a "menace to society." Strode did not burden the reader with scientific data or medical terms. Instead, he relied on the wisdom of everyday life, noting that "human experience has demonstrated that heredity plays an important part in the transmission of insanity, idiocy, imbecility, epilepsy and crime." The assumption that "human experience" demonstrated how important heredity is to the body, mind, and behavior of human beings was not new. Similar language had appeared in the failed "Bill for the Prevention of Idiocy" in Pennsylvania from 1901 and 1905; states like Indiana, New Jersey, and North Dakota had adopted almost the same phrase in their laws. The idea that heredity controlled human defectiveness had also been built into earlier laws that governed the Virginia Colony for Epileptics and Feeble-minded, such as the three different pieces of legislation Strode himself had written in 1916 alluding to the problem of feebleminded women who might bear children.[24] The commonsense assertion linking mental disorders and social problems with inheritance was highly questionable, but it was made repeatedly in sterilization laws. The presumption that the ills of society could be cured through a selective control of the mechanism of heredity was a major tenet of eugenic dogma.

The details of administering the sterilization program filled six lengthy sections. Authority for initiating the sterilization procedures was vested in superintendents of state hospitals or colonies for the mentally deficient and was made contingent on a diagnosis of hereditary defect. The law then enumerated the rights of patients. A superintendent wishing to operate on a patient must first petition the special board of directors, the governing body of the institution. Then the patient and the guardians were to be notified of the proceedings in writing; if no legal guardian was present, one would be appointed by the local court. Patients could attend their sterilization hearings, put witnesses under oath, and receive a written record of the evidence. They also had the right to be represented by a lawyer.

Sterilization could be ordered if the board found that the patient was "insane, idiotic, imbecile, feeble-minded, or epileptic, and by the laws of heredity is the probably potential parent of socially inadequate offspring likewise afflicted." The board must also find that the inmate could be sterilized "without detriment to his or her general health" and that both the welfare of the patient and society would be promoted by the operation. Following a board order for sterilization, patients were given the right to appeal to the local circuit court and to the State Supreme Court of Appeals.

The two final sections of the law made physicians who performed sterilizations immune from civil or criminal liability and also reaffirmed the prerogative of physicians to sterilize patients for therapeutic (rather than eugenic) motives.

While some parts of the law simply repeated provisions in existing statutes, the language of "social inadequacy" was entirely new to the Virginia legal code. The phrase "socially inadequate offspring" and a segment giving therapeutic latitude to physicians were taken directly from Laughlin's 1922

Model Sterilization Law.[25] Strode credited Laughlin's text as the source of several references in an article he wrote in 1924 entitled "Sterilization of Defectives."[26]

Social inadequacy was, like feeblemindedness, somewhat hard to define clinically. But from the vast numbers covered by Laughlin's Model Law to the somewhat smaller group identified in Strode's law, the socially inadequate always included the mentally and physically afflicted, the criminal, and those of morally weak constitution. According to Laughlin, this was the group most likely to reproduce in great numbers and to pose the greatest cost to state programs of social welfare and criminal justice.

Strode took the time to lay the political groundwork that would ensure the passage of his sterilization law. He had the bill introduced by an old friend, who as a Senate committee chairman could guarantee a favorable endorsement. A Senate committee spent two hours on the bill. After testimony by Priddy and other hospital superintendents, the measure was judged to be "not as drastic as senators at first thought."[27] Virginia's House of Delegates was similarly expeditious. Following additional testimony in committee by Priddy and his friend Dr. Joseph DeJarnette, the bill passed the House with only two votes in opposition. Within two weeks the bill passed the Senate unanimously.[28]

Other matters took up the attention of legislators. Helen Keller addressed the Virginia legislature as the sterilization bill made its way through committee. She did not comment on the bill, nor did the press remark on the irony of her visit.[29] Another eugenic enactment—a law to strengthen prohibitions on interracial marriage and "preserve racial integrity"—became law the same day as the sterilization bill. That new legislation attracted considerable press attention both before and after passage.[30] In contrast, Strode's bill attracted only the most cursory mention.[31] Following a record-setting session, during which nearly 150 bills were passed by the House of Delegates in a single day, the sterilization bill arrived at Governor Trinkle's office buried in a pile of 250 enactments awaiting signature.[32] Quietly and without ceremony, "An Act to provide for the sexual sterilization of inmates of State institutions in certain cases" became Virginia law.[33]

About that same time in Washington, D.C., the finishing touches were being put on the new Academy Research Building. Dubbed "America's Hall of Fame for Science," the block-long marble structure facing the Lincoln Memorial was built to hold the National Academy of Sciences and the National Research Council. A bas-relief gracing its facade depicted a parade of thirty-seven historically significant heroes of science. Among them was Francis Galton, the father of eugenics.[34]

Planning the "Test Case"

Priddy waited patiently at the Virginia Colony while Strode ushered the sterilization law through the legislature in Richmond. But he was anxious to get on with his work and prepared to begin operating again as soon as the law took effect in July 1924. In August, Priddy presented a list of patients to his Board members, asking their approval. There were eighteen names on the list, all women. The new law required each patient to have a guardian, and R. G. Shelton, a local Justice of the Peace, had been named to fill that role for every patient. Following an *en masse* hearing, the Board approved fourteen sterilization petitions, holding another four in abeyance until its next meeting.[35]

Priddy was in constant contact with Strode, and he understood that surgery would have to wait for the next step in the lawyer's plan regardless of the Board's actions. Haunted by the humiliation of the *Mallory* lawsuits, Priddy followed Strode's advice and suggested to the Board that "as a matter of precautionery [sic] safety . . . a test case of the constitutionality of the Sterilization Law be made before any operation is performed." Priddy was dispatched to ask the General Board of State Hospitals to approve the lawsuit so that the Colony's legal bills would be paid.[36]

The next month, Strode was formally retained for the litigation. He negotiated payment of $250 for the trial and $250 for each appeal of the case, plus expenses. Retainer in place, he explained his strategy to the Board. The constitutionality of sterilization laws in other states had been challenged for two reasons, and his law might also be vulnerable. The first issue was "due process." Sterilization laws unlawfully impaired life and liberty if they failed to include specific procedural safeguards like the right of the patient to testify or call witnesses in her own behalf. Strode dismissed this concern, noting that he "had so carefully prepared the Virginia Act to meet this objection that he considered the question of its [un]constitutionality on this ground negligible." The second and principal issue, according to Strode, was equal protection. Because the Virginia law applied only to institutionalized patients and not to feebleminded people in the community, an objection could be made that Colony patients did not receive the "equal protection" of the law—a potentially serious constitutional problem. Because unconstitutional laws had no legal force, any sterilization performed before these issues were settled might again expose Priddy and the institution to liability. Strode advised that all sterilizations be postponed until the Virginia law had been validated by the Court of Appeals of Virginia and "possibly the Supreme Court of the United States."

The Board completed its remaining business, considering sterilization petitions for sixteen more women. It set aside fifteen of them indefinitely, pending resolution of the constitutionality of the new law, but one petition took up the rest of the meeting. It involved a young girl from Charlottesville, Virginia, named Carrie Buck.[37]

8
Choosing Carrie Buck

In the fall of 1923, Alice Dobbs discovered that she had a serious problem. Her seventeen-year-old foster child Carrie Buck was pregnant. By Thanksgiving the condition was apparent; by Christmas it could no longer be hidden. By New Years Day, Mrs. Dobbs knew that she must do with Carrie what was always done for girls of "that type"—send her away. On January 23, 1924, only a few days before Aubrey Strode's sterilization bill was introduced in the state Senate, Mrs. Dobbs's husband John was temporarily deputized as a "special constable" and directed to deliver notices and subpoenas to his wife, Carrie Buck's parents, and two doctors and also to bring Carrie to the Juvenile and Domestic Relations Court in Charlottesville, Virginia. There, he and his wife would argue that Carrie should be sent to the Virginia Colony for Epileptics and Feeble-minded.

Mr. and Mrs. Dobbs described what they knew about Carrie in papers they filed with the court and in their testimony to the judge. She was born July 2, 1906, the daughter of Frank and Emma Buck. Emma was already a resident of the State Colony near Lynchburg, but Frank's whereabouts were unknown. Carrie had lived at the Dobbses' home since she was three or four years old and was in generally good health. She spent her time helping Mrs. Dobbs with chores around the house.[1]

The Dobbses claimed that Carrie was subject to "some hallucinations and some outbreaks of temper" and that she was dishonest. They said that she had been born with an unusual mental condition that had been demonstrated by certain "peculiar actions." She had attended school five years and reached the sixth grade and had experienced no problems with liquor or drugs but was guilty of "moral delinquency."[2]

The testimony of the Dobbses was somewhat confused on the subject of Carrie's supposed physical disability. At one point they agreed that she had never been "subject to epilepsy, headaches, nervousness, fits or convulsions," and they confirmed that she had had no "fits or spasms of any kind."

Yet to the direct question "At what age did epilepsy first appear?" they responded, "Since childhood." They said that it was impossible for them to control Carrie and that they could not afford to support her.[3]

The Dobbses' family physician, J. C. Coulter, and another Charlottesville doctor, J. F. Williams, reported to Justice of the Peace C. D. Shackleford that in their judgment Carrie was "feebleminded within the meaning of the law" and ought to be sent to an institution.[4] Shackleford signed the order, directing the Colony to admit her when space was available.[5]

The social worker managing Carrie's case sent the paperwork to Priddy, but the commitment law required that a notice of commitment be delivered to Carrie Buck before the judge's hearing. Seeing no copy of the notice, Priddy promptly sent the papers back for correction. Nearly seven weeks later, the papers were still not ready, and Charlottesville lawyer Homer Ritchie wrote to explain the delay. Speaking on behalf of the welfare authorities, Ritchie assured Priddy that the paperwork was unnecessary, since Carrie had in fact appeared at the hearing and her presence was noted in the record. But Priddy was adamant that his own "severe lessons" learned in the *Mallory* case counseled against any legal irregularities, no matter how technical.[6]

The next day, Priddy received a letter from a Caroline Wilhelm, a Red Cross nurse helping with the case, who said that Mrs. Dobbs would soon need to be available to help with the birth of her own daughter's child. By then Carrie Buck was also very pregnant, and Dobbs was afraid to leave her alone. Wilhelm wanted the paperwork problem to be settled quickly so that Carrie would be at the Colony when her baby was born.[7]

From Priddy's perspective, too many children had been delivered at the Colony already, and even though he had made it clear that expectant mothers would no longer be admitted, he did agree to take Carrie in as soon as the baby was born and all legal requirements had been met.[8] Carrie was sent briefly to a home on the other side of Charlottesville for the duration of her pregnancy.[9] Her baby, Vivian Buck, was born on March 28, 1924.

As a patient on her way to the Colony, Carrie could not keep the baby with her. But Caroline Wilhelm was reluctant to put the child up for adoption since both Carrie and her mother had been declared "feebleminded." Proper homes were difficult to locate for such children, she indicated. After some discussion, Priddy's only suggestion was to place the baby in the city poorhouse. The Dobbs family agreed to care for the girl on the condition that she would be sent to the Colony if in the future she continued to be feebleminded.[10] Carrie Buck left Charlottesville barely two months later, her brief career as a mother finished.

Coming to the Colony

Caroline Wilhelm took Carrie to the Colony on the afternoon of June 4, 1924, along with a young man who had also been committed. The one o'clock train to Lynchburg took only ninety minutes, and after the short buggy ride across the James River, she was admitted as Colony inmate no. 1692. Dr. John Bell examined her, noting that she was well nourished, clean, and free of infection. From the looks of her abdomen, he said, Carrie had recently had a child, but she was in good general health.[11]

Priddy conducted a follow-up examination. Carrie's father, Frank Buck, had died accidentally, but no details about her two living siblings or other family members were available. Priddy thought the Dobbses' report concerning Carrie's "temper and hallucinations" was inaccurate; he found no evidence of psychosis. He noted that she could read and write and "keeps herself in a tidy condition."[12] The mental examiner who performed the intelligence testing wrote inaccurately that Carrie had spent nine years in school and had repeated one grade. Her IQ was listed as 56.[13] In fact, Carrie spent five years in school and repeated no grades.[14] A few days later, her Wasserman test for syphilis came back from the laboratory negative.

Before long, Carrie saw her mother Emma, who had been committed to the Colony four years earlier. Emma Adeline Harlowe was born in Charlottesville, Virginia, on November 28, 1872. She never knew her mother, Adeline Dudley Harlowe, who died during childbirth. Her father, Richard Harlowe, was a farmer in Albemarle County, just outside of Charlottesville. Though family fortunes were modest, Emma was able to finish five years at school. There is no record of her early years.

Emma married Frank Buck in 1896, just before her twenty-fourth birthday. Her first child, Carrie, was born ten years later. At the age of thirty-five, again pregnant, "housewife" Emma entered the hospital for a ten-week stay. She left alone, probably after a miscarriage. When she came to the hospital, pregnant again, the ledger listed her age as forty-three and her occupation as "servant." She left twenty-three days later, accompanied by her husband Frank and her new daughter Doris. At every hospital admission, Emma was designated as a married woman.[15]

Emma Buck was taken to the Colony from Charlottesville in 1920, at the age of forty-eight. She was described in Colony records as a widow who "lacked moral sense and responsibility." The doctor who examined her on arrival described Emma as "well nourished, fat and pale, nervous and restless." She had suffered from pneumonia, rheumatism, and syphilis and was in generally poor health. Her arms showed scars from intravenous injections, suggesting illicit drug use. A reputation as "notoriously untruthful" had followed her from her school days; her record noted that she had been arrested for prostitution and given birth to illegitimate children. She could do housework under supervision but was "untidy." She had four and a half dollars in cash, and her clothes were "in very bad condition." Her IQ test score was 50. The initial diagnosis was: "Mental Deficiency, Familial: Moron."[16]

The summer of 1924 she became reacquainted with her daughter Carrie, and they lived near each other at the Colony for four-and-a-half years, the longest single period of contact in their lives.

Building the Sterilization Case: The Hearing

By the time Aubrey Strode met with the Colony Board to urge that it test the constitutionality of the sterilization law, Carrie Buck had been at the institution for over three months. Upon her arrival, Priddy immediately made the connection between mother and daughter: both poor, both judged feebleminded, both accused of sexual misconduct. He quickly began collecting information to demonstrate the hereditary defects he was certain linked Emma and Carrie.

Priddy checked off every detail of the law as he prepared. He had R. G. Shelton named as Carrie's guardian on July 21, 1924. On July 24 he prepared a petition requesting approval to operate on Carrie Buck. He set the date for the September meeting of the Board and made sure that Emma Buck received written notice of the sterilization hearing.[17] Then Priddy appeared before the assembled Colony Board as a witness to detail his opinions of Carrie's mental condition.[18] Leaving no detail to chance, Priddy even dictated a note for Carrie's medical chart that day, saying that she was "desirous of taking advantage of the sterilization law" and describing the legal procedures that were followed at the hearing.[19]

According to Priddy, although Carrie had a chronological age of eighteen, mentally she was nine years old. He classified her as "feebleminded of the lowest grade Moron class," also noting that Carrie had borne "one illegitimate mentally defective child" and was "a moral delinquent." The value of sterilization to her would consist of the freedom to "leave the institution, enjoy her liberty and life and become self sustaining." The value to society would be not having to maintain her in custody for her child-bearing period, potentially thirty years. The sterilizing operation of salpingectomy was "simple and comparatively harmless," said Priddy, who admitted to operating on about one hundred women without "a single death." Guardian R. G. Shelton questioned Priddy concerning the potential to rehabilitate Carrie through a course of training in Priddy's institution. Could she not, Shelton asked, be restored to society without the operation? No, responded Priddy, reform was impossible because she was "congenitally and incurably defective."

Carrie Buck was present at the hearing as the law required, and Strode posed one question directly to her. "Do you care to say anything about having this operation performed on you?" "No sir, I have not," Carrie replied. "It is up to my people."[20]

The Board voted to sterilize Carrie Buck, casting its resolution in the language of the statute Strode had adapted from Laughlin's Model Law: "Carrie Buck is a feebleminded inmate of this institution and by the laws of heredity is the probable potential parent of socially inadequate offspring, likewise afflicted, that she may be sexually sterilized without detriment to her general health, and that the welfare of the said Carrie Buck and of society will be promoted by such sterilization."[21]

To complete the required procedures, the Board had only to appoint an attorney to appeal the sterilization order and defend Carrie. Shelton was directed to hire "some competent lawyer" to oppose Strode in the litigation. He chose Irving Whitehead, confidant of Priddy, boyhood friend to Aubrey Strode, former Colony director, and sterilization advocate.

Following the Board meeting, Priddy wrote to Caroline Wilhelm in Charlottesville, seeking more information on the Buck family. He asked her to prepare a list of "all defectives" connected to either side of Carrie's family and to bring it to assist her testimony at the coming sterilization trial. The superintendent of the public welfare office in Charlottesville also offered to assist with Priddy's field study of the Buck family.[22]

Then Priddy contacted Edith M. Furbush of the National Committee on Mental Hygiene in New York. He explained that the Virginia sterilization law had been written by "one of the ablest lawyers in Virginia" but nevertheless would require a test case to be "on the absolutely safe side." Priddy asked Furbush to send a field agent down to assist with the investigation of the Buck family. But Furbush offered no help.[23]

As part of his own trial preparation, Aubrey Strode wrote to Harry Laughlin at the Eugenics Record Office. He asked for assistance in assembling evidence for the case and requested a deposition that could be presented to the court. According to Strode, the case involved a feebleminded mother and daughter, both Colony residents. The daughter had a child of her own, also feebleminded.[24] Strode thought that analysis of a "nationally recognized expert" like Laughlin would be especially valuable when the case reached the higher courts, since Laughlin's book had been an important aid in structuring the case to that point.[25]

Laughlin alerted Strode to a Michigan sterilization case that was under way but noted that he was more interested in the Virginia statute, which was more recent and had incorporated the lessons of previous legal challenges. Laughlin believed that the U.S. Supreme Court was ready to "sustain the essential elements" of his Model Law. He pointed to the importance of preparing a family history or pedigree study in the *Buck* case to demonstrate the nature of hereditary feeblemindedness. He congratulated Strode on selecting Carrie Buck as the subject of this case because both her immediate ancestor (Emma) and her immediate offspring (Vivian) were feebleminded. In light of the hundreds of cases he had reviewed, Laughlin could "not recall a single instance in which feeblemindedness appeared in the grandmother, the mother . . . and the child (three generations), by environmental or accidental causes." He sent forms for Priddy to use in collecting information on the relatives of Carrie Buck and offered to analyze it as part of his deposition for the court.[26]

But Priddy was hard pressed to keep up with the pace of work Strode was generating for him. He was very ill, and his tonsils had recently been removed to alleviate what was thought to be an infection of the neck. He was unable to speak because of the surgery, and time in the hospital put him a week behind in trial preparation. On October 14 he responded directly to Laughlin. When the decision was made to pursue a sterilization law, Priddy said, he gave Strode all the information he had on the topic, including his personal copy of Laughlin's book *Eugenical Sterilization in the United States*. He conceded that there were limits to what any sterilization law could do, but he emphasized that the Virginia law was "strictly a humanitarian and economic provision." Without "a greatly increased tax rate," it would not be financially feasible to take in more than a few of the many potential patients in the state, particularly the "anti-social women and girls . . . who bear illegitimate children and increase the population of mental defectives." Sterilization would provide both moral and fiscal medicine. Though both Priddy and his colleague Joseph DeJarnette were initially doubtful that the legislature would actually pass the new law or that the courts had reached a level of "progressive enlightenment" about eugenics that would make them willing to cast off their "constitutional morrings [sic]" and "meet this pressing need," they had been happy to see the Virginia General Assembly rise to the challenge with the new enactment.[27]

Priddy planned to use the Colony as a "kind of clearing house" where inmates would be trained, sterilized, and sent out to earn their own living and relieve the state of "this enormous financial burden." He apologized for his inability to make out a genealogical tree of the type that Laughlin might construct, but, he wrote, "this girl comes from a shiftless, ignorant and moving class of people, and it is impossible to get intelligent and satisfactory data."

Priddy's frustration was apparent. The information collected by Caroline Wilhelm called into question most of Carrie's kinship links. On one hand, people named Harlow who lived at the Colony denied that Emma (Harlowe) Buck was related to them. To complicate the picture even more, there was "considerable doubt" that Carrie's father was actually a Buck. Rumor had it that Frank Buck was not Carrie's father in any event, and there was little else that could be proven on this score, as he was dead. Carrie's pedigree was unclear indeed. Disregarding these details, Priddy judged that the "line of baneful heredity" came through Emma; since all of the Bucks and Harlows who lived at the Colony came from Charlottesville and Albemarle County, he concluded that "they are of the same stock."

Priddy had little else to say about the supposedly "hereditary" nature of Carrie's problems, but he completed the forms that Laughlin had supplied to collect information on her relatives. He did describe her appearance, saying "she is well grown, [and] has a rather badly formed face." Priddy urged Laughlin to prepare a deposition soon since the trial was scheduled for October 20, only a week later.[28]

The next day a gaping hole appeared in the case when Wilhelm refuted a critical assumption of the case that Priddy, Strode, and Laughlin had thought well settled: there was no firm evidence that Carrie's baby Vivian was mentally defective. "I do not recall," Wilhelm said, "and am unable to find any mention in our files of having said that Carrie Buck's baby was mentally defective." She reminded Priddy that in an earlier letter she had stated that placing the child for adoption could be problematic; people might presume the baby was impaired considering that both her grandmother and mother were at the Colony.[29]

In an effort to buy time and find more evidence, Strode had the trial date delayed. With so little clear information from Priddy's informants in Charlottesville, he took Laughlin's advice to solicit the services of his ERO colleague Arthur Estabrook to testify as an expert witness at the Buck trial. Estabrook, who was a friend of Strode's wife, had already volunteered to help with the case before Strode thought he would be needed.[30] Now, with hard evidence lacking and time running out, Strode asked Estabrook to appear in person as an expert witness. The lawyer knew that authoritative testimony from a published expert in eugenics would be powerful at trial and would complement the firsthand information from people who knew the Bucks in Charlottesville.[31]

As the trial date approached, Priddy wrote his friend Joseph DeJarnette, declaring his presence at the trial "absolutely necessary." Priddy urged DeJarnette "to read up all you can on heredity of the jukes, callikaks [sic] and other noted families of that stripe." Priddy's illness prevented him from traveling, but he asked DeJarnette to go to Charlottesville to do a mental test of Carrie Buck's baby. The test would be "the usual one," as outlined on the test sheet Laughlin had provided. "We are leaving nothing undone in evidence."[32]

Priddy, becoming more ill by the day and fatigued from receiving x-ray treatment for his mysterious disease, agreed with Strode about the importance of Estabrook's testimony. But Priddy was unsure how much commentary on the infamous Jukes and Kallikaks would be allowed into evidence to make his argument concerning the costs of defective heredity.[33] Having the author of the modern follow-up study of the Jukes on hand to testify as an expert would clearly help.

Estabrook's fieldwork in the backwoods of Kentucky made him impossible to contact for days at a time, so for ten days Strode's letter to him went unread. As he waited for a response, Strode did not know that Laughlin had already asked Estabrook's wife to send a telegram to her husband. Describing himself "exceedingly anxious" because of the impending trial, Strode asked Laughlin for help again.[34] Laughlin completed his sworn deposition on November 6 and sent it to Strode later that week, but before he had time to answer Strode's request, Estabrook confirmed that he could travel to Virginia to complete the Buck family study and testify at the trial on November 18.[35] Strode wanted him in Lynchburg by November 13, to spend time at the Colony talking to Priddy and examine Carrie, Emma, and their records. He could then go to Charlottesville to fill in the blanks with "additional pertinent facts" not yet provided by the social workers there who had investigated the Buck pedigree. Strode agreed to pay Estabrook his fee of twenty dollars a day plus expenses.[36]

Estabrook was unsure of his arrival time, since he had to ride a full day on horseback just to reach the rail line. But by November 14 he had made the trip, seen Carrie and Emma, and taken a snapshot of them for his photo album collection. He detailed his opinion of the Bucks to Strode in a letter, then took the train to Charlottesville. Strode sent Estabrook a batch of subpoenas for potential witnesses he had found, along with instructions for paying travel expenses to the trial on November 18. Strode also let Estabrook know that a copy of his book *The Jukes in 1915* had arrived in the Saturday mail.[37] Having completed his examination of the infant Vivian Buck and interviews with Buck family members, Estabrook returned to Amherst on November 17 to meet with the lawyer for a final conference before the trial.

9
Carrie Buck versus Dr. Priddy

The morning of the *Buck* trial was unusually cold. Snow in mid-November was rare enough, but getting up before dawn for a train ride to court was not a prospect that most of the witnesses relished. As they made their way up the brick walkway to the County Courthouse, each visitor saw the obelisk placed by the Daughters of the Confederacy to honor the "sons of Amherst County" who had fallen in the Civil War a generation earlier. The plaque stood as a reminder of the sacrifices citizens are sometimes asked to make for the good of the community. The inscription also remembered the survivors, whose later lives were lived "ever proud that they had done their part in the noble cause." Saving society from a threat some considered as dangerous as a foreign army would be a theme in the trial the visitors were about to attend. At least some of the witnesses in that trial regarded eugenics as a similarly noble cause, a kind of biological patriotism.

Aubrey Strode was a careful lawyer. He orchestrated the passage of the sterilization law and constructed a legal test that he could shepherd through the courts. Though last-minute surprises threatened to ruin his lawsuit, Strode quickly recovered to assemble a case he had every intention of winning. It was built on several kinds of testimony from what he hoped were uniquely qualified witnesses. There were teachers who had observed the Buck family in school, and there were social workers from welfare agencies who had monitored similar problem families in the community. Strode even called several neighbors of the Buck family to show how ordinary people viewed the Bucks. But the most important witnesses were the experts, each with the title of "Doctor," all well versed in eugenic theory. Two medical doctors who ran asylums for the defective took the stand, and two eugenic scientists—authorities from out of state—added their opinions on the workings of heredity and the threat posed by girls like Carrie Buck.

Strode began his presentation with a touch of courtesy to put the witnesses at ease. People from Charlottesville had traveled some distance, he said. Though it would not be the usual practice, he planned to question them in the beginning so that they could return home early. Mrs. Anne Harris, a Charlottesville nurse, was first.[1]

Harris claimed to have known Carrie Buck for twelve years. She described Emma Buck as someone who was "on the charity list for a number of years, off and on—mostly on." She lived in the "worst neighborhoods" and could not—or would not—take care of her children. They were "on the streets more or less."

Strode pressed Mrs. Harris for more detail. The Bucks were, she said, "absolutely irresponsible" and dependent on "numerous charity organizations."

Rumor had it that Doris Buck was only half-sister to Carrie and was a "very stormy individual, . . . incorrigible." Emma Buck had children even though she wasn't living with her husband. The nurse knew the phrase "the socially inadequate person" and believed that it defined someone who

Carrie and Emma Buck at the Virginia Colony just before the Buck trial, 1924. Courtesy Arthur Estabrook Papers, University at Albany Libraries' Special Collections.

Carrie Buck versus Dr. Priddy | 113

Aubrey Strode, lawyer for the Virginia Colony in the *Buck* case.
Papers of Aubrey E. Strode, MSS 3014, Special Collections, University of Virginia
Library.

was "irresponsible mentally" and unable to care for herself. In response to
Strode's direct question about "feeblemindedness," Harris said that Emma
functioned like a twelve-year-old, the children no better.

Under cross-examination by Irving Whitehead, Harris repeated what
she had said earlier: she did not know much about them, "very little" about
Carrie past the age of four. Mrs. Harris had little recollection of Carrie except for an incident during her grammar school years:

> *Harris:* The Superintendent called me and said she was having trouble with
> Carrie. She told me that Carrie was writing notes, and that sort of thing. . . .

Whitehead: Writing notes to boys, I suppose?

Harris: Yes, sir.

Whitehead: Is writing notes to boys in school, nine or ten years old, considered anti-social?

Harris: It depends upon the character of the note.

Whitehead: If the child had been sixteen years old, would it still have been anti-social?

A girl like that, suggested Harris, should be sent away to the juvenile home.

Whitehead could easily have challenged Harris, who three times admitted that she had no firsthand information about Carrie. Instead he drew her out on the note-writing incident. He knew that Priddy had sterilized one girl for similar behavior, and as a member of the Colony Board, he had voted for the operation.

Harris's testimony added little to the court's knowledge of Carrie. She repeated secondhand observations that Carrie was a poor and at times unruly girl. Her mother had a questionable reputation, and Carrie was linked to both a sister and a vaguely identified cousin who also had some history of less-than-polite behavior.

Three teachers followed Harris on the witness stand. Beginning in 1918, Virginia law required school teachers to take a course in preventive medicine, so that they would be "prepared for public health work" and could perform medical inspections to identify "defective" children.[2] Searching for evidence of defect, Strode asked each teacher to recall Carrie's behavior and that of her sister and a boy thought to be her half-brother. One teacher described Carrie's younger sister Doris as "dull in her books," but none of the teachers knew Carrie.[3]

The next witness was John W. Hopkins, superintendent of an Albemarle County orphanage. He knew no more about Carrie's relatives than her teachers did. He did offer the opinion that two of her supposed half brothers appeared to be "right peculiar." Pressed by Strode for details, Hopkins admitted that he had told Estabrook that the one of the boys was "mentally defective" but could recall no specific observations upon which to base that opinion. He didn't know Carrie at all.[4]

Nor did Samuel Dudley, a man who identified himself as brother-in-law to Emma Buck's father. He also described two men allegedly related to Carrie as "a little peculiar" but seemed irritated at having been asked so many questions by Estabrook the previous day.[5]

At that point, onlookers in the courtroom were probably quite confused about Buck's relatives. None of the first six Charlottesville witnesses provided firsthand information about Carrie; the best they could do was repeat rumors and anecdotes like the episode of passing notes. When time to prepare for the trial ran out, Strode had to depend on Arthur Estabrook, aided by county welfare workers, to identify useful witnesses and subpoena them for trial. The lawyer had interviewed none of these witnesses himself before the trial, and as the lunch hour approached, he must have been wondering if anyone from Charlottesville really knew Carrie Buck.

The next witness did not allay that concern. Red Cross nurse Caroline Wilhelm was new to Charlottesville, having moved to town the previous February to become county administrator of public welfare. She had little firsthand experience with Carrie apart from bringing her to Lynchburg on the train early in the summer. Her first comments clearly revealed the real reason that Carrie was sent to the Virginia Colony. Wilhelm explained that Mr. Dobbs had reported to the welfare office that Carrie was pregnant and that "he wanted her committed somewhere—to have her sent to some institution."

Carrie's pregnancy was a threatening prospect for the Dobbses. John T. Dobbs did maintenance work on the city streetcar line, and his wife took in foster children. Their modest house backed up to the railroad tracks, where poor whites lived in a racially mixed neighborhood. Most people of means in Charlottesville lived on the other side of the tracks—where the University of Virginia cast its grand shadow. News of a pregnant girl in the Dobbs house would not sit well with the welfare officials, particularly since the boy who got her pregnant was Mrs. Dobbs's nephew. It was hardly surprising that the Dobbses wanted to have Carrie sent away.

Wilhelm spoke confidently, her opinion backed by her training and experience as a social worker. She thought that girls like Carrie were "more or less at the mercy of other people." Though she had never met Emma Buck, Wilhelm repeated the rumor that she had several illegitimate children. That was basis enough for her opinion that Carrie was also "very likely to have illegitimate children." Was Carrie "obviously feebleminded?" asked Strode. Proud of her expertise, Wilhelm replied "I should say so, as a social worker."

Then Strode asked about the Buck baby, who had been left with the Dobbs family. How old was she? "Not quite eight months old," Wilhelm responded.

Strode: Have you any impression about the child?

Wilhelm: It is difficult to judge the probabilities of a child as young as that, but it seems to me not quite a normal baby.

Strode: You don't regard her child as a normal baby?

Wilhelm: In its appearance—I should say that perhaps my knowledge of the mother may prejudice me in that regard, but I saw the child at the same time as Mrs. Dobbs's daughter's baby, which is only three days older than this one, and there is a very decided difference in the development of the babies. That was about two weeks ago.

Strode: You would not judge the child as a normal baby?

Wilhelm: There is a look about it that is not quite normal, but just what it is, I can't tell.

Caroline Wilhelm obviously had a change of heart. On October 15, she had notified Strode that there was no evidence of "defect" in the Buck child. Her reluctance to comment left Strode near desperation, his case crumbling. But in the following month, she left her office in the Charlottesville business district and crossed the tracks herself. There, two weeks before the trial, she found no new evidence but did reach a new opinion. Of two babies crawling on the floor, she found one "not quite normal." She had done no tests nor any extended observations; she merely noticed "a look about it" that she could not explain.

This was a perfect opportunity for Irving Whitehead to begin dismantling Strode's case, and for a moment it seemed as if he would. The morning was gone; seven witnesses into the case, Strode had produced almost no solid evidence to support the sterilization of Carrie Buck. The commentary about her mother was almost entirely hearsay and rumor. Of the seven people who had testified, only two had even met Carrie, one of whom admitted that the last contact had been more than fourteen years earlier. Caroline Wilhelm's conclusion about the Buck baby gave Whitehead the chance to focus on how superficial and weak the previous testimony had been.

"Who was this other baby you examined?" Whitehead asked. "Mrs. Dobbs's new granddaughter," Wilhelm replied.

Whitehead: Have you any impression about the child?

Wilhelm: Mrs. Dobbs's daughter's baby is a very responsive baby. When you play with it, or try to attract its attention—it is a baby that you can play with. The other baby is not. It seems very apathetic and not responsive.

Whitehead did not pursue the question of whether a baby taken prematurely from its mother under stressful conditions might not thrive immediately, nor did he ask how Wilhelm could be so certain when neither child could walk or talk. Instead he probed Wilhelm's knowledge of Carrie. Her file at the welfare department had been opened January 17, 1924, only a week before the commitment hearing, which Wilhelm did not attend. The only thing Wilhelm really knew was that Carrie had been pregnant. Was that the basis of her opinion that Carrie was "feeble-minded" and "antisocial"? Did the fact that she "made a miss-step [*sic*]—went wrong" prove feeblemindedness? Speaking once again "as a social worker," Wilhelm said that she had known girls of "that type"; while that might not prove they were feebleminded, "a feebleminded girl is much more likely to go wrong."

Wilhelm's testimony ended with an exchange about Carrie's ability to work (only under "very careful supervision") and whether Carrie had an "immoral tendency" ("certainly"). At that point, Whitehead seemed to be testifying for her, leading her with questions that could have been taken directly from Harry Laughlin's book.

Whitehead: Judging by the fact that she has already given birth to an illegitimate child, and has an immoral tendency, is it your opinion that by sterilization she would be made less of a liability and more of an asset to the State?

Wilhelm: I think it would at least prevent propagation of her kind.

Whitehead: It would prevent the propagation of her kind, undoubtedly, but is it your opinion that it would have a deterrent effect in that it would make her less immoral?

Wilhelm: I am afraid I am not competent to judge of that.

Whitehead: Your idea is, while she would never become an asset, she would become less of a liability by sterilization, and your idea is that she could be turned over to somebody and under careful supervision be made self-supporting? Is that your idea?

Wilhelm: I think so, yes, sir.[6]

Mary Duke was the eighth witness Aubrey Strode called to testify against Carrie Buck. Duke had been the temporary head of the welfare office in Charlottesville before Caroline Wilhelm took the position. Duke claimed to have heard of Emma Buck, and she had seen Doris Buck as "a little baby." She learned from Mrs. Dobbs that Carrie had been left with someone while Mrs. Dobbs was away in the summer of 1923. Unfortunately, "they didn't watch her closely enough" and Carrie became pregnant. Duke knew of Carrie's case because of the paperwork mixup she had to manage after Carrie's commitment hearing, but she only saw her once briefly and "never had any dealings with her." This lack of direct contact did not prevent her from venturing the opinion that Carrie was following in her mother's footsteps and "didn't seem to be a bright girl."[7]

Like most of the witnesses before her, Miss Duke's testimony shed no new light on Carrie's alleged mental condition. Irving Whitehead challenged nothing Mary Duke said, asking no questions in cross-examination.

Irving Whitehead, lawyer for Carrie Buck in the Buck case. Courtesy Virginia Historical Society, Richmond, Virginia.

All of the witnesses from Charlottesville left a generally negative impression about Carrie Buck and her family. Carrie's background was poor, she was poorly behaved and poor in looks. Though not demonstrably mentally ill, like some of her relatives she seemed somehow "peculiar." Her foster parents had wanted her sent away because of the shame of her pregnancy. Even her baby was presented as less than normal. And while none of the teachers, social workers, and nurses from Carrie's hometown could specify clearly the extent or the source of the Buck family's problems, they all seemed to agree that Carrie was typical of the group and likely to come to little good.

Irving Whitehead questioned each of Strode's witnesses in his turn. The manner of Whitehead's questioning was polite, with a gentle probing of the witnesses' conclusions regarding Carrie. Except questioning the hasty conclusion reached by the Red Cross social worker about the abnormality of Carrie's baby, Whitehead raised no strong challenge to any of the testimony. Whitehead made the first recorded comment in the trial and that comment, like his later objections, focused on a procedural formality. As if reading from the rules laid down in Strode's law, Whitehead announced: "Have the record show that the appellant [Mr. Shelton] is present. I want the record to show that the parties are all here."[8] With similar formality, Whitehead raised objections about the testimony solicited by Strode. Only moments into the testimony of the first witness, he interrupted Strode, who had posed a question about Carrie's mother:

> *Whitehead:* Wait a minute, right there is what I think is one of the important features, I am not objecting right now, but I think I will ask later to strike that out because I think that question violates the constitutional right of the defendant.

Whitehead lodged this kind of objection again after several witnesses had spoken. As the testimony continued, he seemed to tire of challenging evidence about the Buck family history and simply noted his repetitive objection as "the same thing I said before."[9]

Eugenics Experts

After sending the witnesses from Charlottesville home, Strode turned to his more prominent witnesses, the experts in medicine and eugenic science.

Court appearances by medical experts in the first quarter of the twentieth century were very common, and trials about sterilization laws in other states had made use of local as well as national experts. In order to portray Carrie as a person whose defective characteristics cried out for preventive surgery, Strode needed the voices of experts who could deliver eugenics to the courtroom.

First among these was Dr. Joseph S. DeJarnette, who began his service at the Western Lunatic Asylum in Staunton, Virginia, in 1895. He had presided for more than fifty years as a fiscal disciplinarian and iron-fisted manager over a facility that at one point housed more than three thousand patients. His role was one part medical patriarch, one part public scientist, and one part political moralist. An imposingly large, stout, bald man, DeJarnette reveled in his position as father figure to his mentally disordered patients and proudly counted himself an early "alienist"—a forensic scientist in the field of mental health—whose skills included the ability to navigate the muddy waters between madness and crime, illness and evil. An advocate for the prohibition of alcohol his entire career, DeJarnette waded into other political controversies without hesitation.

Barely two years into his tenure as head of Western State, he entered the public debate over restricting marriage of the mentally ill, arguing that weddings among "the unfit" should be prohibited in order to eliminate "defectives and weaklings." His reach included not only people with "dipsomania [alcoholism], insanity, epilepsy [and] feeblemindedness" but also those with tuberculosis and syphilis. Some had mocked the anti-spitting laws designed to prevent tuberculosis, he said, but now they were universally observed. He prophetically declared that sterilization would also eventually be adopted as a public health measure.[10] He would later cite this as the first official public statement in favor of a Virginia sterilization law, justifying his self-proclaimed nickname: "Sterilization" DeJarnette.[11]

Reproduction among the "unfit," said DeJarnette, was "a crime against their offspring and a burden to their state," and he demanded their sterilization. He decried those who constantly invoked "the so-called inalienable rights of man" to oppose surgery and compared accepted rules of stock breeding with the deference paid to "the syphilitic, epileptic, imbecile, drunkard and unfit." In contrast to the political sentimentalists, DeJarnette believed in a Darwinian scheme where "only the fit survive." A good farmer, he said, "breeding his hogs, horses, cows, sheep, . . . selects a thor-

Dr. Joseph "Sterilization" DeJarnette, expert witness for the Virginia Colony at the Buck trial.

oughbred"; even crops are grown from the "best seed." Yet "when it comes to our own race any sort of seed seems good enough."[12]

The economic burden shouldered by taxpayers could be lifted by sterilization, he said. Care of "the defective quota of our population" cost nearly 15 percent of the state's income; sterilization was "cheap and effective." Moreover, DeJarnette's own experience had taught him that "in many instances the patient can be sterilized without his knowledge."[13] DeJarnette set his policy preferences to verse, repeatedly publishing his doggerel in official reports to the Virginia legislature.

He was especially proud of the poem "Mendel's Law: A Plea for a Better Race of Man," in which he railed against policies that allowed "the fools, the weaklings, and crazy [to] Keep breeding and breeding again." DeJarnette's understanding of the Mendelian laws of heredity was simplistically reduced to the formula of "like begets like," where "Defectives will breed defectives, And the insane breed insane." The solution to this social burden was equally simple to him: "Sterilize the misfits promptly."[14]

Strode had no difficulty establishing DeJarnette's credibility as an expert in the field of mental health. He was superintendent of Virginia's Western State Hospital, where he had been a specialist in the care of the insane for thirty-six years. During that time, he had treated over eleven thousand patients. He had testified in scores of cases, and he was intimately familiar with the new sterilization law.[15]

DeJarnette's opinion included the common observation that "feeble-mindedness runs in families." If parents and children in the same family are affected, there is every reason "to believe it is from inheritance." The doctor then began a long discourse on the operation of Mendel's laws in the hereditary scheme, using the red and white flowering pea plants of the enterprising monk's classic experiments to describe the idea of dominant and recessive traits. He claimed that Mendel also understood that "in breeding people, a man that has a certain quality of mind will breed one-fourth a certain way."

Asked by Strode to discuss how Mendel's rules applied to humans, DeJarnette was less certain. "I have never worked the law out, but it seems from the history of the cases I have had that they work out pretty much that way, but I have no accurate knowledge of it because inheritance is such a complicated thing that unless a man devotes himself particularly to it. . . ." Then, despite his admitted ignorance on the topic, DeJarnette interrupted his own comments to claim that it was "practically certain" that for every four children a feebleminded woman had, one would also be feebleminded.

Reading from notes, he then described the case of the infamous Kallikak family whose more than one hundred defective offspring were an example of a genetic disaster that had produced great expense for the state. Apparently Strode was more prepared than DeJarnette on this topic. When he asked whether the "normal" side of the Kallikak line produced any citizens of "eminence," DeJarnette could not remember. Strode teased him: "You don't know about that, Doctor DeJarnette?" Strode knew that Henry Goddard himself had explained that one of the founders of Princeton University was related to Martin Kallikak, while a collateral branch of the family included a signer of the Declaration of Independence.[16] It was a standard feature of instruction in "heredity" to recall positive facts about the "untainted" side of the Kallikak pedigree and emphasize the contrast with the "defective" side, spawned by the feebleminded tavern wench who once lured Martin Kallikak into sin. Strode also knew that Priddy had alerted DeJarnette before the trial to study the famous texts of *The Jukes* and *The Kallikak Family* as preparation for his testimony—apparently to no avail.

Strode pressed for details of the Virginia law. DeJarnette had trouble pronouncing "salpingectomy," the surgery for women prescribed in the statute. "The operation must be by sectomy—sylgectomy," he stated. "Now, sylgectomy is the operation that the hospital ordered." It was a "very safe" procedure that allowed a woman to keep her "normal health." And it was a necessary measure because the "unfit and unsafe of our state cost us approximately one-seventh of our income . . . [and] . . . reproduction among the unfit is just as much a manufactured article as plows and threshing machines, because the feeble-minded produce their kind."

Strode reminded the doctor that according to the new law, the inmate's welfare must be promoted by surgery. How was that accomplished? Sterilized patients would benefit by being "liberated" from institutions, allowed to marry and have sexual gratification without the fear of producing children, said DeJarnette. Then Strode asked how society was to benefit from this surgery. DeJarnette was completely confident explaining how the operation would promote social welfare. "The standard of general intelligence would be lifted . . . [and] . . . it would lower the number of our criminals."

Strode completed his questioning of DeJarnette with a repetition of the evidence as it related to the new law. The girl Carrie Buck was feebleminded, her mother was also feebleminded, and her eight-month-old daughter did "not appear to be normal." In light of those facts, asked Strode, "Is she the probable potential parent of socially inadequate offspring, by the laws of heredity?" "I think so," DeJarnette responded.

Irving Whitehead's cross-examination of DeJarnette began with a formal objection that the doctor's comments contained too much legal commentary—not a topic within DeJarnette's expertise—and thus should not be allowed. The judge turned Whitehead's suggestion aside, and the conversation quickly returned to a discussion of heredity. Like Strode, Whitehead focused on DeJarnette's experience.

> *Whitehead:* Doctor, what is, in your opinion as a physician and from your experience as superintendent of that hospital, what in your opinion is the greatest cause of insanity?

DeJarnette: Inheritance.

Whitehead: Have you ever had occasion to trace back along the lines of heredity to find out what was the beginning of the thing?

DeJarnette: No, sir. Adam, I think, was a little off himself on some things.

There was no follow-up to this joking evasion of Whitehead's question, nor any attempt to probe the level of DeJarnette's actual ignorance about Mendelian theory, made clear in his stumbling attempts to explain heredity during Strode's questioning. Mendel had never suggested that his pea plant experiments should be applied to humans, as DeJarnette had claimed, but Whitehead did not challenge the statement. Nor did he question the doctor's supposed expertise in the surgical technique he had such difficulty pronouncing.

Whitehead turned instead to the topic of sexually transmitted infections, asking DeJarnette how often defects in children were caused by such disease. DeJarnette produced statistics from the state's hospitals, agreeing that a sterilized woman could very well be sexually active if she were allowed to leave an institution. "How would society be benefited" by such a practice, asked Whitehead. "By not producing any more of that kind," DeJarnette answered. As if to emphasize his agreement with the eugenical strategy, Whitehead went on: "It is a question of selective breeding, in other words. You are cutting out the unfit by breaking up the source?" "Yes, sir," DeJarnette concurred, "and you are raising the standard of intelligence in the state."

DeJarnette also conceded that many prostitutes were feebleminded and infected. Many such women had come to his asylum, and, in his words, "they come there, having had venereal diseases and having had children, and, brother, worse than all, white women come there having negro children."

Whitehead then posed a "hypothetical question" that admitted all the charges lodged against Carrie Buck. She was feebleminded and had an "immoral tendency" demonstrated by her illegitimate child, he said. If she were sterilized and released, it would be more likely that she would contract a disease. What benefit then, Whitehead asked, would come from releasing sterilized women from institutions while they could still be infected with syphilis?

Men who associated with such women got what they deserved, DeJarnette suggested, and in any event, sterilization would insure that only one generation would suffer from their promiscuity. Operating on girls like Carrie "benefits society by not taking care of them, and by the work they do. They are the hewers of wood and drawers of water, and there is not very much more likelihood that they would spread venereal disease if sterilized than if they were not."

The state could use an "in and out method," and the institution could become a sterilization "clearing house," performing operations, releasing patients, replacing them with others who were still liable to reproduce, all of which would save the state money in the long run. In completing this phase of questioning, Whitehead walked DeJarnette through another argument against sterilization—that it constituted bodily mutilation:

> *Whitehead:* Now Doctor, one more question I want to ask you: this law here says that you shall not take sound organs out of the body. Do you consider the cutting of that fallopian tube—
>
> *DeJarnette:* You don't take out a thing.
>
> *Whitehead:* You just cut it?
>
> *DeJarnette:* Yes, sir.
>
> *Whitehead:* Now that destroys it?
>
> *DeJarnette:* No, sir, it shunts the egg off from its destination where it would develop.
>
> *Whitehead:* It merely prevents reproduction?
>
> *DeJarnette:* Yes.

By the end of DeJarnette's testimony, Whitehead had taken him through several criticisms that had already been leveled against a sterilization program. It would be unfairly applied only to those in institutions, not defectives in the community. Since it would free women who could become infected with sexually transmitted diseases, it could exacerbate a public health problem. It involved a drastic intervention—abdominal surgery in women—that many considered a mutilating operation. Finally, surgery had no direct medical benefit to the patients who would endure it. Whitehead raised all these issues carefully with DeJarnette but did not challenge the answers, ending his questions with this witness as he did with others, by reiterating points that supported the validity of the sterilization law.

Far from being a vigorous advocate for Carrie Buck, Whitehead showed extraordinary deference to Strode's witnesses, at one point even conceding the very factual claims—that Carrie was feebleminded, promiscuous, and hereditarily defective—that were the basis of Strode's case. Whitehead knew that Carrie was uninfected with disease when she entered the Colony but failed to use that information to confront the false implications of DeJarnette's testimony concerning venereal disease. A bystander might reasonably have reached the conclusion that there were two lawyers working for Dr. Priddy and none for Carrie Buck.

Arthur H. Estabrook, Eugenics Record Office field worker and expert witness at the Buck trial. Courtesy American Philosophical Society.

Strode structured his case carefully to include a description of Carrie Buck and the rest of her family that would comport with an orthodox version of eugenic theory. Much eugenic propaganda had originated at the Eugenics Record Office, and there were few experts with as much credibility on the topic of problematic heredity than the staff members there. Strode's next witness was Dr. Arthur Estabrook, whose goal was to demonstrate how Carrie Buck was exactly the type of person eugenic laws could eradicate. Strode knew Estabrook long before *Buck v. Priddy*. Research for a study of mixed racial groups, entitled *Mongrel Virginians*, had brought him to Amherst, Virginia, numerous times.[17] Estabrook already knew Strode's wife Louisa Hubbard. He had encouraged her to study social work in Philadelphia, and they maintained a friendly correspondence. After she married Strode, Hubbard arranged for the two men to meet during one of Estabrook's field trips to the South.[18]

At the *Buck* trial, Estabrook's testimony began with a recitation of his expert qualifications, including books describing his studies of hereditary mental defect. Strode paid particular attention to Estabrook's book *The Jukes in 1915,* which brought the famous Dugdale study up to date. It was written, Estabrook testified, "to show that certain definite laws of heredity" explained how the "mental and social defectives" in the Jukes clan were replicated. Feeblemindedness was predictably inherited and "the basis of the antisocial conduct, showing up in criminality and pauperism."[19]

Estabrook went on to describe how a person's traits are inherited in pairs, using a person with six fingers as an example of a dominant trait. Strode interrupted: "Doctor, we are not interested in fingers." He then asked specifically about the transmission of feeblemindedness. When Estabrook calculated the odds of two apparently normal people having a feebleminded child, Strode asked him to draw a diagram. Estabrook produced a pedigree with N symbolizing "normal" and F standing for "feebleminded," explaining how cells contained "little chemical bodies" that could be seen under a microscope. Estabrook's explanation of the heredity of feeblemindedness was lengthy and confusing. By the time court adjourned for lunch at half past one, he had said nothing specific about sterilization and had not even mentioned the name of Carrie Buck.

Forty-five minutes later, after the court reconvened, Strode wasted no time getting to the point. Emphasizing the need for simplicity, he asked for a report on Estabrook's investigation of Buck and her ancestry in light of the rules of heredity. Estabrook recounted how he visited the Colony, saw Carrie and her mother, and read their files. He told how he had traveled to the Virginia Colony to examine Carrie Buck, then to Charlottesville and Albemarle County to question other members of the Buck family. He concluded that the "Dudley germ plasm" that was inherited via Emma Buck's mother carried "a defective strain." He also speculated that Emma's father Richard Harlowe (whom Estabrook called "Harley") was "of a defective makeup." As for the other Buck children, Estabrook had seen Carrie and Doris but had not met Roy. He nevertheless gave his opinion that the boy was feebleminded based on a "study of his school behavior and his general behavioral reactions."

Strode asked Estabrook for his understanding of the term *feebleminded:*

Estabrook: A feebleminded person is a person who is so weak mentally that he or she is unable to maintain himself or herself in the ordinary community at large.

Strode: Now, what is a socially inadequate person?

Estabrook: That is anybody who by reason of any sort of defect or condition is unable to maintain themselves according to the accepted rules of society.

Estabrook agreed that based on his examination, Carrie Buck fit both of these definitions and that they were compatible with the language of the Virginia sterilization law. He also described how he had given "the regular mental test" to Carrie's baby Vivian and found her to have performed "below the average."

Strode then had Estabrook supplement DeJarnette's testimony, explaining the college presidents and governors who had descended from Martin Kallikak and the similarities between the "defective" strains of both the Jukes and Kallikaks. Strode displayed Estabrook's book, *The Jukes in 1915,* to the court and asked Estabrook what influence the environment might have on feeblemindedness. According to Estabrook, environment might affect someone's behavior, but it would not change his or her "inherent mental ability."

Irving Whitehead might have disputed a good deal of Estabrook's testimony. As he began cross-examination, it again appeared that he might make a challenge. Whitehead asked Estabrook if he recommended the sterilization of people who were not feebleminded but might carry "the taint" in their germ plasm. When Estabrook said no, Whitehead did not pursue the questioning. He could easily have attacked the effectiveness of sterilization when even eugenic theory itself posited that most defects were hidden as recessive traits that might remain submerged for generations. Instead, Whitehead once again provided the answer he wanted from Estabrook: "The idea would be to sterilize them as the feeblemindedness breaks out in the offspring? Sterilizing them as it appears?" Estabrook agreed.

Estabrook's testimony concerning Carrie's brother was also suspect. While he claimed that he had reviewed records and judged Roy feebleminded because of his "general behavioral reactions," the very people he had interviewed concerning Roy had already testified in court. The most they were able to contribute was to say that Roy had not kept up in school and that he was "a little peculiar." But Whitehead did not revisit that line of questioning.

Irving Whitehead's only challenge to Estabrook's testimony concerned how society might benefit by Carrie's sterilization. Like DeJarnette, Estabrook agreed that destroying Carrie's ability to have any more defective children would be one benefit and that decreasing the cost of state care would be another. Asked where Carrie would live after discharge from the Colony, Estabrook surmised that "she would probably land in the lower-class area in the neighborhood in which she lives." Whitehead concluded as he had with other witnesses, by having Estabrook repeat the language of the new law as it applied to the Bucks. Whitehead reminded Estabrook about Emma Buck's family: "You say both of those strains have feeblemindedness in them" and "are socially inadequate?" Estabrook responded dutifully, "Yes, sir."

Estabrook's one-day survey of four generations of the Buck family in Charlottesville yielded conclusions about people who were dead, people he had never met. But little hard data about Carrie Buck's immediate family emerged in his testimony. He had done a "brief study" of Carrie and Emma Buck, based primarily on their records. He apparently gave no test to Carrie Buck at all and only gave the short version of the IQ test to Emma—not even the standard test. As for the eight-month-old infant Vivian, he gave the "the regular mental test" and concluded that she was below average. But based on this cursory investigation, he was able to conclude that Carrie should be sterilized.

Alice Dobbs holding Carrie's baby Vivian Buck in Charlottesville days
before the *Buck* trial, 1924. Courtesy Arthur Estabrook Papers, University at Albany
Libraries' Special Collections.

The final person to take the stand that day was Dr. Albert Priddy. Though
he could not know it at the time, this was to be the last time he testified
at any trial; exactly eight weeks later he would be dead. But Priddy had
worked to make his unchecked medical power over patients a part of state

law since taking over the Colony fourteen years earlier. Even during illness, his efforts to make eugenic sterilization a reality had been unrelenting.

Strode drew out Priddy's expert qualifications. He had worked in the mental health world for twenty-one years, and had run the Virginia Colony since its inception. He claimed to have cared for nearly five thousand patients.

Strode immediately turned to the point of the trial. Why was "this girl" Carrie Buck chosen for sterilization? She was, said Priddy, "a highly proper case for the benefit of the Sterilization Act," since her family history and personal condition met the provisions of the law. She was eighteen years old, said Priddy, and it was likely that her childbearing period would span thirty years. She would therefore be in state custody at a cost of two hundred dollars per year for that entire period, deprived of "outdoor life and liberty," a "burden on the State."[20]

Strode then explored Priddy's previous experience with sterilization. The doctor admitted that he had operated on many patients, asserting that there was no law prohibiting such surgery. In fact, he claimed "a right to do whatever is best for the mental and physical advantage of the patient." He recalled the *Mallory* case, boasting that he had successfully defended a lawsuit on the grounds that it was "accepted medical practice" to operate on diseased patients.

The discussion moved to Carrie Buck, whom Priddy linked to her mother Emma. That made "two direct generations of feebleminded," he said. There were also "eight Bucks and Harlowes, all coming from the Albemarle stock" at the Colony. These inmates came from several different counties in the state; Priddy's admission that he did not "know anything about their kinship" and could not "vouch for their relationship" undercut his statement linking them to Carrie.

Priddy then summarized Carrie's history. Contradicting his earlier assertion at the hearing before the Colony Board that Carrie was "of the lowest grade of the Moron class," Priddy at the trial declared Carrie to be a "middle grade moron" who could be sterilized without any damage to her general health. She could then be released and earn a living as a housekeeper. Because "the demand for domestics in housework is so great," probably "half of our young women of average intelligence" could be placed in jobs, but that practice was discontinued because of the "constant chance of them becoming mothers," he said.

Priddy explained how from 1916 to 1917 he had sterilized some eighty women and placed sixty of them as housemaids; he also conceded that Irving Whitehead knew some eight or ten of those cases. Priddy related the story of a boy, an "incorrigible . . . imbecile" son of a Baptist minister, whom he had sterilized "by the complete method" of castration. The boy later married a girl who had also been sterilized. Irving Whitehead knew these patients too, Priddy said. Whitehead interrupted to clarify the record: "Yes, put in there that I know them through being a member of the Special Board of Directors." In fact, only a few months before Carrie Buck's trial, a building had been completed at the Colony and named in Whitehead's honor.[21]

There were other male patients who would qualify for the operation, but because of their "anti-social tendencies" they could not be released, said Priddy. "They rank below the tramps and hoboes." Seventy-five women could easily be returned to the community—but for the likelihood that they would "have illegitimate children." Sterilization was "an absolutely safe and harmless operation" that would allow women to be trained to work and live outside the institution. Otherwise, they must remain at the Colony, where "they become helpless and lose confidence in themselves." Priddy said that John and Alice Dobbs wanted Carrie back; if she was sterilized, the Dobbs home would be open for her to return.

On cross-examination, Whitehead asked about the "grades of feeble-mindedness," eliciting the response from Priddy that Carrie was a "middle grade moron" who could "earn her living under proper supervision." He suggested that sterilization of girls like Carrie might "tame them down some," but Priddy, while certain there was supposed to be no effect on libido, did surmise that "it seems to make them better." Whitehead repeated his final questions to Priddy as he had to DeJarnette, establishing that salpingectomy involved only "cutting that fallopian tube and tying it back" and that ovaries were not taken out. His final question clarified that Carrie Buck had no diseases.

Priddy supplied the final testimony in court, but Strode had one witness who did not travel to Amherst. Harry Laughlin's deposition was taken in New York twelve days before the trial.[22] The written statement responded to five questions that required Laughlin to review a set of "facts" set out by Priddy, then analyze them in light of his eugenical expertise. Thus a major portion of the deposition was merely a repetition of what Priddy had written about the Bucks.

The deposition began with a recitation of Laughlin's qualifications as an expert witness. He was the assistant director of the Eugenics Record Office, and since 1910 he had trained over 250 "scientific field workers in eugenics." He had appeared as "Expert Eugenics Agent" for the Committee on Immigration and Naturalization of the House of Representatives and was also a "Eugenics Associate" for Judge Harry Olson's Psychopathic Laboratory of the Municipal Court of Chicago. He had published a book of 502 pages, entitled *Eugenical Sterilization in the United States.* The clear implication was that these experiences qualified him to submit a "short analysis of the heredity nature of Carrie Buck."

Priddy's "facts" about Carrie Buck pointed to evidence "of social and economic inadequacy" and a record "of immorality, prostitution and untruthfulness." Carrie's illegitimate child was "supposed to be mentally defective." Priddy supplied a similar record about Carrie's mother Emma. She was "divorced from her husband on account of infidelity" and had a record of prostitution and syphilis. She had "one illegitimate child and probably two others inclusive of Carrie Buck." Under Carrie's "family history" Priddy had written, "These people belong to the shiftless, ignorant, and worthless class of antisocial whites of the South."

The Bucks were "a moving class of people" about whom it was "impossible to get intelligent and satisfactory data." Priddy had Miss Wilhelm "try to work out their line" of kinship, but since none of the Bucks at the Colony admitted to being related to the Harlowes, there was "considerable doubt" whether Carrie was actually a Buck. Priddy falsely suggested that Miss Wilhelm had declared Carrie's baby "feebleminded" at Carrie's commitment hearing, even though Wilhelm had not been present on that day. Priddy described Carrie as "well grown, has rather badly formed face; of a sensual emotional reaction; . . . is incapable of self-support and restraint except under strict supervision."[23]

These supposed "facts" were Laughlin's starting point. He began by noting that his analysis was based entirely on Priddy's comments; he had assumed that what Priddy said was true. His conclusions followed. Since Carrie had more than one feebleminded sibling and had turned out to be sexually immoral despite having been raised away from her mother in the superior environment of the Dobbs household, Laughlin judged her defects to be caused by heredity.

To Laughlin, Carrie's history was typical of "a low grade moron." The major question was whether her children would be "assets or debits to the future population of the State." Laughlin answered this question using the language of his Model Law as echoed in the Virginia statute: "The family history record and the individual case histories . . . demonstrate the hereditary nature of the feeblemindedness and moral delinquency described in Carrie Buck. She is therefore a potential parent of socially inadequate or defective offspring."

The rest of Laughlin's deposition included lengthy excerpts of case histories from his book *Eugenical Sterilization* along with two articles scheduled to be printed in the *Eugenical News* entitled "Segregation Versus Sterilization" and "Purging the Race." These publications supported "the right of the State to limit human reproduction in the interests of race betterment." He judged the surgery allowed by the Virginia law to be effective at interrupting fertility and its eugenical effect to be equally salutary. Since socially inadequate people could not obey laws concerning marriage or reproduction, surgery made their voluntary obedience to law irrelevant. Laughlin's deposition concluded with an exhortation: "The administrative and institutional forces in any state, . . . require authority to segregate or to sterilize inadequate individuals, under proper legal regulations, if the State is to prevent race degeneracy." The Virginia statute was, Laughlin added, the best law that had been enacted to bring about these eugenical reforms.

Strode finished reading the Laughlin deposition for the court. Whitehead made one final objection asking that it be struck from the trial record, but he was overruled. It would normally have been the moment for Whitehead to put on evidence in opposition to the sterilization of Carrie Buck. But he offered no witnesses, and produced no evidence. Less than five hours after it began, the trial ended.

10 Defenseless

Aubrey Strode's presentation had followed a simple pattern. He asked for details of Carrie's life and the lives of Carrie's family members. Fastening upon the most useful details such as her mental slowness, her bad temper, or her moral weakness, Strode led the witnesses to a conclusion about Carrie. Would these negative traits, he asked, lead a person of professional experience (for example, a teacher, a nurse, or a social worker) to conclude that Carrie was "anti-social" or "socially inadequate"? With a mother like Carrie's, one could hardly expect otherwise, Strode would imply. And with Carrie as her mother, how normal could we expect a child to be? Carrie "simply didn't seem to be a bright girl."

Strode asked both Dr. DeJarnette and Dr. Priddy to comment on the efficiency of the surgery in reducing the cost of maintaining "defectives" in institutions. As superintendents of two Virginia facilities, the doctors spoke with the wisdom of locally verified experience. Finally, Strode attempted to explain how Carrie's condition fit within eugenical theory via testimony from nationally prominent eugenicists. The major premise of sterilization advocates in the *Buck* case—that Carrie's genealogy demonstrated hereditary deficiencies—could be traced to the earliest work of the Eugenics Record Office and its publications on eugenical field studies. Those publications suggested a particular way of thinking about people from the lower rungs of the social ladder, people whom the eugenicists described as socially inadequate: the feebleminded, the criminals, the misfits, and the paupers.

The picture Strode painted of the Buck family was uniformly negative. At this remove, portions of the testimony sound like mere name-calling, but the language Strode elicited from witnesses to describe Carrie's family was clearly part of the specialized vocabulary of eugenics. This testimony linked Carrie to her supposedly promiscuous mother, who was similarly tainted by the inherited trait of "moral degeneracy." Evidence about Carrie was assembled to fit neatly within eugenical theory and the label of "feebleminded." That term, particularly as it was applied to women, suggested weak intelligence, lack of inhibition, and a predisposition to sexual misbehavior.

While many eugenicists agreed that feeblemindedness was difficult to diagnose, they often claimed that being an illegitimate child or having illegitimate children was highly correlated with being feebleminded.[1] Thus, the most damning evidence against Carrie came from two witnesses who had examined her child. In a logically circular argument, the illegitimate infant was used as confirmation that Carrie was feebleminded, just as Carrie's alleged illegitimacy was used to confirm other behaviors (such as sexual license) that were theoretically passed down through a family line from Carrie to Vivian. Illegitimacy begot feeblemindedness; feeblemindedness begot illegitimacy.

Henry Goddard himself had declared that any person regularly described by neighbors as "'not quite right' is certain to have been decidedly defective."[2] So other witnesses were prompted to say that Carrie's sister, brother, and several cousins were "peculiar." This vague term appeared regularly in the eugenics literature as a description of the eccentric, mentally unstable, or socially unorthodox, and the witnesses referred to general "peculiarity" as cumulative evidence of inherited defects in the Buck family.[3]

Shiftlessness was another term familiar to eugenical advocates. Field workers trained at the Eugenics Record Office learned to prepare a pedigree chart and analysis of the trait of "shiftlessness" as a characteristic feature of the Jukes family.[4] Specific directions for charting this trait were listed in ERO manuals.[5] Charles Davenport described one of his subjects portrayed in a pedigree chart as a "shiftless, worthless epileptic."[6] Albert Wiggam, one of the most successful popularizers of eugenics, declared that "blood does tell," particularly if you descend from a line of people who are "lazy, shiftless and worthless."[7] When Harry Laughlin testified before Congress on the "Biological Aspects of Immigration," he stated that family histories would reveal "what sort of material the individual is made of" and clarify if someone came from "an industrious or a shiftless family." He described the infamous Jukes and the Kallikaks as examples of "worthless, mentally backward" families. He read his definition of the "socially inadequate" into the *Congressional Record* as a standard to use in purging "degenerating qualities" from the country's gene pool.[8] As prominent a person as Teddy Roosevelt used similar language, describing the consequences flowing from "the unrestricted breeding of a feebleminded, utterly shiftless and worthless family."[9]

Thus, when Laughlin echoed Priddy's description of the Buck family as "mentally defective" and part of the "shiftless, ignorant, and worthless class of antisocial whites of the South," his language was typical of the jargon that eugenicists used to describe the problem families who were their research subjects. These terms, seemingly so hostile and contemptuous, were actually meant to be disinterested labels of hereditary analysis based on well-understood scientific categories.

We will never know how Amherst County Judge Bennett Gordon might have ruled in light of other evidence, but certainly by the time the trial ended, he had ample reason to believe that surgery could remedy the multiplication of families like the Bucks. Tracking the work of eugenic theorists from the Eugenics Record Office, Strode had used the testimony of DeJarnette and Priddy, Estabrook and Laughlin to place Carrie Buck alongside the Jukes and the Kallikaks as Virginia's prime example of a defective problem family. If Virginia's law was upheld, surgery would empty the hospitals and asylums, taxes would be lowered, and doctors would no longer need to fear the embarrassment and expense of patient lawsuits.

But there was no second half to the *Buck* lawsuit, and Judge Gordon never heard the evidence that Irving Whitehead should have presented. His questioning of Strode's witnesses was so weak that it was often unclear which side he was representing. He repeated and reemphasized evidence that Strode had elicited in support of sterilization and never seriously disputed the allegations against Carrie Buck or the eugenic theory on which the sterilization law was founded.

Whitehead could have exploited the almost empty testimony of several witnesses from Carrie Buck's hometown. None of them had recent, direct contact with Carrie. None had examined her, taught her, or done a firsthand diagnosis of her. They supplied no school records to review and no documents to support their judgments. Apart from the rumors and secondhand stories they repeated, Carrie was a stranger to the first eight people who testified. Even the least competent advocate would have pointed out this obvious vulnerability in Strode's case. Instead, Whitehead spent his cross-examinations reemphasizing claims that Carrie had "an immoral tendency" and was an unwed mother. For two witnesses, he had no questions at all.

Whitehead also let inaccurate statements about Carrie's background go undisputed. Several witnesses said that she and her siblings were illegitimate, but no sound evidence was introduced to prove that contention. Frank and Emma Buck were married in 1896 and when Carrie was born in 1906, they were still married.[10] They remained married until Frank Buck died. When Emma Buck was admitted to the hospital for childbirth, she was listed as a married woman, and several times her husband accompanied her. Absent other evidence, according to Virginia law Carrie and her siblings were born legitimately.[11]

Was Carrie feebleminded? Several of Carrie's own teachers could have attested, with documents, that Carrie was not mentally deficient. Carrie's school records indicate that she was a normal child: She attended school for five years and was promoted to the sixth grade. The year before she left school, her teacher recorded the comment "very good—deportment and lessons" and recommended her for promotion.[12] That teacher and her records could have contradicted the testimony of the single witness who claimed that Carrie was "anti-social" because she had written notes to boys in school.

The circumstances of Carrie's commitment and the contradictions in the testimony of her foster parents should have alerted any conscientious attorney to probe further. As Caroline Wilhelm had testified, Carrie's foster parents were anxious to have her sent away. Their responses on the petition for commitment and later in court were often inconsistent. At one point they denied that Carrie was epileptic but later said that she was. They claimed in their petition that Carrie's symptoms of feeblemindedness did not emerge until she was ten or eleven. Later, they described a "mental peculiarity" they had noticed at her birth, even though they did not have custody of Carrie until she was three or four years old. The Dobbses claimed to be unable to afford to support Carrie but later told Priddy they would be happy to have her back, if she were sterilized.

Had Whitehead called the Dobbses as witnesses, he could have revealed both their contradictory statements and their personal interest in having Carrie committed. Bringing the Dobbs family to court might have identified Clarence Garland as the father of Carrie's baby, and it could have exposed that the maiden name of Mrs. Dobbs was Dudley, the very family Priddy had identified as being the source of the "tainted germ-plasm." If Mr. and Mrs. Dobbs had taken the witness stand, the record could have shown that

Carrie had become pregnant not because of her much-discussed "immoral behavior" but rather because she had been raped.

Whitehead did not call John or Alice Dobbs or any other witnesses; there is no record that he even met with Carrie Buck before the trial. Had he taken the time to confer with his client, he might have heard the story she repeated numerous times to others right up until the time of her death. While Strode's witnesses focused on "moral degeneracy," Carrie's own recollection was that she had been assaulted.[13]

During the summer of 1923, when Alice Dobbs was out of town, her nephew Clarence Garland visited Carrie, then sixteen years old, at the Dobbses' home. Carrie and Clarence had been childhood schoolmates.[14] Carrie recalled the episode later, saying, "He forced himself on me" and "He took advantage of me." She also noted, "He said he was going to marry me, but he didn't."[15]

If we take Carrie's account seriously, what does it say about the quality of Whitehead's representation? We can speculate how she might have been treated if she had been Strode's client. Strode became famous in 1907 for successfully defending a Virginia judge who had mistakenly suspected a man of assaulting his daughter. Without hesitating to learn the facts, the judge took his shotgun and killed the suspected rapist.[16] Strode's defense consisted of invoking the "unwritten law" that allowed for such vengeance killings of rapists or other men who sullied the purity of Virginia maidens.[17]

Strode did not, however, apply those rules to his own sexual misadventures. Before Carrie Buck was even born, Strode had had a liaison with a young woman whom he subsequently erased from his social calendar, but not before she beseeched him for help in the face of what she thought was an impending pregnancy. There was no talk of "moral delinquency" in that girl's case, and unlike Clarence Garland, young Strode was not punished or sent away. In a dozen letters sent while she waited to determine her condition, Strode's young correspondent decried the many "liberties" Strode took and how he had accomplished her "ruin" and settled her destiny "never again to be a pure woman." In response, he suggested that she go to a hospital, or in the event she found herself "in a certain condition," find a man who might marry her. Fortunately for him, the girl did not become pregnant.[18]

Carrie did become pregnant, but a defense in her case would have been even simpler. It need not rely on mere tradition but the full force of Virginia law. Clarence Garland could have been charged with rape, which Virginia punished with the death penalty.[19] But with no witnesses during a summer afternoon's meeting between a girl and her "boyfriend," such a charge would have been difficult to sustain. At least sixteen at the time of the incident, Carrie was by law old enough to consent, thus too old to make Garland liable for statutory rape. But there were other legal options. Virginia also punished men guilty of "seduction." A man who, "under the promise of marriage," had an "illicit connection with any unmarried female" could be sentenced for up to ten years in prison, as long as the woman was of "previous chaste character." And in the absence of contrary evidence, the chastity of the female was presumed.[20]

Thus Clarence Garland could have been prosecuted for either rape or seduction under a promise of marriage. Yet the Dobbses made no criminal complaint against their nephew, and Whitehead never called them as witnesses. He did not explore these charges with Carrie, nor did he even raise the possibility that her "moral delinquency" and subsequent pregnancy was the result of an assault. Setting aside his own youthful indiscretions, Aubrey Strode defended those who avenged the honor of pure Virginia maidens; in contrast, Irving Whitehead took it for granted that his client from the wrong side of the tracks was promiscuous.

Whitehead's efforts in confronting Strode's experts were similarly deficient. DeJarnette's testimony had been almost laughable. He had admitted his ignorance of theories of heredity and confusion concerning Mendelian laws. He couldn't even pronounce the name of the sterilizing operation planned for Carrie Buck. But Whitehead never seriously questioned his expertise.

Whitehead's approach to Arthur Estabrook was also puzzling. By the time he testified against Carrie Buck, Estabrook understood how a test case works. He was thoroughly familiar with *Smith v. Williams,* the lawsuit that challenged the very first sterilization law adopted in the United States, Indiana's 1907 statute. Irving Whitehead could have used the record Estabrook left from that case as the basis for vigorous cross-examination.

In that case, Warren Wallace Smith was an inmate at the Indiana Reformatory labeled a "moral pervert" and imprisoned for the crime of incest.

During his sterilization hearing, physicians concluded that Smith's "pro-creation was inadvisable" since it was unlikely that his mental condition, which arguably had made him a criminal, would improve.[21] Smith had no opportunity to respond to the doctors' findings or retain a lawyer to pro-test the planned surgery; he could not present witnesses on his own behalf to challenge the state's experts, and none of the proceedings were public. Because of these procedural deficiencies, the Indiana Supreme Court de-clared that the Indiana law violated the constitutional guarantee of due process.[22]

After his research in Indiana schools and on Indiana's Tribe of Ishmael Estabrook knew many people in that state's correctional system.[23] After the *Smith* case was decided, state officials contacted him and at their request Estabrook offered a complete analysis of Warren Smith's family. He said that Smith was the wrong person to sterilize because prison officials had so little evidence of his family background. It was hard to determine whether acts such as incest were the result of feeblemindedness, and it was difficult to distinguish them from similar criminal acts that might be the result of a hereditary propensity to crime. The Indiana law stated that "heredity plays a most important part in the transmission of crime, idiocy and imbecility," yet no demonstration of the hereditary nature of Smith's offense had been offered. Too few "mental defectives" had been produced in Smith's extend-ed family to justify sterilization. To Estabrook, a complete family study carried out under "present day methods of eugenics" was mandatory.[24]

Estabrook's commentary, subsequently published in Laughlin's book *Eugenical Sterilization in the United States,* was accompanied by a pedigree chart of Smith's extended family, showing a smattering of alcoholism, epi-lepsy, mental defect, tuberculosis, and sexual offenses. He described the group as "below the average intelligence, unambitious, [and] shiftless" but mostly "hard working laborers" who had little education. He concluded with the observation that sterilization was appropriate only when "a family is producing practically all mental defectives."

Harry Laughlin provided his own editorial comments on the *Smith* case, criticizing Estabrook's "extreme conservatism" and suggesting that as few as one in four "hereditary degenerates" born to a family would demonstrate ad-equate defective potentiality to justify sterilization under his Model Law.[25] It became clear that there was no unanimity even among the Eugenics Re-cord Office staff about how to prove the hereditary basis of some "defects."

Anyone who read Laughlin's book would have known of these disagreements among the eugenics experts. The Virginia law Carrie Buck confronted also declared that "heredity plays an important part in the transmission of insanity, idiocy, imbecility, epilepsy and crime." But Whitehead never questioned the flimsy evidence of "hereditary defect," nor the guesswork genealogy used to prove her familial taint. And he never pointed out how much Estabrook's testimony in the *Buck* case contradicted his more conservative opinion about the *Smith* case. Attention to previous sterilization cases, readily available through minimal legal research and discussed in exhaustive detail in Laughlin's book, would have provided Whitehead with the opinions of other eugenicists to counter the testimony that surfaced at the *Buck* trial.

For example, Charles Davenport testified in the New York case of *Osborn v. Thomson,* a case filed by the state Board of Examiners, a eugenic body, to test New York's 1912 sterilization law.[26] Called as an expert witness in *Osborn,* Davenport said that *feeblemindedness* was not a biological or medical term, but "is rather a social term."[27] Once willing to argue in scientific publications that "unit characters" were the "essence of the individuals," in the *Osborn* trial Davenport denied that feeblemindedness was a unit character and concluded that Mendel's laws did not describe how it was inherited.[28] Frank Osborn's lawyer used Davenport's book *Heredity in Relation to Eugenics* and Goddard's *Feeblemindedness* as well as the critiques that had appeared in the *Journal of Criminal Law and Criminology* to refute the idea that feeblemindedness would be wiped out by sterilization.[29]

Massachusetts asylum superintendent Walter Fernald also testified at the *Osborn* trial. Though he had been a strong supporter of eugenic ideals, more than eight years before the *Buck* trial Fernald turned on sterilization advocates and declared that "the laws of heredity are not as simple as some of these black charts which the eugenic publicists show us would seem to indicate." Fernald said that understanding the workings of inheritance was "infinitely a more complicated problem than the believers in the potency of sterilization would have us believe."[30]

There were many similarities between the way officials described Frank Osborn and Carrie Buck. In both cases testimony focused on feeblemindedness, shiftlessness, sexual misbehavior, and the expense of state institutions. It was clear that Frank Osborn was chosen to test sterilization law because of the costs his family had generated for the taxpayers of New York.[31]

Complete details on the *Osborn* case were published in Laughlin's book, but Whitehead did not refer to it during his representation of Carrie Buck.[32] With competent and well-prepared counsel, Frank Osborn won his case, and the New York law was declared unconstitutional.[33]

Whitehead might also have questioned the use of mental tests as a foolproof measure of mental ability. Several witnesses argued that Carrie Buck was deficient because of her performance on IQ tests, though most had not administered a test themselves. By 1924, the weaknesses of "mental measurement" had been widely debated, and charges had been leveled against eugenic field workers and other amateur "Binet testers" who made superficial "diagnoses" of feeblemindedness. The use of IQ tests to evaluate feeblemindedness in women had been questioned extensively after one study showed 98 percent of unmarried mothers to be mentally deficient. The habit of attributing most cases of juvenile crime to mental problems had also been criticized.[34]

In the early years of mental testing, Henry Goddard claimed that a person with "considerable experience" could diagnose feeblemindedness from a distance and "pick out the feeble-minded without the aid of the Binet test at all."[35] Arthur Estabrook's halting testimony suggested that he might have claimed similar skills of clairvoyance, because there is no evidence that he gave any test to Carrie Buck. His cursory "mental test" of baby Vivian Buck led to the conclusion that she was likely to be "socially inadequate" as well.

Whitehead never asked what test was given to Vivian or what it included, though it is likely Estabrook followed the forms used at the Virginia Colony supplied by Priddy. They include directions for testing infants at ages four months, six months, and one year. These assessment tools included several simple tasks, among them holding the head without support and following the source of a sound and a light.[36]

It was Estabrook's habit to photograph the subjects of his eugenical family studies, and one surviving photo shows Alice Dobbs holding Carrie's baby.[37] It appears that Mrs. Dobbs is holding a coin in front of Vivian's face in an attempt to catch her attention. The baby looks past her, staring into the distance, apparently failing the test. Estabrook described that moment during his testimony at trial a few days later: "I gave the child the regular mental test for a child of the age of six months, and judging from her reaction to the tests I gave her, I decided she was below the average."[38] The untimely responses of an infant became the unquestioned evidence that supported a theory of inheritance linking "feeblemindedness" with future moral failings.

Strode's other experts—Priddy and Laughlin—were also vulnerable. In the material he supplied for the Laughlin deposition, Priddy repeated the unsubstantiated comments of other witnesses concerning Emma Buck's marital status and her children's supposed illegitimacy. Priddy claimed that Carrie was guilty of "prostitution" even though there was absolutely no evidence introduced to support this charge.[39]

Laughlin's deposition relied entirely on Priddy's "facts." Potentially undercutting his own case, Priddy told Laughlin that "it is impossible to get intelligent and satisfactory data" about the Bucks.[40] How would it have been possible to so firmly conclude that Carrie was suffering from hereditary defects, if you couldn't even be sure who her parents were? If it was impossible to get satisfactory records about the Bucks, how could anyone be sure all the people described at trial were even related? Whitehead and Strode had agreed on the questions that would be asked of Laughlin well in advance of the trial, yet other than the blanket objection to Laughlin's deposition on broad constitutional grounds, Whitehead never challenged the "facts"—provided by Priddy—on which the entire deposition was based.

Priddy's opinions about sterilization were contradicted by his own words. In 1913, as he urged the legislature to expand facilities at the Colony and increase his funding, Priddy cited many arguments against surgery and agreed with those who believed that a general scheme for sterilization "will never be practicable."[41] Whitehead had been a member of the Board that issued that report, but he did not cite it at the trial.

There were other problems with the evidence Strode elicited. For example, the Virginia law required that surgery must be beneficial both for the patient and society, and the preamble of the law made it clear that the "health of the patient" must be promoted. Yet for more than ten years the therapeutic value of sterilization had been denied, and the lack of evidence for its benefit to patients had been extensively publicized in professional journals for lawyers.[42] Even local sterilization enthusiasts like Harvey Earnest Jordan of the University of Virginia Medical School doubted that there was any therapeutic benefit to patients from surgery.[43]

Other experts didn't believe that feeblemindedness was clearly hereditary. For example, Walter Fernald began to voice public doubts on this issue during the *Osborn* trial and made public comments as early as 1919.[44]

His 1924 presidential address to the American Association for the Study of the Feeble-Minded presented a strong critique of the "hereditarian" posture that ignored the environmental causes of mental defect.[45]

Strode's experts emphasized how Carrie Buck's family history was an example of the hereditary transmission of defective traits in a predictable pattern that mirrored Mendel's laws. At many points they relied on the thoroughly outdated idea that complex characteristics like criminality, feeblemindedness, or pauperism were inherited as single "unit characters." Estabrook specifically pointed to the "little chemical bodies" that carried the stuff of heredity, suggesting that feeblemindedness was transmitted in these "particles" of the germ plasm. But by the time of the *Buck* trial, those theories had been rejected by some of the most prominent scientists in America.

Johns Hopkins University geneticist H. S. Jennings, who was never completely opposed to sterilization, attacked the "unit character" theory of feeblemindedness directly. "This theory of representative particles is gone, clean gone. Advance in the knowledge of genetics has demonstrated its falsity. Its prevalence was an illustration of the adage that a little knowledge is a dangerous thing. . . . It is not true that particular characteristics are in any sense represented or condensed or contained in particular unit genes. Neither eye color nor tallness nor feeblemindedness, nor any other characteristic, is a unit character in any such sense."[46]

The "unit character" idea was not the only weak spot in the version of eugenics that became part of the *Buck* trial record. The economic argument in favor of eugenic sterilization rested on the assumption that operating would wipe out defects after a few generations. But even if the unit character theory had been correct, the Hardy-Weinberg principle had disproved the effectiveness of sterilization more than ten years before Carrie Buck's trial. It had "helped to demolish the major assumptions of eugenicists."[47] Geneticist R. C. Punnett calculated that it would take eight thousand years to eliminate the supposed genetic character for feeblemindedness by segregating or sterilizing all known victims of that condition.[48] And in 1919 geneticist Raymond Pearl criticized the effectiveness of sterilization laws, arguing that they did not address the many "normal" people who carried hidden defects that would appear in later generations. Unless both obvious "defectives" and their seemingly innocent kin were sterilized, a regimen of legally mandated surgery would be of little value. And even in the unlikely event such a law could be widely adopted, wrote Pearl, "it is clear that unless a sterilization programme is thoroughly comprehensive in its scope and carried out rigorously and unremittingly for a period of probably at least a hundred years, no significant eugenic results will come of it."[49]

The "evidence" generated by field workers about the Buck family history was also seriously suspect. Genetic experts had questioned the value of hearsay reports regarding missing family members since the appearance of Goddard's *Feeble-Mindedness* more than twenty years before the trial. The critique reappeared in 1922 in the comments of University of Maine president and longtime ally of the eugenics movement, Clarence Cook Little. Little called the bulk of information recorded by field workers concerning people absent or even dead "too uncertain to use in the finer methods of genetic analysis." Since about 1910, he said, eugenics had made "little or no progress towards a truly experimental attitude."[50] Little was widely recognized as a "biologist and eugenist" and someone who favored sterilization years before the *Buck* case.[51] He nevertheless believed that pedigree study as practiced by the eugenicists should be abandoned.

By the 1920s, the utopian vision of eugenics advanced early in the twentieth century had given way to a more critical perspective and more modest claims from many supporters. Describing the field in his 1924 *Bibliography of Eugenics*, University of California Professor Samuel J. Holmes declared: "There is a great deal of rubbish written on this as upon most other topics, and I have perhaps included too much of it . . . many of the most important problems are still very puzzling."[52] Others saw danger in the political uses of eugenics and voiced concern about how "some of the most ferocious enemies of human freedom" sought to enshrine eugenics in U.S. law.[53]

But at Carrie Buck's trial, Irving Whitehead seemed unaware of any of these critics. Some of Whitehead's performance could be attributed to carelessness, inattention, or lack of preparation. But Whitehead had more than nine weeks to prepare his case to counter the experts. With a similarly busy schedule and the same budget, Aubrey Strode assembled a dozen witnesses. Whitehead offered not a single one.

How should Whitehead have countered the expertise marshaled in behalf of sterilization? Lawyers who made challenges to sterilization laws in other states had no difficulty finding experts to speak in opposition to eugenic theory. Irving Whitehead could have hired his own experts to speak out against sterilization, and the General Board of State Hospitals had authorized the expense of litigation. There were experts like geneticist H. S. Jennings living in Baltimore, where Whitehead worked, and dozens of other knowledgeable scientists and physicians were available in Washington, D.C., several of whom had critiqued ERO sterilization arguments. Arthur Estabrook traveled more than five hundred miles for almost two days by horse and train to reach the trial in Amherst, and he was paid for his time and his expenses. For the cost of a four-hour train trip, Whitehead could have brought experts to Amherst to counter Estabrook, but there is no record that he ever even contacted any.

As for Whitehead's own preparation, in one afternoon any competent attorney could have collected the arguments against sterilization and prepared an effective cross-examination of every one of Strode's witnesses. At the very least, a survey of previous cases on sterilization would have revealed the several important objections that should have been raised against the Virginia statute. Whitehead had access to legal journals containing that material near his Baltimore office, or he could have reviewed them at the University of Virginia in Charlottesville, a short walk from Carrie Buck's home.[54]

Whitehead knew of Priddy's desire to enact a scheme of "legalized eugenics" and sterilization and his constant lobbying in favor of the surgical policy beginning in 1911.[55] Whitehead served for fourteen years on the Colony Board. He supported Priddy's extra-legal sterilizations, describing them as "therapeutic" as early as 1916. He was entirely in concert with sterilizing the Mallory women and, even after protracted litigation stemming from Priddy's sterilization policy, continued to stand behind Priddy's most questionable practices. At the close of evidence in the trial of *Buck v. Priddy*, the record contained only evidence in favor of sterilization. At many points that record was bolstered by Whitehead's own words. He did not fail in his advocacy of Carrie Buck simply because he was incompetent; Whitehead failed because he intended to fail.

11

On Appeal: *Buck v. Bell*

The trial had gone well for Albert Priddy. The doctor's witnesses testified as expected, and Judge Gordon made encouraging comments after the trial ended, promising a written opinion as soon as his schedule allowed. Priddy told Caroline Wilhelm how satisfied he was "with the way in which we presented our case," noting his expectation that he would "get a favorable decision."[1] Aubrey Strode was also hopeful. He thanked Arthur Estabrook for coming to Amherst, noting that less than two weeks after the Buck trial, Judge Gordon "intimated that he would probably decide it in our favor."[2] Estabrook asked for a copy of the decision as soon as it was delivered; he planned to write a magazine article analyzing the case.[3]

Priddy's tone with Strode was formal, thanking the lawyer on behalf of the Colony Board for his "painstaking care and legal ability . . . regardless of what the result may be."[4] Priddy was more direct in responding to a letter from his friend Irving Whitehead, in which he voiced his hopes that the judge would uphold the law, writing, "I am inclined to believe he will."

As encouraged as he was by the progress of the *Buck* case, Priddy confided to Whitehead that he was "awfully depressed and discouraged" by his own physical condition. His health had not improved despite medical attention, and he planned to go to Philadelphia to see a specialist, promising to meet with Whitehead when he returned. Priddy sent Whitehead $250 as the first installment on legal fees in the case and, as though racing against time, immediately began preparation for an appeal. He asked Caroline Wilhelm to send a clean copy of Emma and Carrie Buck's commitment papers, which had apparently been misplaced during the trial in Amherst. They would be needed as part of the case record, as would the printed trial transcript Priddy had ordered for submission to the Virginia Court of Appeal.[5]

While he waited for the decision, Strode completed work on an article entitled "Sterilization of Defectives" that he was invited to write for the *Law Review* at his alma mater, the University of Virginia.[6] The article emphasized that no sterilization would be performed under the new Virginia law unless the "welfare of the inmate would also be promoted." But Strode also conceded that his topic involved "elemental personal rights" and credited Harry Laughlin's book as the source of much of his insight into legal trends.[7]

Priddy had less than two months to live after the trial, and his death from Hodgkin's disease happened before Judge Gordon formally endorsed the sterilization of Carrie Buck. Out of town on business, Strode heard only belatedly of Priddy's death and funeral.[8] Even though Irving Whitehead was supposedly the legal adversary to the State Board of Hospitals and the Colony Board, he appeared at their joint meeting in February 1925 to report the trial decision upholding the sterilization law and remind the Board that the matter would eventually be taken to the U.S. Supreme Court for a final decision.[9] Upon his return, Strode was faced with the task of identifying someone to replace the deceased Dr. Priddy as the named party in the sterilization lawsuit. The likely candidate was Dr. John Bell, Priddy's successor as Colony superintendent.

John H. Bell was born in rural Virginia in 1883, near the institution known originally as the Western Lunatic Asylum. After his medical education, Bell worked as a physician for several coal-mining companies. Returning to his home county in 1916, he spent a brief turn in private practice, also serving as assistant physician to Dr. Joseph DeJarnette at the asylum. The next year he moved to the State Colony for Epileptics and Feebleminded near Lynchburg, leaving the Colony briefly during World War I for service as an Army physician. As a former student at the University of Virginia, Bell knew Charlottesville well; as Priddy's right-hand man at the Colony, he was the first to examine Carrie Buck when she arrived there on the train in 1924.[10]

Strode asked Bell to continue as named defendant as the case proceeded through the courts. Bell replaced Priddy in the lawsuit that became *Buck v. Bell*, declaring himself "in entire sympathy with the effort being made to reach a final conclusion as to the legality of this sterilization procedure."[11] While some objected to the idea of sterilization, others who spoke up for eugenics as the case was pending on appeal shared Bell's sympathies. DeJarnette asked for a "eugenic society in every county and city in the State" and suggested that the wages of workers should be "regulated according to the number of children and the mental quality of parents." He demanded support for sterilization laws to "sever the black thread of inheritance" and "save untold suffering, crime and expense."[12] Invited by eugenic enthusiasts in Virginia, Judge Harry Olson lectured the Virginia State Health Convention about inherited criminality.[13] In Richmond a Unitarian minister gave a series of lectures highlighting the need to "purify the human blood stream": "The people of the slums make the slums," he said, bemoaning the "appalling increase of the defective classes."[14] But the critical decisions in the public debate over sterilization would not be made by doctors, newspapers, or the clergy. Judges, deliberating in the quiet of their chambers, would decide Carrie Buck's fate.

Virginia Court of Appeals

From the beginning, Aubrey Strode designed *Buck* to create a legal record that would be reviewed by appellate judges. Only they could provide a lasting endorsement for his sterilization law. His choice of witnesses reflected that plan, including people who could give a firsthand factual report on Carrie Buck and her family and experts who could explain the sometimes complex theories of eugenics that would be used to justify her sterilization. His questions to them were crafted to ensure that the information offered to the court was clearly relevant to the sterilization law the case was meant to validate.

Strode's brief for the appellate court made good use of the record he had generated at the trial. He stated the facts of the case in detail, drawing from the records of Carrie Buck's commitment hearing, the sterilization proceedings before the Colony Board, and the trial transcript. He quoted expert commentary from Laughlin, Priddy, DeJarnette, and Estabrook at length to show how Carrie's "heredity defects" had been proven. His most important argument focused upon the prerogative of states to use the inherent "police power" and enact legislation that protected the public health and safety. How else could society protect itself and future generations against the "multiplication of socially inadequate defectives?"[15]

Strode dismissed the suggestion that the Virginia sterilization law violated the Constitution's Eighth Amendment prohibition of "cruel and unusual punishment," declaring that the law was not punitive. It did not apply to prisons, and Carrie Buck had never been charged as a criminal. Eugenics was the point of the Virginia law, Strode argued, not punishment. He then listed all the provisions of the Virginia law that protected a patient's rights. Patients must be notified that surgery was planned and had the right to appear before the Colony Board to challenge the sterilization order. They also had the right to a guardian and an attorney and the opportunity to appeal a sterilization order. It was clear, Strode declared, that the law provided due process to Carrie Buck.[16]

With Laughlin's book as a road map, Strode cited all the state courts that favored a broad application of the police power in sterilization cases, paying particular attention to the 1905 Supreme Court decision in *Jacobson v. Massachusetts,* a case that endorsed the state power to require vaccination against smallpox.[17]

Strode argued that the constitutional guarantee of equal protection was met by Virginia's more general plan for the care of the feebleminded. Echoing DeJarnette's "clearing house" idea, Strode wrote that if all people to whom the law applied within Virginia institutions were sterilized and released, there would be plenty of room to admit others. Thus, the state could use other Virginia laws to get "defectives" into institutions and by serial admission eventually sterilize everyone covered by the statute. Eventually, if not immediately, all of the feebleminded would be treated equally.[18]

Strode's brief went on for more than forty pages, citing the precedent cases on sterilization from other states and addressing issues Whitehead had not even raised. The brief concluded with praise for the humanitarian motives behind the Virginia sterilization law and the legal protections for patients it included. Arguments about the deficiencies in other sterilization laws, Strode concluded, should not be used as an excuse "to block the path of progress in the light of scientific advance toward a better day both for the afflicted and for society whose wards they are."[19]

Appellate courts are constrained to a great extent by the record they are given to review. They can look at transcripts of testimony, and they can question lawyers who file written briefs and argue before them in person, but they have no access to the people who bring the lawsuit, nor do they see or hear the witnesses who provide testimony at trial. And if a lawyer does not challenge the evidence presented by an opponent there, as a rule, an appellate court may not supplement that record with arguments or facts not heard by the trial court but must decide the case based on the evidence before it.

Having presented almost no factual evidence in defense of Carrie Buck during her trial, Irving Whitehead left little that might have been used by an appellate court that disagreed with Judge Gordon's decision. Whitehead did not contest the claim that Carrie was feebleminded, nor did he raise questions that might change a court's view of her infant child. He often repeated information introduced by Strode in support of the argument that she was a proper subject for the sterilization law. Having failed to challenge those "factual" assertions during the trial, Whitehead could only attack the law itself when he took the case to the appellate level.

The brief Whitehead submitted to the Virginia Supreme Court of Appeals on September 8, 1925, was consistent with his poor showing at the trial. It covered barely five pages, glossing over the question of how Carrie Buck might benefit from involuntary surgery as the law prescribed and focusing entirely on the question of due process.[20] Even though he had failed to question witnesses who invoked the names of the Jukes and the Kallikaks and had said nothing to counter Priddy's description of the Buck family on which Harry Laughlin's testimony was based, Whitehead claimed that he was given no opportunity at the trial to challenge statements about the infamous problem families or Laughlin's "hearsay" evidence. Courts in New Jersey, Iowa, Michigan, New York, and Nevada had invalidated sterilization laws, but Whitehead referred to only one of those cases to support his argument.[21] Legal commentators had pointed out that no sterilization case had examined the importance of the public welfare as weighed against individual rights in light of "the doctrines of science."[22] But Whitehead had passed up the opportunity to question those "doctrines" at the trial.

The decision in Virginia's appellate tribunal seemed a foregone conclusion. The Virginia Supreme Court of Appeals affirmed the order for Carrie Buck's sterilization on November 12, 1925. Justice John West said that there had been no controversy as to the "legality or regularity of the proceedings" by which Carrie was found to be feebleminded and sent to the Colony, and the evidence had shown that the operation of salpingectomy was "harmless and 100 per cent safe." No argument had been presented that questioned the general police power of states to deprive "defective" citizens of many liberties otherwise guaranteed to other citizens. The Virginia law was different than the laws struck down in other states, said the Court, because in those instances the state's power to sterilize did not require that the operation would "promote the welfare" of the patient. Virginia's law, in contrast, demanded that the state prove how surgery would benefit the patient, and evidence in the trial record spoke to that proof.[23] Since neither the "feeblemindedness" nor the "hereditary defects" attributed to Carrie Buck had been contested, West's colleagues on Virginia's highest court all agreed that Virginia's law was valid.

The *Virginia Law Register* celebrated the decision, congratulating the Court for having "happily decided a question of great public interest." Full castration was not an option, said the *Register,* since Virginians are "not sprung from those races who allow the making of eunuchs." Lifelong segregation of the unfit is "impracticable"—therefore "the race must protect itself with sterilization."[24]

There is no record of how Carrie Buck received the Court's decision or whether Whitehead ever reported it to her, but both he and Aubrey Strode met with the Colony Board to give them the news. The minutes of the Board's December 7, 1925, meeting includes this description of their report:

> Colonel Aubrey E. Strode and Mr. I. P. Whitehead appeared before the Board and outlined the present status of the sterilization test case and presented conclusive argument for its prosecution through the Supreme Court of the United States, their advice being that this particular case was in admirable shape to go to the court of last resort, and that we could not hope to have a more favorable situation than this one.[25]

This was Whitehead's report, even though he had failed to call any witnesses of his own, neglected properly to cross examine witnesses of his opponent, and appeared ignorant of widely published legal and scientific arguments condemning the operation Carrie Buck faced. One might be tempted to excuse Whitehead as a reluctant and inexperienced lawyer working pro bono, contributing to a philanthropic enterprise out of civic duty. But Colony Board records show that Whitehead eventually collected fees and expenses totaling over $1,150 for representing Carrie Buck—more even than his supposed adversary Aubrey Strode.[26] Whitehead's poor performance cannot be blamed on inadequate funding.

Whitehead practiced law more than twenty-eight years before the *Buck* case, and he was no novice in the courtroom. Given this level of experience and the compensation he received, it is only fair to evaluate his performance at Carrie Buck's trial as grossly negligent, but that description does not adequately portray the fraud he inflicted on his client and the court. Whitehead was not merely incompetent; his failure to represent Carrie Buck's interests was nothing less than betrayal. His poor showing in court was damning enough, but if the letter Aubrey Strode wrote recommending Whitehead for advancement to the position of general counsel to the Federal Land Bank at Baltimore the week before the *Buck* trial had become publicly known at the time, it might well have raised serious suspicions about Whitehead's loyalties.[27] Whitehead's duty to Carrie Buck was clear. Despite his friendship with Priddy, his connection to the Colony, and his personal support for sterilization, he had a duty to oppose Priddy's plan to sterilize her. Virginia cases long established every lawyer's obligation of loyalty to a client.[28] Reporting regularly to your opponent's client and declaring your pleasure upon losing a case, as Whitehead did, violated every norm of legal ethics. Whitehead acted as if his real client was not Carrie Buck but his now-deceased friend Albert Priddy, whose sterilization program he had supported for over a decade.

Sterilization under Attack

The public critique of eugenics increased in the months that *Buck* was in the courts. Famous lawyer Clarence Darrow attacked the movement in a series of articles in the popular *American Mercury* magazine. He mocked the work of Arthur Estabrook and other eugenic popularizers who attempted to "diagnose" feeblemindedness in families long dead and the "utter absurdity of tracing out any given germ-plasm or part thereof for nine generations, or five, or three."[29] Darrow went on to say that any "talk about breeding for intellect, in the present state of scientific knowledge and data, is nothing short of absurd" and declared himself "alarmed at the conceit and sureness of the advocates of this new dream. I shudder at their ruthlessness in meddling with life. I resent their egoistic and stern righteousness." He condemned governmental intrusions into mate selection, declaring that "it requires unlimited faith, unbounded hope, and a complete absence of charity to believe that the human race . . . would actually profit by placing the control of breeding in the hands of the state." Darrow denounced the eugenic plan to remake society as "the most senseless and impudent that has ever been put forward by irresponsible fanatics."[30]

In his book *The Misbehaviorists: Pseudo-Science and the Modern Temper*, written just before the *Buck* decision was announced, Harvey Wickham argued that the Mendelian belief in unit characters was outdated, even as an explanation for something as simple as eye color. Said Wickham: "This greatly increases the difficulty of even guessing what hereditary possibilities the children of any given couple will receive." In light of known scientific facts, Wickham wondered, "what is the biological warrant for this worship of heredity? We are face to face with a political and social, not a scientific movement."[31]

As Strode and Whitehead prepared to present the *Buck* case to the U.S. Supreme Court, Henry Goddard recanted almost everything that he had to say about the treatment of the feebleminded. Explaining how he coined the term *moron,* Goddard bemoaned that after twenty-five years of using the term *feebleminded,* there was still no accurate definition for this "unscientific and unsatisfactory" word. "There was a time to be sure when we rather thoughtlessly concluded that all people who measured twelve years or less on the Binet-Simon scale were feebleminded. However, we had already begun to discover our error" in 1917 when large groups of soldiers were tested as part of the mobilization for World War I, noted Goddard. To call all who tested poorly "feebleminded" was "an absurdity of the highest degree." Goddard, who had been quoted at the *Buck* trial as the author of *The Kallikak Family,* abandoned the position upon which the *Buck* trial would turn—that the feebleminded were a danger to propagate and must be sterilized—a full ten years before the *Buck* decision. By 1927, he said that childbearing in this group should no longer be feared, and he even raised questions about segregating the "feebleminded" into colonies.[32]

Other voices countered. Speaking on "Heredity in the Feebleminded" at San Francisco's Commonwealth Club, psychologist George Ordahl explained the economic motive for sterilization. If it were possible to calculate clearly "the economic burden of the moron" borne by every taxpayer, said Ordahl, "every patriotic citizen would become a eugenicist in search of methods of prevention even more drastic than any now known." Unfortunately, the "only effective remedy" available, according to Ordahl, was sterilization.[33] On the other coast, addressing the students of Vassar College, Margaret Sanger proposed that government should offer a bonus payment to all "obviously unfit parents" who opt for surgery and thereby take "the burdens of the insane and feebleminded from your backs." The remedy to being "taxed, heavily taxed, to maintain an increasing race of morons," she said, is sterilization.[34] That message was echoed by the press in headlines that shouted: "Sees in Eugenics Way to Cut Cost of Government."[35]

12

In the Supreme Court

As Strode predicted, the final act in the *Buck* legal drama played out before the U.S. Supreme Court. The case was accepted for review in September 1926, and both lawyers prepared new briefs. Strode repeated the arguments he had made previously, paying particular attention to the use of state police power to enforce eugenical laws. Again he cited the 1905 Supreme Court decision of *Jacobson v. Massachusetts,* one of the few federal precedents to support direct medical interventions undertaken for public, rather than private, benefit.[1] *Jacobson* involved a compulsory smallpox vaccination law. Reverend Henning Jacobson, a Swedish Lutheran minister in Cambridge, Massachusetts, had been vaccinated as a child and suffered a severe reaction. He maintained a strong objection to vaccination and refused to be revaccinated. A local court imposed a fine on Jacobson, but he appealed, arguing that medical procedures undertaken against the will of a patient were unconstitutional. The Supreme Court allowed the fine, saying that measures like vaccination, designed to protect public health and safety, were justified under a state's police power.[2]

Whitehead raised a new objection in his brief for the high court. Reframing his argument for "due process," Whitehead stated that the Fourteenth Amendment's protection of a person's "life, liberty, and property" guaranteed people like Carrie Buck "the inherent right to go through life with full bodily integrity." Regardless of how many procedural steps might be involved, an intrusion into the personal prerogative of having children was always a violation of this guarantee, he urged. Repeating his concerns about "equal protection," Whitehead also focused more clearly on an issue he had only hinted at earlier. What benefit accrued to patients who were sterilized under the law? Benefit was impossible to prove and was "at best . . . a mere guess" inserted into the Virginia law to provide an excuse for a court sympathetic to the eugenic scheme. If the Virginia law was upheld, "the reign of doctors" and the "worst kind of tyranny" would ensue. If the law was held to be constitutional, Whitehead wrote, the "power of the state (which in the end is nothing more than the faction in control of the government) to rid itself of those citizens deemed undesirable" was unlimited.[3] But his arguments were sketchy at best, failing to emphasize the usual reluctance among courts to interfere with what the Supreme Court itself had termed "sacred . . . the right of every individual to the possession and control of his own person."[4]

As the Court waited to hear arguments in the *Buck* case, eugenics remained a controversial topic. Eugenic marriage laws proliferated: Kansas was considering a bill to require a financial bond of all candidates for marriage; it had already prohibited marriages between "members of the Ethiopian race with those of the Caucasian race."[5] Indiana, the pioneer in sterilization law, reenacted a sterilization statute that applied to institutional inmates, while the governor of Colorado vetoed a similar bill passed by that state's legislature.[6] *Time* magazine observed that the majority of U.S. physicians still opposed the procedure; a German scientist called U.S. sterilization methods "senseless."[7]

The Justices

The men who would ultimately determine Carrie Buck's fate were led by Chief Justice William Howard Taft, the only former president of the United States who also served on the nation's highest tribunal. Taft was no bigot, and he rejected "the unsanity [sic] of race prejudice" that characterized eugenic laws mandating racial separation.[8] He also rejected some methods of immigration restriction and, as president, had vetoed the literacy test for immigrants adopted by Congress.[9] But like Theodore Roosevelt, who preceded him as president, and Woodrow Wilson, Warren Harding, Calvin Coolidge, and Herbert Hoover to follow, Taft supported some eugenic reforms. And for more than thirty years, Taft lent his public support to some of the most prominent leaders of the eugenics movement.

As a student of William Graham Sumner at Yale University, Taft was exposed to important precursors to eugenic thought in America. Sumner's use of Herbert Spencer's book *The Study of Sociology* created a controversy that drew national attention soon after Taft graduated.[10] The Yale administration challenged the book for its unorthodox treatment of religion, though little was made of Spencer's attitude toward poverty and social dependence—a perspective captured in the phrase "social Darwinism."

Spencer, who coined the phrase "survival of the fittest," declared that the "quality of a society is physically lowered by the artificial preservation of its feeblest members. . . [and] . . . by the artificial preservation of those who are least able to take care of themselves."[11]

Irving Fisher, whose doctoral dissertation was supervised by his "revered master" Sumner, was Taft's most direct contact in the U.S. eugenics movement.[12] Fisher took Yale's first Ph.D. in economics and became a prominent academic there while simultaneously developing a very public profile. He was quoted regularly in the press and published "Fisher's Weekly Index"—a number that incorporated commodity prices and the relative purchasing power of money—in the *Wall Street Journal*.[13] Fisher invented and held the patent on the "visible card index" commonly known as the Rolodex. Merger with a competitor created the company known as Remington Rand, making him a multimillionaire.[14]

Fisher was one of the original directors of the Eugenics Record Office, and he sent his own family's genealogical records there for preservation.[15] He was a longtime colleague of Harry Laughlin, with whose help he organized the American Eugenics Society as the propaganda arm of the eugenics movement.[16] In 1921, as president of the Eugenics Research Association, a related organization, Fisher deplored the "great load of degeneracy" carried by America and argued, as Francis Galton and Charles Davenport had, that "eugenics must be a religion."[17] He also belonged to the Galton Society, founded by Madison Grant, one of the most prominent supporters of every brand of negative eugenic policy, and another Sumner student at Yale. Grant graduated the year before Fisher.[18]

Fisher founded the Committee of One Hundred in 1907 to lobby for a national health department and to collect support for a federal role in public health and eugenic interventions; Taft lent his assistance.[19] In 1908, President Theodore Roosevelt created the National Conservation Commission, charging it with studying ways to preserve the nation's natural resources so as to prevent national "degeneration." That study concluded as Taft's presidency began. In the midst of the commission's sprawling three-volume report was Fisher's section on "National Vitality," which linked prevention of disease and a lengthened life span to increased economic productivity. Fisher argued for the "money value of increased vitality" and devoted a major part of his discussion to the impact of eugenics.[20] "If the aims of eugenists are carried out," Fisher said, "an obviously unhygienic marriage will be frowned upon." Quoting liberally from Galton and his followers and citing famous eugenic family studies like *The Jukes* and "The Tribe of Ishmael," Fisher urged "governmental interference with the birth rate" among the less fit. He suggested that the prohibition of marriage among "criminals, paupers and the feeble-minded" was in order. He praised the new Indiana sterilization law, predicting that laws "of gradually increasing severity" would eventually proliferate and that the public would accept "communal restriction of the right to multiply." "We insure," he said, that "certain of these classes are not permitted to propagate their kind" and claimed that "marriage and 'sterilization' laws will reduce the number of marriages of degenerates."[21] Ironically, Fisher himself suffered from tuberculosis and his father had died from it; many of his colleagues in the eugenics movement would have found in these facts alone ample justification to sterilize Fisher himself.[22]

Fisher emphasized the need to study and "gradually put into practice" eugenic measures to protect future generations. Rapists, criminals, idiots, and "degenerates generally" should be "unsexed." Marriage should be prohibited among the "unfit," including syphilitics, the insane, feebleminded, epileptics, paupers, and criminals, and public opinion should be molded so as to inform marriage choices for future generations.[23] Fisher later published his report as a book entitled *National Vitality: Its Conservation and Preservation.*[24] In 1913, when his Committee of One Hundred ceased to function, its remaining funds were donated to the Eugenics Committee of the United States of America.[25]

In 1913 Taft agreed to become a director—later chairman of the board—of Fisher's Life Extension Institute, which promoted healthful living and preventive medicine.[26] By the 1920s the institute took over the work—previously done by the Eugenics Record Office—of providing eugenic advice to families contemplating marriage or children.[27] Fisher set up a "Hygiene Reference Board" as part of the institute, containing nearly a hundred experts who could advise government and the private sector on health and policy issues. The original board included a "Division of Race and Social Hygiene," which was soon renamed the "Division of Eugenics."[28] The most prominent names in eugenics were included for their expertise; with Fisher, Charles Davenport, Lewellys Barker, William Welch, Elmer E. Southard, and Alexander Graham Bell, the list eventually contained six of the eight original directors of the Eugenics Record Office.

HOW TO LIVE

RULES FOR HEALTHFUL LIVING
BASED ON MODERN SCIENCE

*AUTHORIZED BY AND PREPARED IN COLLABO-
RATION WITH THE HYGIENE REFERENCE
BOARD OF THE LIFE EXTENSION
INSTITUTE, INC.*

BY

IRVING FISHER, *Chairman,*
PROFESSOR OF POLITICAL ECONOMY, YALE UNIVERSITY

AND

EUGENE LYMAN FISK, M.D.,
DIRECTOR OF HYGIENE OF THE INSTITUTE

EIGHTH REVISED EDITION

FUNK & WAGNALLS COMPANY
NEW YORK AND LONDON
1916

William Howard Taft, chairman of the board of the Life Extension Institute, from Irving Fisher's *How To Live*, 1916.

In 1915 Fisher and Dr. Eugene Lyman Fisk published *How to Live,* an enormously popular book that combined recommendations for healthful living and proper hygiene with a liberal dose of eugenic ideology.[29] Its final section on eugenics, written by Charles Davenport, filled over thirty pages.[30] Taft wrote an introduction, printed in each edition of the book, and every reader encountered his photograph in the book's introductory pages. *How to Live* recommended eugenics boards for every state to extend the work of the Eugenics Record Office. Harry Laughlin's *Eugenical News* listed it along with other "Eugenic Publications."[31] It listed "prevention of reproduction by the markedly unfit" and sterilization of "gross and hopeless defectives" as important social goals.[32] Book reviews paid attention to the "special significance" of eugenics in the book as well as Taft's contribution.[33] The book went through sixteen editions between 1915 and 1922—ninety in all—selling more than 135,000 copies. It demonstrated Taft's regular and intimate involvement with the leaders of the U.S. eugenics movement.

As open as he was in endorsing the regimen of good hygiene, proper diet, and straightforward eugenic attitudes that made up the message of the popular self-help text, Taft was also wary of being too openly associated with the organized eugenics movement. He was particularly cautious when his friend Fisher attempted to involve him in moneymaking schemes disguised as philanthropy. Soon after the Supreme Court decided to hear the *Buck* case, Fisher wrote to Taft on American Eugenics Society stationery that listed Charles Davenport, Harry Laughlin, and Harry Olson as officers. Fisher offered "an investment opportunity" involving "important philanthropic work" of the society. Taft, who over years of public life was besieged with solicitations, was wary of another request for an endorsement. He was also aware of the political danger of openly taking sides as an advocate of

Oliver Wendell Holmes, 1926. Courtesy Virginia Historical Society, Richmond, Virginia.

eugenic schemes launched for profit. He responded brusquely, rejecting Fisher's invitation, noting simply, "I have to keep out of it."[34]

But Taft's aversion to entrepreneurial eugenics did not prevent him from endorsing eugenics as public policy; as one of the recipients of Harry Laughlin's book, perhaps he understood the Virginia plan better than most. He cast his vote in favor of the Virginia law and assigned the job of writing the opinion in the *Buck* case to Associate Justice Oliver Wendell Holmes Jr.

"A Burning Theme"

The Holmes name was well known among eugenicists, who often quoted an aphorism of famous physician and poet Oliver Wendell Holmes Sr.: "A man's education should begin with his grandparents."[35] Even Charles Davenport's eugenics textbook echoed the elder Holmes. According to Davenport, Holmes said that the prescription for longevity was heredity—"first to select long-lived grandparents."[36] Renown as a judge with a way with words seemed to insure that Holmes Jr., like his father, would also become a darling of the eugenics movement.

After nearly twenty-five years on the Supreme Court, Holmes was the most celebrated judge in America. In March 1926, his portrait appeared on the cover of *Time* magazine as a tribute to his eighty-fifth birthday. The *New Republic* regaled him as a "tender, wise, beautiful being" who "redeems the whole legal profession" and, a few months later, applauded him as the wise "Yankee, strayed from Olympus."[37]

Holmes had embraced the most radical ideas for social improvement when the formal eugenics movement was only in its infancy. Describing the typical criminal as "a degenerate," Holmes despaired of any potential for improvement or reformation; such people, he said, simply "must be got rid of."[38] Science, he said, could "take control of life, and condemn at once with instant execution what is now left to nature to destroy."[39]

As a young man Holmes had read Herbert Spencer's *Social Statics* and *First Principles of Evolutionary Philosophy.*[40] He quoted Spencer freely, declaring that legislation "must tend in the long run to aid the survival of the fittest."[41] But while Spencer's prescriptions were harsh, condemning both government sponsored welfare schemes and the "injudicious charity" that made future generations dependent on philanthropy, he did not dispute the need for private beneficence at times. Holmes trumped Spencer, declaring that philanthropy was "the worst abuse of private ownership."[42] At least in *Social Statics,* Spencer never called for the execution of the socially deficient, as Holmes later would. In fact, Spencer condemned those like Holmes who delighted in "passing harsh sentence on his poor, hard-worked, heavily burdened fellow countrymen." Herbert Spencer also believed in a radical libertarianism and freedom from state control; Holmes did not.[43]

Holmes' famous dissent in the 1905 Supreme Court case of *Lochner v. New York* criticized the decision striking down a New York law limiting the work week of bakers to sixty hours. The Court's majority held that the law violated the constitutionally protected right to contract, under which any baker should be allowed to work for whatever hours he wished. Holmes' dissent pointedly rejected that analysis, with the declaration that "the Fourteenth Amendment does not enact Mr. Herbert Spencer's *Social Statics.*" In the same dissent he clarified another definition, saying that "the word liberty in the Fourteenth Amendment is perverted when it is held to prevent the natural outcome of a dominant opinion" except when a law might "infringe fundamental principles."[44] In the first decade of the twentieth century, the definition of "fundamental rights" was quite limited; it was clear to Holmes that "liberty" did not mean freedom from government coercion or control.

While Spencer's social scheme found the strong surviving on their own, Holmes was keen to help evolution along. He found the idea that "any rearrangement of property, while any part of the world propagates freely, will prevent civilization from killing its weaker members" to be "absurd."[45] And he was equally cynical about plans for social amelioration. In 1910 he wrote legal scholar John Wigmore, asking, "Doesn't this squashy sentimentality of a big minority of our people about human life make you puke? [That minority includes people] who believe there is an upward and onward—who talk of uplift—who think that something in particular has happened and that the universe is no longer predatory. Oh bring in a basin."[46]

Few of Holmes' friends were as outspoken on eugenics as British socialist Harold Laski. Despite their many political differences, they remained longtime correspondents and often discussed the importance of eugenics in social planning. Laski decried the habit of "fostering the weaker part of mankind, until its numbers have become a positive danger to the community" and condemned a "kindly but thoughtless humanitarianism." Eugenics was available, he said, with the "scientific means" to deal with social problems. Laski pointed to family studies like *The Jukes* as the basis for laying "the foundations of our science." National demise followed "the fostering of the unfit at the expense of the fit, and their consequent over-propagation." The solution was to bolster "natural selection" with "reproductive selection" and, by using genetics, "build a strong political superstructure."[47]

Holmes' own thoughts on eugenics combined Malthusian concerns with overpopulation and a similar faith in science to yield social solutions. There might be a way, he noted to Laski, to help the Malthusian scheme along: "As I have said, no doubt, often, it seems to me that all society rests on the death of men. If you don't kill 'em one way you kill 'em another—or prevent their being born." He was impatient with welfare schemes of socialist reformers, writing, "One can change institutions by a fiat but populations only by slow degrees . . . and . . . while propagation is free and we do all we can to keep the products, however bad, alive, I listen with some skepticism to plans for fundamental amelioration. I should expect more from systematic prevention of the survival of the unfit."[48]

Earlier in his career, Holmes supported substituting "artificial selection for natural by putting to death the inadequate."[49] But by the time of *Buck*, the tide had turned against euthanasia and other radical proposals of the eugenicists. Scientists and public figures had begun to speak out against eugenic sterilization. In contrast, Holmes still clung to the positions held by Harry Laughlin and his colleagues. He had no compunctions about "restricting propagation by the undesirables and putting to death infants that didn't pass the examination."[50] Holmes was clearly locked into earlier iterations of eugenics and had not kept up with the science that in the meantime had undercut the old positions of Laughlin and his ilk.

Strode's brief to the Supreme Court seemed perfectly fitted to Holmes' beliefs. The eugenical emphasis was clear; the argument in favor of a broad police power was equally compatible with Holmes' personal and public opinions. While Holmes and the other Justices reviewed the records and briefs in *Buck v. Bell*, Strode and Whitehead prepared for their appearances in Washington. Strode had appeared before the Court on two previous occasions. He lost both times; both times Holmes delivered the Court's opinion. Strode went to great lengths to assure that he would not miss his third attempt at a Supreme Court victory before Holmes. Whitehead seemed somewhat less certain that he would find time away from his banking concerns in Baltimore, but he too finally was scheduled to argue before the Court.[51]

It was Strode's good fortune that Taft assigned Holmes the task of writing the opinion in the *Buck* case. It would soon become clear that Holmes agreed completely with the Strode position that the state could intrude on almost any personal liberties when the public welfare was at stake. Holmes believed that the Malthusian calamity of population growth could be delayed or at least averted by preventing some births from occurring. And he would not be reticent in prescribing, in a Supreme Court opinion, which births those should be.

Taft warned Holmes that he could avoid controversy by following a strategy that focused on the Buck family's hereditary defects. "Some of the brethren [the Justices] are troubled about the case, especially [Justice Pierce] Butler. May I suggest that you make a little full [the explanation of] the care Virginia has taken in guarding against undue or hasty action, proven absence of danger to the patient, and other circumstances tending to lessen the shock that many feel over the remedy? The strength of the facts in three generations of course is the strongest argument."[52]

On April 25, 1927, Holmes sent his opinion to be printed. He had followed Taft's lead but, as was his habit, gave only an abbreviated justification for the Court's decision and let rhetorical flourishes define the opinion's structure. His first draft triggered criticism from other Justices who found the abrupt style jarring. But as Holmes confided to his friend Laski, style was important:

> I have had some rather interesting cases—the present one, as I believe I mentioned, on the Constitutionality of a Virginia act for the sterilization of imbeciles, which I believe is a burning theme. In most cases the difficulty is rather with the writing than with the thinking. To put the case well and from time to time hint at a vista is the job. I am amused (between ourselves) at some of the rhetorical changes suggested, when I purposely used short and rather brutal words for an antithesis, polysyllables that made them mad. I am pretty accommodating in cutting out even thought that I think important, but a man must be allowed his own style.[53]

Holmes had complained that in other cases, if he was left alone, his original compositions emerged in a fully masculine style, sometimes seeming "to have a tiny pair of testicles." But the sensitivities of his colleagues occasionally forced him to neuter his opinions. He tired of writing opinions that sang only "in a very soft voice."[54] But in the days before the *Buck* opinion was formally announced, Holmes seemed satisfied with his work and pleased to have had the occasion to speak his mind. Time and his many years on the Court were on his side, as he later told Laski, because "sooner or later one gets a chance to say what one thinks."[55]

The Opinion

The Supreme Court had never dealt with the merits of government-mandated surgery before, much less a law prescribing operations primarily for state benefit. But Holmes disposed of the issue in an opinion of shocking brevity. The text ran just under three full pages in the official Supreme Court Reports, and it contained little original thought. Holmes' first paragraph introduced the case as a challenge to the constitutionality of the Virginia law that called for "the operation of salpingectomy upon Carrie Buck . . . for the purpose of making her sterile." Following Taft's suggestion, Holmes immediately described Carrie as "a feeble-minded white woman" living at Virginia's State Colony for Epileptics and Feebleminded. Eighteen years of age at the time of her trial, she was the "daughter of a feeble-minded mother in the same institution, and the mother of an illegitimate feeble-minded child." He also repeated language from the preamble of the law, declaring that "the health of the patient and the welfare of society may be promoted in certain cases by the sterilization of mental defectives," surgical sterilization could be effected "without serious pain or substantial danger," and that "experience has shown that heredity plays an important part in the transmission of insanity, imbecility, etc."

By a recitation of the many procedural steps included in the law, Holmes summarily dismissed Whitehead's first contention—that Carrie had not received "due process." The state had followed "the very careful provisions" of the law, which "protects the patients from possible abuse." After outlining the state procedures, Holmes continued, "There can be no doubt that so far as procedure is concerned the rights of the patient are most carefully considered." Virginia officials had maintained "scrupulous compliance" with "every step" prescribed in the statute, and Carrie Buck had been subjected to "months of observation." There was no doubt that she had received "due process at law."

Holmes knew that Buck's case was not about procedural protections but rather the substantive scope of the law. Whitehead claimed a right of "bodily integrity" for Carrie, contending that it was impossible to justify a sterilization that would extinguish that right. But the Virginia legislature had specified certain grounds under which sterilization could take place. State officials had found that Buck was "the probable potential parent of socially inadequate offspring" and could endure an operation that would benefit her as well as society more generally. The trial court confirmed that proper grounds existed in Carrie's case—as Holmes wrote, "if they exist, they justify the result."[56]

In the Civil War, Holmes' 20th Massachusetts Regiment lost more men than any other unit in the Union Army. Holmes himself was wounded three times during the conflict, and the experience irrevocably marked his life thereafter. Harkening back to his military service, Holmes invoked the memory of his dead comrades at arms, the "best and noblest of our generation" who had fallen.[57] "We have seen more than once that the public welfare may call upon the best citizens for their lives," he recalled. But the Carrie Bucks of the world were not to be counted in their ranks. She lived, in contrast, among those he had years earlier labeled the "thick-fingered clowns we call the people . . . vulgar, selfish and base."[58] Their dues to society were less costly. Though Holmes was later praised as the "champion of the Common Man," it is clear that if he ever had any sympathies for the lower classes, they were long gone by the time he wrote the *Buck* opinion.[59]

The people called "unfit" were locked in prisons, poorhouses, and asylums, where they would not be called upon by their country to make the ultimate sacrifice. Yet "it would be strange," said Holmes, if the public "could not call upon those who already sap the strength of the State for these lesser sacrifices," such as yielding to surgery and forgoing children. Though sterilization might not seem a "lesser sacrifice" to those who faced it, it was nevertheless justified "in order to prevent our being swamped with incompetence." To a man who had taken a minié ball in the chest at Ball's Ridge and returned to the field to suffer a second wound at Antietam and a third at Chancellorsville, the inconvenience of surgery was the least that those less-than-best citizens should be asked to endure to protect society from their profligate spawn. Life was short and early death even more likely for the unfit, suggested Holmes. It would be preferable for them and the rest of society if they had never been born at all. He offered a simple solution: "It is better for all the world, if instead of waiting to execute degenerate offspring for crime, or to let them starve for their imbecility, society can prevent those who are manifestly unfit from continuing their kind."

Most judicial opinions are full of citations from similar cases, precedents from which principles can be drawn to apply to the case at hand. Despite the radical new ground this opinion broke, Holmes cited only one precedent case in *Buck*. It was *Jacobson v. Massachusetts*, the 1905 case that Harry Laughlin had recommended in his book, that Aubrey Strode had included in his briefing, and that Holmes had participated in deciding, early in his Supreme Court career.

In that case, Reverend Henning Jacobson paid a five-dollar fine for refusing to be vaccinated for smallpox. The public health principle of requiring vaccination and enforcing the greater medical good for those who refused via fines was close enough for Holmes to the eugenic benefit of forestalling promiscuous reproduction of defectives by coercive surgery. Referring to *Jacobson*, Holmes wrote: "The principle that sustains compulsory vaccination is broad enough to cover cutting the fallopian tubes." At this remove, it is hard to gauge the power of plague imagery; a generation that had seen the devastation wrought by both smallpox and influenza would not need to be convinced that the usual rules must be set aside when epidemics threatened. But Holmes was not satisfied with invoking the vaccination analogy; as if to silence any objection to its accuracy, he then wrote his boldest and most caustic line. Carrie Buck was accused of being a "high grade moron"; the charge that her daughter was abnormal rested on the slimmest of evidence, but Holmes ignored these details. He had learned the dangers of faulty heredity from the same father who had taught him a penchant for phrase-making. (According to popular report, the master of the celebrated Holmes breakfast table rewarded clever lines with an extra dollop of marmalade.)[60] Never one to disappoint his audience, Holmes called upon this famous talent to give the eugenics movement a popular and unforgettable slogan that would remain the rallying cry for the forces of sterilization for years to come. Invoking the image of the Buck family, with its generations of hereditary defect cascading through grandmother, mother, and daughter, Holmes proclaimed: "Three generations of imbeciles are enough."

The decision's final paragraph returned to an important point from Whitehead's argument. Some state courts, like New Jersey, had voided sterilization laws that were applied only to people in institutions. Using similar logic, Whitehead had argued that the Virginia law violated the equal protection clause of the Fourteenth Amendment to the Constitution by imposing an unfair, unequal treatment of one small group that the much larger numbers of similarly situated people were allowed to escape merely because they remained in the community. Mocking Whitehead's assertion, Holmes discarded the claim, asserting that "it is the usual last resort of constitutional arguments to point out shortcomings of this sort." It was not necessary for such laws to cover all peoples at all times, he wrote, "the law does all that is needed when it does all that it can." By a policy of serial admission such as Strode suggested, many could be sterilized and released and then many more brought to institutions for surgery. In that way, Holmes concluded, "the equality aimed at will be more nearly reached."

As brief as it was, most of the Holmes opinion appeared to be lifted from earlier writers. Many of its most memorable phrases were commonplace among eugenicists, perhaps recalled from his younger years. And the main arguments had all been made before, everywhere from obscure journals to the *New York Times.*

For example, Holmes ranked the Bucks among the "manifestly unfit." Twenty-five years earlier, that phrase had been used to argue for institutionalization of the "feebleminded" who were lacking in self control.[61]

In 1904, "asexualization of those who are manifestly unfit" was supported in a journal on ethics.[62] In 1912 Lewellys Barker, as president of the National Committee for Mental Hygiene and Board Member of the Eugenics Record Office, said that "denying . . . the privilege of parenthood to the manifestly unfit . . . [was] the problem of eugenics," and a writer in the *New York Times* judged the feebleminded "manifestly unfit" to attend the public schools.[63] By 1915 the *Washington Post* reported that some clergy had endorsed a prohibition on parenthood for those deemed "manifestly unfit to produce."[64] To condemn society's least favored members as "manifestly unfit" was merely to echo a eugenic cliché.

Holmes' assertion that sterilization was a "lesser sacrifice" in comparison to giving a life for one's country was also a well-worn argument. Eugenicists repeatedly claimed that since the state had power over life, it also had power over birth. Because soldiers could be drafted and convicts executed, states could mark some people for sterilization as part of the police power, protecting the general welfare just as public health laws did. Charles Davenport even included the argument in his popular 1913 textbook: "Concerning the power of the state to operate upon selected persons there can be little doubt, . . . the right to the greater deprivation—that of life—includes the right to the lesser deprivation—that of reproduction."[65] But Holmes' language had an even older pedigree than that. Holmes' lines from *Buck* sound the same themes, with similar language, as a 1909 essay from a medical journal by Dr. J. Ewing Mears, an early advocate of sterilization:

> *Holmes:* We have seen more than once that the public welfare may call upon the best citizens for their lives. It would be strange if it could not call upon those who already sap the strength of the State for these lesser sacrifices, often not felt to be such by those concerned, in order to prevent our being swamped with incompetence.

> *Mears:* The state when attacked sends forth its best citizens to battle in its defense. Is it asking too much . . . when the state, through carefully considered legislation, seeks to protect itself against the degrading influences of the continually flowing stream of transmitted pollution, which saps the mental, moral and physical vitality of its citizens . . . ?[66]

Joseph Mayer, a German Roman Catholic scholar, had also voiced similar sentiments:

> *Mayer:* For its protection the State may sacrifice millions of its best sons in battle; for its protection, namely for the prevention of epidemics, it may make vaccination obligatory and inoculate innocent children. . . . By the same token, the State must also have the right for its own protection to deprive mental defectives and the criminally insane of the power of generation by a relatively trivial operation.[67]

Original or not, Holmes' language resonated with the public, as it did with most of his colleagues on the Supreme Court.

Following the Holmes opinion, the Supreme Court decision was printed with only one line more: "Mr. Justice Butler dissents." There had been pressure on the Court to avoid dissents, and even when he felt compelled to register his disagreement with the majority of his colleagues, Justice Pierce Butler was not in the habit of amplifying discord in writing.[68] It has been widely speculated that Butler dissented from the *Buck* decision because of his Roman Catholic background. While there was no Roman Catholic dogma forbidding sterilization in 1924, there was much sentiment characterizing any eugenical surgery as "mutilation." In fact, as early as 1912, sterilization advocate Bleecker Van Wagenen claimed that defeat of early Pennsylvania eugenic legislation had been the work of one particularly strident opponent in the Assembly, a Roman Catholic.[69]

Similarly, some thought that Butler's regular contact with the Catholic hierarchy may have influenced his vote, and he was said to rank only the Church among institutions he defended as zealously as "his corporate friends." In one account he was even called the "Papal Delegate to the Supreme Court." Satisfied that his opinion was an accurate interpretation of the law in the *Buck* case, Justice Holmes commented that Butler was "afraid of the Church. I'll lay you a bet the Church beats the law."[70] But claiming a religious motive for Butler's dissent is mere speculation. He left no opinion, and no other evidence has surfaced.

Perhaps other motives were important to Butler. He became embroiled in a dispute over a will he had written for his older brother John at exactly the time the *Buck* case was under consideration at the Court. John Butler's only child was an illegitimate daughter, born of a servant girl. The daughter later claimed that he had abandoned her in a church pew and subsequently arranged for her to be adopted by a family in a distant state. When John Butler died, she discovered that she had been cut out of his will, while all eight children of Pierce Butler and the children of John's other siblings were generously provided for via the $2.7 million estate. The forgotten daughter of John Butler sued to challenge the will, which had been written by Justice Butler, claiming "fraud, misrepresentation, and conspiracy" and "undue influence" on the part of the Justice and his sons, who together had acted as executors of the estate. Butler paid his illegitimate niece what one account described as an "immediate and handsome settlement" and was spared the public embarrassment of litigation over unseemly family matters when the suit was dismissed.[71] How these events may have colored Butler's sympathies for a Virginia girl described as poor and illegitimate cannot be known.

Except for Butler, every other member of the Court sided with Holmes. Justice James McReynolds was a predictable vote for the majority. A legendary racist and anti-Semite, McReynolds's refusal to sit next to Louis Brandeis, the first Jewish Supreme Court Justice, left the 1924 Court with no official portrait. McReynolds also had a reputation for being generally intolerant of women, and he rarely voted to strike down state enactments. His attitude toward disability was later demonstrated when McReynolds declared himself unwilling to "resign as long as that crippled son-of-a-bitch [Franklin D. Roosevelt] is in the White House."[72] So his hesitance to overrule a law that would sterilize a "defective" girl from the town in which he had attended law school was unremarkable.

Eminent jurists Harlan Stone and Louis Brandeis also favored the Holmes position, as did former president Taft. Taft had reason to be irritated by Holmes earlier in his career, when Holmes wrote an opinion summarily invalidating millions in taxes Taft had levied as governor of the Philippines.[73] On eugenics, however, they had no such disagreement. Taft wrote to his wife after the *Buck* opinion was announced to praise the eighty-three-year-old Justice: "Holmes is wonderful. I gave him three cases this week . . . and today he sends me three good opinions. His quickness and his powers of catching and stating the point succinctly are marvelous. His brilliance does not seem to abate at all."[74]

Justice Brandeis is usually identified with the idea of privacy, and in that light his vote in favor of Carrie Buck's sterilization may seem an anomaly. But the concept of privacy has expanded considerably in the past hundred years, particularly in the realm of reproduction. It is important not to identify Brandeis' early comments with our later understanding. Only a year later, in a dissent that included his famous invocation of the "right to be left alone," Brandeis cited *Buck* for the proposition that limitations on governmental power contained in the Fourteenth Amendment do not stand in the way of governments "meeting modern conditions" through laws that might previously have been dismissed as "arbitrary and oppressive."[75]

On several occasions in the weeks following *Buck v. Bell,* Holmes wrote to friends to comment on the case. In one letter, he declared that "sterilizing imbeciles . . . was getting near the first principle of real reform."[76] To another correspondent, Holmes boasted of his satisfaction in the outcome of the case: "One decision that I wrote gave me pleasure, establishing the constitutionality of a law permitting the sterilization of imbeciles."[77] Holmes' final surviving comments on eugenical sterilization concerned a letter he had received, calling him a "monster" who should expect the "judgment of an outraged God" for his opinion in *Buck v. Bell.* Holmes' response was characteristically glib: "Cranks as usual do not fail."[78]

13

Reactions and Repercussions

There was strong public reaction to Holmes' opinion. Some, like Holmes' "crank" correspondent, saw blasphemous intent in a law that would negate the divine command to "be fruitful and multiply." Others welcomed the decision as a powerful stroke in favor of social progress. News of the Holmes opinion made its way into dozens of papers in every corner of the country. Press reaction was overwhelmingly positive.

An editorial in Carrie Buck's hometown newspaper, the *Charlottesville Daily Progress,* declared that Holmes was "in sympathy with the most progressive tendencies in our social machine" and called his opinion "a genuine classic." Virginia was deemed "fortunate in having this eminently sane and beneficial law . . . permanently enrolled on its statute books."[1] The *New York Times* covered the *Buck* decision, quoting liberally from the Holmes opinion, which had given states the "right to protect society."[2] The *Los Angeles Times* and the *Chicago Tribune* repeated the *New York Times* story, thought to be important because of "agitation for similar legislation in other states."[3]

The *Baltimore Evening Sun* noted that twenty-year-old Carrie Buck had the mental age of a nine-year-old, and its daily counterpart, the *Baltimore Sun,* described the "vigorous opinion" by Justice Holmes.[4] The *Boston Daily Globe* reported that "fifteen other states have similar laws."[5] "Eugenists cheered" the Holmes decision, and "sentimentalists were vexed" by it, according to *Time* magazine.[6]

In praising the opinion of the U.S. Supreme Court, the *Louisville Courier-Journal* commented that sterilization would be used "much more frequently in the cases of women than of men."[7] The *Atlanta Constitution* highlighted the Holmes opinion: "'Men must die for the state,' said Justice Oliver Wendell Holmes, who was dangerously wounded in the civil war, and the lesser sacrifice called for in sterilization may properly be required for protection of the public welfare."[8] The Associated Press picked up the news, relaying the *Buck* result to the *Washington Post.*[9] Virginia's flagship newspaper, the

Richmond Times-Dispatch, summarized the state's argument, including the claim that the law "would result to the advantage" of Carrie Buck and allow her to be released from custody.[10] North Carolinians learned that *Buck* provided the "remedy for imbecility," while Iowans found the decision to "protect the world against the morons" to be a "notable step in the right direction."[11] One Tennessee paper trumped the Holmes formula, saying that even "one generation of imbeciles is enough."[12] In Montana the decision was applauded as a lesson that "we do not dispose of enough human weeds."[13] A wire service reporter quoted the U.S. surgeon general, who called the decision a "step toward a super-race."[14] A paper in Alabama saw the decision as an opportunity "to convince open-minded folk that such legislation is wise."[15]

A brief note summarizing news accounts of the sterilization case appeared in the *Journal of the American Medical Association.*[16] The *American Journal of Public Health* praised Holmes, "that great jurist," and a decision that "opens future possibilities of vast importance in the field of eugenics and public health."[17] Even the *Literary Digest* weighed in, proclaiming that the Holmes opinion was designed, as the headline trumpeted, "to halt the imbecile's perilous line."[18]

The *Buck* decision provided an opportunity for commentary from pundits and eugenic partisans. For example, Mrs. Victoria Woodhull Martin had been the first female presidential candidate, running under the banner of the Equal Rights Party in 1872. She wrote and lectured on eugenics before the turn of the twentieth century, calling asylums the "over-crowded receptacles of human misery" and urging the "scientific propagation of the race." After all, she said, the "first principle of the breeder's art is to weed out the inferior animals." Contacted for a comment, she noted her pleasure at the *Buck* decision. As for sterilization, said Martin, "I advocated that fifty years ago in my book, 'Marriage of the Unfit.'" Time, she said, had proven her right.[19]

Other commentators were more cautious. Readers in Illinois learned that just as Virginia was embracing sterilization, Wisconsin physicians were questioning the role of heredity in the birth of "defectives." Sterilization, wrote the *Decatur Herald,* was "not a panacea."[20] While not disputing the accuracy of eugenic theory, a New York paper warned of the "danger of error" in the administration of any sterilization program, such that the public must be on guard "to prevent abuse."[21] The *Philadelphia Evening Bul-*

letin warned that "eugenic enthusiasts are apt to go to extremes."[22] But some countered this concern over abuse with dire predictions that without sterilization, in two hundred years, "morons would hold the upper hand."[23]

Supporters of the eugenics cause were disheartened when in the same week as the *Buck* decision, the *New York Times* reported on its front page that Harvard University had turned down a sixty thousand dollar bequest in the will of sterilization advocate J. Ewing Mears. When eugenic sterilization laws were first introduced, writers at the *Harvard Law Review* were convinced that states could constitutionally protect themselves from "the birth of undesirable citizens" such as criminals and the insane, particularly since the "insanity of lunatics is generally inherited" and a "large number of criminals have an inborn and hereditary tendency to crime."[24] When Dr. Mears died in 1919, he left the money to Harvard with confidence that it would be put to good use for the eugenics cause. But by the time the estate finally cleared probate eight years later, the Harvard trustees were apparently much less convinced of eugenic theories, and declined to commit the university "to teach that the treatment of defective and criminal classes by surgical procedures was a sound doctrine."[25] That opinion infuriated some Laughlin correspondents, of which one quipped: "Wouldn't this make a minister strike his father?"[26]

Despite such minor setbacks, supporters of sterilization perceived a shift in the momentum of public opinion and forged ahead. Harry Laughlin received a flurry of correspondence from California eugenicists E. S. Gosney and Paul Popenoe of the Human Betterment Foundation. The day after the Supreme Court decision, their local newspaper published a lengthy account of Gosney's research on the results of eugenic sterilization in California. He later wrote to Laughlin: "We were successful in heading off an attempt to repeal the California law at the past session of the legislature, and the federal supreme court's decision will strengthen our hands in the future."[27] America's northern neighbors also paid attention to *Buck*. The *Vancouver Sun* praised the Holmes opinion, urging Canada to heed the U.S. example "the sooner the better."[28] The next year Alberta passed its first sterilization law.

Back at the Virginia Colony, Dr. John Bell received a postcard mailed from New York City that said:

> May God protect Miss Carrie Buck
> ~~from feebleminded justice~~ from injustice.[29]

Oblivious to critics, Bell's *Annual Report* declared: "Virginia has placed herself in the forefront of civilization" through the sterilization law.[30]

Petition for Rehearing

Aubrey Strode was pleased at how much attention the decision had attracted, boasting that he had received "a number of letters about it from over the country."[31] He also made it known that he reported to the Colony Board that "an appeal had been made to the Supreme Court by some anti-eugenic society" and that the organization was pressing for a rehearing of the case. He advised the Board that surgeries should be postponed until the Court acted on the petition.[32]

The petition originated with a Roman Catholic men's group. In the early years of the twentieth century, Catholic opinion on the moral appropriateness of sterilization was far from unanimous. No official Church teaching was proclaimed on the issue until the 1930s, though ancient prohibitions against grave bodily mutilation had moved many to question surgery—particularly castration—that had only questionable medical benefit. Some writers entertained the idea that vasectomy could be directed toward "weak-willed degenerates and naturally defective criminals."[33] Others questioned the "materialistic principles" embedded in popular hereditarian arguments, arguing that degeneracy has many sources.[34] A more severe stand was taken by those who felt that people "with a supposedly inherited and incurable bent to evil" were better off not being born.[35] The general opinion of Catholic theologians seemed to accommodate several conclusions, with at least a respectable minority judging "the right of society to protect itself against moral and social evil" a sufficient motive for approving of surgical vasectomy.[36]

As the *Buck* case made its way to Washington, several lengthy essays containing an ethical analysis of sterilization filled the pages of a religious periodical called the *Homiletic and Pastoral Review.* Charles Bruehl argued that while there may actually be a "menace of the feebleminded," as eugenicists insisted, the "invasion of degenerates" predicted by some was still a distant danger, even in the worst-case scenario. A "panicky fear" of this deluge could lead "society to adopt measures of self-protection of which in soberer moments it will feel ashamed." Bruehl denied the "state of emergency" rhetoric that compared the danger of mental deficiency to war and asserted that the "present state of knowledge" of heredity was still inadequate as a basis for laws that would deprive large groups of "a basic human right."[37]

Bruehl attacked the expansiveness of Laughlin's Model Law, with its broad definition of the "socially inadequate." He described the powers granted to the states in such laws as "simply appalling." Eugenical education was valuable—in some sense "eugenical morality" was not even a new thing but merely "an ethos of social responsibility in marriage." Emphasis on chastity, healthful living, and social justice completed his vision of "Christian eugenics."[38] Critics like Bruehl were left with no specific teaching from the Catholic hierarchy on the practice of eugenic sterilization.

After the *Buck* case was decided, Bruehl completed a detailed critical analysis of sterilization and predicted that the *Buck* decision would provide "a new impetus" for eugenics and the revival of moribund laws that had fallen out of use as well as "other unsavory schemes for racial improvement." He admitted that sterilization might be justified, but only in the face of an imminent demographic emergency—a situation that was particularly unlikely. More pointedly, he emphasized the "colossal, unlimited contempt" shown by leaders of the eugenics movement for anyone they might consider unfit.[39]

Others in the Catholic press had criticized sterilization laws without specific reference to *Buck* even as the case moved through the courts. Quoting scientific publications, one critic concluded that the sterilization movement had been based on a faulty theory that moral qualities were transmitted via heredity, an idea "which most recent scientists, philosophers and criminologists" had begun to discard. Nevertheless, he said, the laws were still endorsed "for economic reasons" without regard for the defunct theory.[40]

In the absence of a clear doctrinal directive from Rome, the Catholic publication *America* conceded that sterilization of "the mentally defective" by the state might be permissible under some emergency circumstances, but it blasted the Holmes opinion for the unsupported assertion that Virginia was in immediate danger of "being swamped with incompetence."[41] In another critique of *Buck*, the Catholic periodical *Commonweal* suggested that the breadth of Laughlin's definition of the "socially inadequate" might make sterilization of the "economic incompetent" a standard method to prevent poverty.[42]

The general Catholic opposition to sterilization did not go unnoticed. Even before his opinion was complete, Holmes wrote to Harold Laski that "the religious are astir."[43] One group to which Holmes may have been referring was the National Council of Catholic Men. Even though the lower court decisions in *Buck* were widely publicized, the NCCM only belatedly learned of the progress of the *Buck* case to the Supreme Court.[44] Immediately after the decision became public, they went into action to see whether some strategy could forestall the decision's impact.

Charles F. Dolle, executive secretary of the NCCM, traveled to Lynchburg to meet with Aubrey Strode and Irving Whitehead. He reported that "both of the attorneys in the case were surprised and rather disappointed that the court did not treat the principle involved more fully in its decision." There was a forty-day window available to file a petition for rehearing, and Dolle succeeded in convincing Whitehead that the effort must be made. Because Whitehead complained that he would not be paid to draft the petition, Dolle promised him one hundred dollars and also agreed to cover the cost of printing the application itself, possibly another hundred dollars.[45]

The lawyer for the NCCM conferred with his colleagues early in June to resolve the question of whether the Catholic group would intervene to urge a rehearing of *Buck*. Even though the considered opinion was that the petition was "practically hopeless"—the decision had after all been 8 to 1—they agreed that the decision could not be allowed to stand without a fight. In the light of existing anti-Catholic sentiment, prudence demanded that neither the group nor its lawyers should be named explicitly in the brief and that Whitehead should present the petition. Dolle met with Whitehead in Baltimore to deliver material he hoped that Whitehead would use. In the unlikely event that the petition was considered favorably, counsel could be associated with Whitehead for the oral argument.[46]

The final form of the petition was not what Dolle had proposed, because Whitehead excluded commentary that would contradict the opinion of Harry Laughlin and the other experts who had testified in favor of sterilization. Whitehead justified the omission by saying he might be criticized for bringing up material on rehearing for the first time, when he had presented no evidence nor elicited testimony at trial that would have challenged "the propriety and utility" of sterilization. Due to the efforts of Dolle and his colleagues, these arguments were in hand for the writing of the petition, but Whitehead omitted them from the final draft.[47] He signed the watered-down petition and Dolle had it printed and filed with the Court.[48]

The petition questioned for the first time the eugenical theories and "scientific" propositions about the hereditary nature of mental illness, crime, and disease that had been the basis for the Virginia sterilization law. It noted that it was "common knowledge that the beneficent effects of sterilization . . . [were] denied by competent medical and sociological authority all over the country."[49] In fact, sterilization and the eugenical argument in its favor had been condemned in a publication of the State Board of Public Welfare in Virginia the same month that Whitehead filed his Supreme Court brief.[50]

The petition also attacked the Court's reliance on *Jacobson v. Massachusetts,* which upheld a law mandating smallpox vaccination, as irrelevant to the question of "permanent bodily mutilation" authorized by the Virginia sterilization statute. The government could abridge personal freedoms when the general public health was seriously endangered, and a smallpox epidemic represented such a danger. However, Carrie Buck's case was different. The Virginia law merely suggested that the patient's health and the welfare of society "*may* be promoted in *certain* cases" by sterilizing defectives. The statute did not explain any "existing or imminent danger" and was, therefore, based on an arbitrary assumption.[51]

Another objection to the *Jacobson* precedent had to do with the result in the case. Reverend Jacobson merely faced a fine for violating the vaccination law; Carrie Buck faced the surgeon's knife. "In no case which has yet reached this Court," the brief declared, "has the right of the State to inflict permanent bodily mutilation upon any of its citizens been upheld or even considered."[52] To force such "mutilation" upon Carrie Buck would set a new and dangerous precedent.[53]

Dismissing the "scientific" foundations of eugenical theory, the brief stated that sterilization was a new, "experimental" procedure that was "not generally practiced." It condemned the eugenic law as a "legalized experiment in sociology." Quoting an opinion of Chief Justice Taft from the 1921 Supreme Court case *Truax v. Corrigan,* the brief attacked the attempt to deprive a person of "fundamental rights." As Taft had written: "The Constitution was intended, its very purpose was, to prevent experimentation with the fundamental rights of the individual."[54]

The brief for rehearing relied extensively on the records of the Michigan case of *Smith v. Command,* a decision recorded in 1925 after Buck's Virginia trial was held. Willie Smith was a sixteen-year-old boy whose parents filed a sterilization petition. Smith's lawyer cited all the authorities that disfavored sterilization noted in the Laughlin book, among others. He scoured the briefs and records of other cases, citing contradictory or otherwise inconsistent commentary on sterilization from eugenicists. He collected opinions against the practice from published commentaries and solicited opinions from noted experts, such as a doctor from St. Elizabeth's Hospital in Washington, D.C. In short, he did the kind of job that should have been expected of a competent attorney. As a result, the judge who dissented in the *Smith* case included a reference to the fundamental right of bodily integrity and a detailed critique of sterilization law.[55]

The Michigan court endorsed the power of the state to perform sterilizations, agreeing that feeblemindedness was hereditary. It found no constitutional barriers to the law except the provision that allowed for sterilization of some who "are unable to support any children they might have" and because those children "probably will become public charges." Because this provision would allow the financially secure to avoid sterilization while the poor were marked for surgery, it represented a defect in the law. The Michigan court even criticized Aubrey Strode's comments in the *Virginia Law Review,* calling his "the stock argument."[56] Though most of the sterilization statute survived court scrutiny, Willie Smith escaped the surgeon's knife because the parts of the law requiring the appointment of a guardian, the convening of a hearing, and the taking of testimony from witnesses had not been followed. The *Smith* case was widely reported, and Harry Laughlin was aware of it even before it reached court.[57]

The petition for rehearing—written not by Irving Whitehead but by lawyers for the Roman Catholic men's group, NCCM—was the finest effort in Carrie Buck's defense. But by then, as Whitehead no doubt knew, there was little chance that the Court would reconsider. As the authors of the petition were aware, at least one of the eight justices who concurred with the Holmes opinion would have to change his mind and vote for rehearing.[58] Supreme Court records for the period show that not even 2 percent of petitions for rehearing were granted.[59] Even before a ruling on the petition, Joseph DeJarnette publicly declared that a "rehearing has been requested, but in all probability the opinion will be sustained."[60] When the petition was denied in October 1927, newspapers immediately announced the result.[61]

Attacks on eugenics persisted. Raymond Pearl, a professor at Johns Hopkins University, described how propaganda and science were always mixed together in the movement until both features of eugenics were "almost inextricably confused." What passed for scholarship on the topic was "a mingled mess of ill-grounded and uncritical sociology, economics, anthropology, and politics, full of emotional appeals to class and race prejudices, solemnly put forth as science, and unfortunately accepted as such by the general public." It was impossible, in the light of genetic research, to assert any longer that the traits of children could be predicted by looking at the traits of their parents, and claims that "inferior people [produce] inferior children" were "contrary to the best established facts of genetical science." Pearl declared that it was "high time that eugenics cleaned house, and threw away the old-fashioned rubbish that has accumulated in the attic."[62] Pearl's critique was echoed even in the Virginia press.[63]

One Eugenics Record Office publication was mocked as "a flagrant example of scientific guessing," but some thought that the ridicule was misplaced. Eugenics might be a "silly fad" but must be taken seriously, particularly in light of the growing respect it was receiving in other scientific areas and by "the average layman."[64] For example, it was reported that famous surgeon and clinic founder Dr. Charles Mayo approved of sterilization.[65] Other doctors spoke out as a wave of state legislation followed the *Buck* decision. Noting that the American Medical Association had done nothing more than "stand idly by" as eugenic laws passed, the *Journal of the American Medical Association* called for a thorough study of the field of "eugenic asexualization."[66]

But the critics spoke in vain, and the time for study had passed. Strode's third visit to the Supreme Court had proved successful, and though to all outward observers Irving Whitehead had lost, his defeat stood as a perverse tribute to his deceased colleague Albert Priddy and the validation of the sterilization law fulfilled their longstanding hopes for the Colony's future.

Estabrook's Demise

Arthur Estabrook's expert testimony provided some of the most important evidence in the 1924 *Buck* trial. Unlike any other witness, Estabrook had actually seen the "three generations" whose problematic heredity became the foundation of the Supreme Court opinion. In court, Estabrook condemned Carrie Buck as a "socially inadequate" person who needed supervision to oversee her "moral welfare." His own career after *Buck* represents a fascinating counterpoint to the pose he struck in the courtroom.

Immediately following the *Buck* trial, Estabrook submitted his Virginia travel expenses to Charles Davenport at the ERO. Davenport chided him for attempting to inflate his expense account, writing, "I am just as surprised to find you at Amherst on a sterilization test case as the auditor would be." Davenport eventually approved the travel payment.[67]

Estabrook then prepared an invoice for Aubrey Strode including fees as an expert witness and travel expenses; he was paid $112.08.[68] It is no small irony that the man who described Carrie Buck's defective ethical sense as an inherited condition double-billed his employer and the state of Virginia for the cost of appearing at the *Buck* trial.[69] And Estabrook, who described Carrie Buck as a "moral degenerate," had a questionable sexual history himself.

It was Estabrook's habit to employ young women from nearby colleges to assist in fieldwork, and he corresponded with them regularly. His notebooks included snapshots taken in the field, often showing the dapper scientist surrounded by his smiling helpers. Aubrey Strode's wife Louise Hubbard filled the role of assistant to Estabrook for a while, and her letters make it clear that his relationships with young women were not hampered by rigid professional formality.

After one visit by Estabrook, young Miss Hubbard complained that she was not feeling well and scolded herself and another young woman for allowing "gentlemen to carouse around in our room so late!" Estabrook himself was one of those "gentlemen." She also told him that a friend had given her a quart of "the most marvelous scuppernong wine" for medicinal purposes. While that must have seemed a humorous admission in the Prohibition era, it could have led to much less comical conclusions for a less privileged woman, particularly one under evaluation by Estabrook for moral weaknesses.[70]

Ivan McDougle was Estabrook's research colleague and coauthor of the book *Mongrel Virginians*.[71] Research for that book, which focused on a mixed racial group, was carried out in central Virginia near Amherst. McDougle recruited students from nearby Sweet Briar College to help collect data, commenting that his latest cohort of twenty college seniors was "the finest bunch of college girls I ever saw." McDougle's agenda for research included "very definite plans for hard work and strenuous play—thanks to our senior bunch."[72]

Estabrook shared McDougle's captivation with the coeds, and his sexual indiscretion was the direct cause of his dismissal from the Eugenics Record

Office staff. Charles Davenport had known that Estabrook struggled with "loneliness" during his long fieldwork assignments. The matter came to a head when Estabrook's wife, having learned of an inappropriate relationship in which her husband was involved, wrote to Davenport. Noting his own distress at the news, Davenport said: "I have not been unaware of the danger involved in sending Dr. E. to a place where his wife was not to accompany him, but fondly hoped the danger might be averted." Davenport himself had difficulty maintaining contact with Estabrook, who often did not reply to letters. Davenport was also frustrated over Estabrook's "wholly unassigned and unauthorized trips" that Davenport found difficult to justify. Davenport recommended that Estabrook move back to Cold Spring Harbor and that the couple find a way to live together in order to remove the temptations that extended separations had yielded.[73]

Just over six months later, Estabrook's career as eugenic field worker came to an end. He completed his work in Kentucky and sold the horse he had ridden during his research there. He returned his camera along with his final expense report and pressed Davenport to tell him what type of recommendation for future employment he could expect. Davenport identified yet another travel cost of Estabrook's that had been rejected by the accountant but assured him that he could expect a positive reference concerning his "highly creditable" work at the ERO.[74] After almost twenty years as a key investigator at the Eugenics Record Office, the man who had argued that Carrie Buck should be sterilized because she could not conform to "the accepted rules of society" was discovered by his employer to be a philanderer and dismissed as a common cheat.

Estabrook eventually found a position at the American Society for the Control of Cancer, where he studied the epidemiology of disease.[75] He was later asked during a news interview whether people like the Jukes would be likely suspects in a recent wave of crime. He judged the infamous paupers unfit for "modern" crime because of their low mentality. He also surmised that the clan had probably dispersed. "Though weak minded," he said, the Juke women tended toward "good looks." Men of "good heredity" would have found them attractive, "with obvious results."[76]

After the Supreme Court

With all appeals in the *Buck* case exhausted, the Supreme Court decision was no longer in doubt, and Dr. John Bell could finally plan for surgery. On the morning of October 19, 1927, Carrie Buck was taken to the infirmary of the Colony; at nine thirty, she was given drugs to prepare her for sterilization. At ten o'clock she entered the operating room fortified with morphine, to blunt the inevitable pain, and atropine, to decrease salivation. There were two doctors in the room, assisted by two nurses. Nurse Roxie Berry gave Carrie ether to render her unconscious. After the anesthetic had taken effect, the operation began.

Bell made a "midline incision," removing a section of each Fallopian tube; he tied off the remaining ends of the tubes, and the nurse cauterized them with carbolic acid and then sterilized them with alcohol. Though the nurse's notes recorded the "immediate postoperative condition; hemorrhage, shock etc.," Bell noted that his patient responded well to the surgery and that all of her organs were normal. He was careful to record that "this was the first case operated on under the sterilization law," and he pointedly summarized in the medical record how authorization to sterilize Carrie Buck had been won through the court system. That unusual notation captured Bell's priorities in the *Buck* case: this was surgery, but its legal significance overwhelmed the importance of any particular patient.[1]

Carrie Buck was back in bed by eleven o'clock, and for the next several days was fed a light diet. Her bandages were changed daily. After two full weeks in the infirmary, Bell declared her to be infection-free, and since recovery had been "uneventful," the patient would be "allowed up today." She left the infirmary on November 3, 1927.

On Parole, 1927–1930

Nine days later, Carrie spent her last night at the Colony. She was discharged on November 12, 1927, on "furlough" to the Coleman family, friends of Bell and owners of a lumber company in the tiny southwestern Virginia village of Belspring. By the time of her release, Carrie had been entangled in the legal system for almost four years. She was twenty-one years old, and her baby Vivian, living back in Charlottesville with the Dobbs family, was a few months shy of her fourth birthday.

The process of "furlough"—also referred to as "parole"—was the logical extension of Dr. Priddy's long-term plan for Colony inmates. After the *Buck* case, there would be no need to house them at the institution for several years. They could be identified, scheduled for surgery, and, as Carrie was, sent back into the community less than a month after an operation. While on parole, young women would work as domestic servants or in similar occupations for a family needing help and willing to take in a Colony girl.

Priddy's sterilization program had been stalled by years of litigation, but just before Carrie left the Colony, its Board of Directors met and voted to reinitiate the surgery schedule.[2] At age thirteen, Carrie's sister Doris Buck was the youngest member of the first group chosen by Bell for surgery.

Welfare workers in Charlottesville had monitored Doris Buck closely since Arthur Estabrook examined her during his preparation for the *Buck* trial. In April 1926, the superintendent of the Board of Public Welfare wrote to Bell asking for advice. Doris had lived for almost nine years in a private home, but anonymous letters and people in the neighborhood reported that she was "running wild." There were also unconfirmed rumors that she was "meeting men." The superintendent concluded that Doris was "undoubtedly feeble-minded and cannot be controlled by the people who have her now."[3]

Bell could not recall Doris's IQ score from Estabrook's report, but he did not hesitate to agree that she was "unquestionably feeble-minded." He recommended that she be brought to the Colony, because "sooner or later she will become the mother of illegitimate children, from what I know of the history of the family."[4]

When twelve-year-old Doris Buck was registered as inmate no. 1968 at the Colony on June 18, 1926, her sister's case was still in the courts. The commitment forms called her "incorrigible . . . untruthful and in danger of being a moral delinquent." She was almost eight years younger than Carrie, and in five years of tearing up books and skipping classes, Doris had finished only two grades in school. Less than a month after Carrie initially left the Colony on parole, the Colony Board issued a sterilization order for

Doris. Board records noted that she was "of ample mind to fully understand the nature and consequences of the operation."[5]

The Buck family was reunited briefly when Carrie returned to the Colony for a visit during Christmas week. Carrie made the round-trip train ride without incident, even though Bell had cautioned that she should not be sent alone, since "a girl of her type" might be "picked up by someone and carried off."[6]

Before long, Carrie was back at the Colony again. A letter sent to Roxie Berry, a nurse at the Colony, describes the incident that led to Carrie's return. Carrie's guardian on parole said that she planned to share the story with a friend "to get a good laugh out of it": "I went in the kitchen this morning and found Carrie Buck using the dishpan in which she had washed dishes, for a chamber[pot]. The picture I saw will never leave my mind. Of course I was furious, but I was too tickled to say much. I have written Dr. Bell to take her back. I just couldn't stand any more of that. . . . I may be calling on you to get me another girl later. . . . I certainly do not want one so feeble-minded. [The last one] had plenty of sense but didn't care to work, & Carrie is the opposite."[7]

Carrie returned to the Colony just before Doris entered the infirmary for a procedure that was medically necessary—so she was told. When the surgeon cut into her abdomen, he found an appendix that was "long and twisted," showing evidence of prior disease, so it was removed. The medical record showed, however, that the real reason for operating was to make her sterile.[8]

Earlier, people in Doris's hometown had offered to take her in as a boarder. One, aware of her "unruly" temperament, said, "I really do not think Doris is feebleminded." But Bell insisted that Doris must be confined at least one year and could not be released until she was sterilized.[9]

While records described Doris "of ample mind" to understand the effect of the operation she endured, the woman to whom she was eventually paroled was uncertain about her own understanding of Bell's actions. To the generation that came of age in the first two decades of the century, the term *sterilization* was commonly understood as a hygienic process—as in, to kill the germs in milk or water, to clean implements for canning food. When the term was used to describe a eugenic procedure, it did not immediately denote sexual surgery to most people.

Doris's guardian asked Bell for clarification. Did the word *sterilize* have some significance beyond its "ordinary meaning"? Bell explained that whatever the "ordinary meaning" of sterilization was, he meant it as "an operation which renders the individual incapable of reproduction." The purpose of the operation was to ensure that she would not become pregnant, and the purpose of her parole placement was not education—from which he said Doris could not benefit—but rather to learn how to work. As an added incentive to take Doris off the Colony rolls after the surgery, Bell said that there would be no need to pay her.[10]

Bell also planned on new work placement for Carrie Buck. He wrote to Mrs. Dobbs in Charlottesville, offering to place Carrie there on parole. Mrs. Dobbs had become attached to the infant Vivian and said that she didn't "think it wise" for both Carrie and baby Vivian to be together. While Dobbs claimed that she thought "a great deal" of Carrie, she said that her husband's advancing age made it impossible for Carrie to return. Contradicting her statements at Carrie's commitment several years earlier, Mrs. Dobbs said that she expected Carrie to stay at the Colony so that she would have "a place to call home" for the rest of her life. Bell was disappointed because Carrie was "very anxious" to return to Charlottesville.[11]

In another attempt to place Carrie on furlough, Bell wrote to a family living in the mountains of southwestern Virginia. He described her as a good worker who was "strong and healthy . . . good tempered and easy to handle," who only needed to be paid five dollars per month. After the placement was complete, the new guardian reported to Bell that she was pleased with Carrie as a worker, even if she found the girl "rather inclined to be stubborn." Carrie apparently had some difficulty getting along with the guardian's elderly mother-in-law. Carrie looked forward to her final release from Bell's legal custody, but he insisted that she was unable to care for herself and so must remain on parole another year.[12] She was twenty-two.

In the late winter of 1928, Carrie wrote Bell in an attempt to preempt bad news: "I am expecting Mrs. Newberry to write you about some trouble I have had. I hope you will not put it against me and have me to come back there as I am trying now to make a good record and get my discharge." Her expectations were confirmed. Newberry told Bell that Carrie was "obedient" and added that she had "never seen a better girl to work." But, she complained, Carrie "is beginning her adultery again. (I say again for I believe she is an old hand at the business) I feel that this has been her downfall before. I advise her as best I know and try to impress upon her the importance of living a clean, pure life." Carrie had promised Newberry that she would reform, as she was "deeply grieved for fear of going back to the Colony. [If discharged] I fear that she will go back to the bad."[13]

Bell was philosophical at the news, saying that Carrie's "sexual delinquency is probably a thing that will have to be contended with for many years, unless she should find a suitable husband and marry and settle down." There was little that could be done with Carrie, Bell felt. She "has a sister who was also delinquent; they come of a long line of mental defectives and delinquents, and due allowance must be made for the same." Bell eventually tired of monitoring Carrie. He authorized Carrie's formal discharge from Colony control on January 1, 1929, but kept in touch with her guardian, as he was certain that Carrie would need to work under supervision in someone's home "indefinitely."[14]

Carrie wrote to her sister Doris regularly.[15] Though she could read and write with proficiency, Doris responded less often and moved so regularly that the mail often failed to catch up with her. After almost a year on parole, Doris was again sent back to the Colony, accused of theft and "terrible immoralities" with boys. Several other placements failed, primarily due to "her determination to get with the men." After escaping from the farm where she had been placed, Doris wrote to Bell complaining of mistreatment. She eventually returned to the people who several years earlier had taken her from the Children's Home at the request of her mother Emma. They invited her to return not "as a servant" but as "one of the family." In contrast to other reports, they described her as always well behaved before being taken to the Colony. Bell promised Doris that she would be free of Colony control if she could show good behavior on parole for a year. When the year was up, Doris reminded Bell of his promise, continuing to ask for her release.[16] She was discharged from the Colony rolls on December 5, 1930.

Carrie maintained a correspondence with Bell and Nurse Roxie Berry at the Colony, asking about her mother and others who remained there. In the spring of 1932, she had special news. "I am married and getting along alright so far. I married Saturday at about 11:30 and went to Ada. West Va. Had a real nice time. I married Mr. Eagle a man I had been going with for three years. . . . He is a good man. I thought it was best for me to marry. . . . Tell my mother I will send her something when I can." She signed the letter, "Carrie Eagle, Bland, Va."[17]

Carrie turned twenty-six years old a few months later. She would remain married to Eagle until his death twenty-five years later. Back in Charlottesville, her eight-year-old daughter Vivian was completing the second grade. She was an average student during her brief school career; at its high point, she earned a spot on her school's honor roll.[18] But this year she would not live through the summer. In late June she came down with the measles; she then developed a secondary intestinal infection and died soon thereafter. She was buried on July 3, 1932. John and Alice Dobbs had treated her like their own child and given her their surname; they were unwilling to identify her on the death certificate as a "Buck." She was buried as adopted child Vivian Alice Elaine Dobbs. Her death was certified by Dr. J. C. Coulter, the same physician who had signed the papers sending her mother Carrie to the Colony.[19]

In March 1933, famed eugenicist Paul Popenoe of California's Human Betterment Foundation wrote to Bell, describing his efforts to assemble "material that would serve future historians on the subject of eugenic sterilization." Popenoe called the *Buck* case a "milestone in this history" and asked for a photograph—preferably of Emma, Carrie, and Vivian—as well as a photo of Bell for his archives. Bell contacted Carrie, explaining Popenoe's request and asking for a picture of her and of Vivian. Carrie arranged for Bell to get a negative of her wedding photo that included her husband, but she had no pictures of her daughter, Vivian. Bell let Popenoe know that the child's "present whereabouts are unknown."[20] Neither Bell nor Carrie Buck knew that Vivian had been dead for nine months.

Popenoe was not satisfied with cropping the Carrie Buck wedding photo for his historical record. He pressed for the photograph of "three generations of imbeciles," saying that he and his colleague Ezra Gosney would not be satisfied with "a studio photograph which will be retouched until it falsifies the facts." He also wanted Bell's personal photo and tempted the doctor with hints of his legacy. "A hundred years from now you will still have a place in this history of which your descendants may well be proud."[21]

Facing page

(Top) Virginia school grade book, 1930–31.

(Bottom) Vivian Buck honor roll grades, 1931.

VIRGINIA

Daily Attendance Register

AND

Record of Class Grades

FOR

Venable _____ School

In the County or City of

Charlottesville _____

For Year 1930-1931

Read the Introduction Before Making any Entries in this Register

A. M.

Lena Hochman _____ Teacher

ISSUED BY
SUPERINTENDENT OF PUBLIC INSTRUCTION
Richmond, Virginia

Form T. No. 15—6-15-30—35M.

Record of Class Grades

For month, semester, year

For Grade __1 B__ with enrollment of ____

	Month	Deportment 5	Reading 6	Spelling 7	Writing 8	English 9	Mathematics 10	Community Study 11	History 12	History 13	Civics 14	Geography 15	Hygiene 16	Phys. Education 17	Music 18	Drawing 19	Home Economics 20	Agriculture 21	Industrial Arts 22	Aver. Scholarship
Name _Dabbs, Vivian_	1																			
Parent or Guardian " _J. J._	2																			
Residence _Fifeville Stone 1501_	3																			
Grade when enrolled _1 B_	4																			
Promoted to Grade _2 a_ (Month) _May_ 1931	5																			
Remarks: _April Honor Roll_	Exam.																			
	Avg.																			
	6	a a		C a B																
	7	a a		B B B																
	8	a a		C B B																
	9	a B		C B B																
	10																			
	Exam.																			
	Avg.																			
	Fin. Gr.	a B		B B C																

Bell eventually sent pictures of Carrie and Emma to Popenoe and renewed his search for Vivian in Charlottesville, writing to the Red Cross in an attempt to contact Caroline Wilhelm, the nurse who had accompanied Carrie to the Colony and testified at her trial. He also wrote to Mrs. Dobbs. Bell was reluctant to disclose how the photo would be used, indicating instead that he wanted to "make our files complete in the *Buck* case." A Red Cross worker had already interviewed Mrs. Dobbs and learned that the child had been "very bright" and had finished the second grade but had died the previous summer.[22]

Apparently oblivious of the significance of this report of Vivian's achievement, Bell sent his regrets at the news of her death but urged the Red Cross representative to secure a picture "for our records," emphasizing how "extremely important" it was. Mrs. Dobbs had a picture of Vivian but "emphatically refused" to send it to Bell. The woman, by now obviously distraught, asked Bell to cease his "investigations," indicating that the child was "considered a member of [her] own family." Bell insisted that his inquiries were in the "interest of scientific advancement" and complained that Mrs. Dobbs "probably has deprived the child of an opportunity to become a permanent figure in eugenic history." A "distinguished scientist" wished to use the photo in writing a history of eugenics, according to Bell, who asked one final time for the Red Cross official to explain to Mrs. Dobbs that use of the photo "could not possibly do any harm to anyone."[23] Bell sent his own photo to Popenoe but repeated that Dobbs was "adamant" about not giving up the photo of young Vivian. Bell had no further contact with the Dobbs family. Both John and Alice Dobbs died in 1935.

Popenoe printed his history of sterilization in the *Journal of Heredity*. The article included photos of Indiana's Dr. Harry Sharp, the "first eugenic sterilizer"; "sterilization pioneer" Dr. F. W. Hatch, who was credited with drafting the first California law; and "student of sterilization" and eugenic philanthropist E. S. Gosney, also from California.[24]

In the article, a picture of Bell was shown alongside Carrie Buck's wedding picture, cropped to exclude her husband. Justice Oliver Wendell Holmes Jr. was shown above his memorable quotation, shortened to "Three Generations Enough," which was followed by an account of the Supreme Court test case and the full *Buck* opinion. The caption under the Holmes picture characterized his opinion as a "fair minded balancing of the somewhat conflicting claims of the individual and society." On June 26, 1933, while Popenoe

Dr. John Bell and Carrie Buck, 1934. Courtesy Oxford University Press.

searched for the *Buck* photographs, the new Nazi regime in Germany passed the Law for the Prevention of Hereditarily Diseased Offspring. Popenoe's article was rounded out with sterilization statistics in the United States and a brief comment about the new German sterilization law.

Laughlin after *Buck*

Harry Laughlin was extremely busy in 1927. He continued to work on immigration restriction, and he launched another project for the Eugenics Research Association to survey the European "racial descent" of every United

States senator. He was also a candidate for the presidency of the University of Arizona.[25] But none of these activities took priority over his interest in the sterilization case. Within days after the Supreme Court heard arguments in *Buck v. Bell,* Harry Laughlin wrote to the Supreme Court clerk. With an eye toward economy and a penchant for self-promotion, Laughlin offered to exchange a copy of his book *Eugenical Sterilization in the United States* for the full court record. In July he renewed his request, noting that he had received no records and had already donated "a book of 500 pages and another pamphlet of 75 pages" to the Supreme Court Library.[26] Laughlin's persistence eventually paid off, and he received copies of the official Supreme Court documents.

Harry Laughlin's hopes of working more closely with Judge Harry Olson were dashed when the Chicago Department of Public Health took over the Psychopathic Laboratory in 1925, preventing Olson from following through with his offer of a full-time position for Laughlin.[27] But their relationship survived. Following the *Buck* decision, Laughlin told Olson that the case represented "a complete victory for the authority of a state to use eugenical sterilization for preventing reproduction of degenerates." Olson credited Laughlin's book with providing the model on which the Virginia sterilization law was based.[28] He was contacted by a group of Canadians who wanted a copy of the book, as they were interested in writing new sterilization laws for Canada on the *Buck* model.[29] Olson distributed more copies of Laughlin's treatise and publicized his own laboratory at meetings such as the Third International Conference on Race Betterment. Using eugenic sterilization to prevent criminals from being born, said Olson, will "reduce our taxes one-half."[30]

Olson subsidized another Laughlin publication, printing his analysis of the *Buck* case as a supplement to the Municipal Court's annual report for 1929. Olson's introduction to the report explained that Laughlin had earned "the thanks of the American people" for managing legislation and litigation during the "experimental period" of U.S. sterilization. "Our group of workers has followed eugenical sterilization in all of its aspects" since 1907, continued Olson. "The road is now open for its much wider application."[31]

Never one for brevity, Laughlin stretched his supplement to more than eighty pages containing his most thorough discussion of the *Buck* case and incorporating large chunks of the Supreme Court transcript. He described the legislative history of the Virginia law and reprinted its text. He sum-

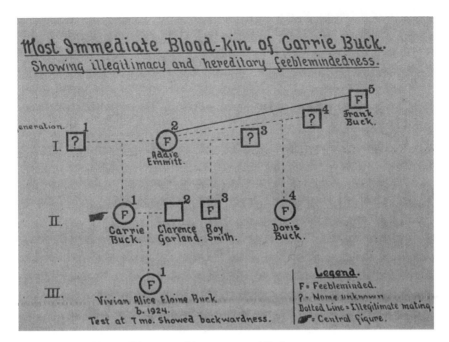

Buck family pedigree slide, created by Harry Laughlin in 1929. Courtesy Truman State University.

marized the testimony at the *Buck* hearing and subsequent trial, repeating generous portions of his own deposition.

Laughlin crafted a special pedigree chart showing evidence presented at the *Buck* trial describing the "Most Immediate Blood-kin of Carrie Buck, Showing illegitimacy and hereditary feeblemindedness." Parts of Whitehead's Supreme Court brief were included, as was the Holmes opinion. Laughlin declared that following *Buck*, the operation of eugenical sterilization would no longer be considered "a wild or radical proposition" but would be seen by most Americans as "a reasonable and conservative matter." The case, he continued, "made use of the best available method of biological analysis of family stocks." Laughlin emphasized that the endorsement of state authority to perform compulsory operations without regard for the consent of the patient or his family was an "outstanding feature" of the *Buck* decision. It constituted an application of the scientific method to statecraft, and employed the "modern sciences" of law and eugenics.[32]

Laughlin drafted a shorter précis of the *Buck* decision and published it in Margaret Sanger's *Birth Control Review.* There he noted that the case had provided "the final legal test" paving the way for any state to prevent "the reproduction of defectives or degenerates."[33] In another discussion of the *Buck* decision, Laughlin recalled how the study of heredity required tools of measurement to quantify "human physical, mental and moral qualities." When those tools were developed, he wrote, "the eugenicist becomes a geneticist."[34]

The *Buck* case was Laughlin's second national policy victory. He had also provided consultation and given congressional testimony on the national origins quotas that were built into the Johnson Immigration Restriction Act of 1924, which placed severe quotas on future immigration by Jews, Italians, Poles, and Hungarians.

In 1928 he spoke to Congress again on the "Eugenical Aspects of Deportation," as his testimony was later titled.[35] Using data such as a breakdown of the Virginia Colony's inmate population by country of birth, Laughlin argued that defective foreign immigrants threatened the purity of the American "race." He provided Congress with eugenical expertise until 1931.[36]

Along with his work on sterilization and immigration restriction, Laughlin's work on state laws to limit "miscegenation," or interracial marriage, was well known. His successes in the arena of public policy and law fed his growing ambitions for eugenics. He revised a plan to revamp the U.S. Census as a vehicle for collecting eugenically important data. He proposed the preparation of a "eugenical code" that would bring together all state laws on reproduction as well as laws on "marriage, divorce, adoption, differential tax in relation to size of family . . . and the like." Part of the plan was a "complete outline of eugenical instruction" for teachers all the way from grammar school through the university level. Laughlin also had plans for using the press, the church, and "organized societies of all sorts" for disseminating the message of eugenics.[37]

Seeing no limit to what eugenics might achieve in a future federal administration, Laughlin in 1929 sketched out a grand plan for a Bureau of Eugenics containing three divisions. The Division of Eugenical Research would consist of a Laboratory of Human Measurements that would keep family records of all Americans. This laboratory would be assisted by a Section of Eugenical Statistics that would track birth, death, marriage, and immigration records and "take stock of the hereditary qualities of the American people." The Section of Eugenical Analysis would investigate the "fundamental nature" of such matters as "mate selection," "race crossing," and "other subjects which bear upon eugenical fortunes."

The Division of Applied Eugenics would contain the Section of Constructive Eugenics to "investigate the nature and control" of factors that influence "fit and fertile matings of those best endowed by natural heredity." Its counterpart, the Section of Social Inadequacy, would be charged with reducing and preventing "social inadequacy" in the United States caused by "defective heredity." It would administer a "Federal Sterilization Statute" that could be applied—consistent with principles of federalism and state control—to all "cacogenic" persons who were beyond the reach of state sterilization laws. That would include the inhabitants of the District of Columbia, U.S. territories, Indian reservations, inmates of federal civil and military prisons, and any potential immigrants to the United States whose heredity was suspect.

Harry H. Laughlin after the Supreme Court's *Buck* decision, 1928. Courtesy American Philosophical Society.

The final office of Laughlin's proposed governmental agency would be the Division of Federal Custodial Institutions. It would control any people in institutions whose conditions appeared on the Model Sterilization Law list of the "socially inadequate," from the incurably ill or chronically diseased (tuberculous, lepers, syphilitics) to the "ruptured and crippled" (deformed) to orphans and "old folks." The chief of Laughlin's Bureau of Eugenics would report directly to the secretary of the Department of Welfare, a member of the president's cabinet. In Laughlin's scheme, eugenics would be a cornerstone of U.S. social policy, the key element in a grand plan of welfare reform.[38]

But at a time when Laughlin's own hopes for eugenics seemed to be reinvigorated, his friend Harry Olson stepped away from public life and retired from the bench in 1930.[39] Despite Olson's work, attempts to pass sterilization laws in Illinois in 1925, 1927, 1929, and 1933 were unsuccessful, just as they had been in 1911.[40] But Olson's support for Harry Laughlin was critical, and the effort had an impact around the world. Laughlin stayed in touch with Olson's staff, asking that copies of the sterilization volume be sent to other dignitaries following the judge's death in 1935. For example, Laughlin wanted one copy mailed to the head of the U.S. Attorney's Advisory Committee on Crime and another to Professor Eugen Fischer, rector at the University of Berlin and director of the Kaiser Wilhelm Institute of Anthropology, Human Heredity, and Eugenics. Laughlin's contact with Fischer was timely, occurring when sterilizations under the German eugenics law accelerated dramatically.[41]

15
Sterilizing Germans

By the 1930s, the links between German and U.S. eugenicists were well established. In the shadow of the Nazi Holocaust, such relationships seem sinister, but as they were developing they were entirely acceptable and even predictable. In the late nineteenth century, German scholars were preeminent in the sciences; it was commonplace for U.S. students early in the twentieth century to travel to the Continent to study with them. The community of doctors and scientists who had a special interest in questions of heredity was small enough that individuals of prominence in that field from both countries would inevitably meet either by chance, as students, or by choice, as colleagues at international meetings. By the first decades of the twentieth century, growing interest in eugenics in Germany throughout the scientific community mirrored similar developments in the United States.

The movement toward a governmental policy emphasizing preventive health interventions had deep roots in Germany, and concerns about the workings of heredity fit firmly within that tradition. The first German treatise specifically dealing with eugenics was Wilhelm Schallmayer's *Concerning the Threatened Physical Degeneration of Civilized Humanity,* published in 1891. Alfred Ploetz, who would become known as "the father of the race hygiene movement," founded the *Archiv für Rassen- und Gesellschafts-Biologie* (Archive for Racial and Social Biology) in 1904. It was the first journal in the world devoted entirely to eugenics.[1] Ploetz was also involved in the 1905 initiation of the Gesellschaft für Rassenhygiene (Society for Race Hygiene). It represented the first professional organization for eugenicists in the world; with participants from other countries, it became international in 1907. As early as 1910, the Internationale Gesellschaft, as it was now called, listed "marriage restriction" and other measures designed to prevent the "reproduction of the inferior" among its goals.[2] A translation of *The Kallikaks (Die Familie Kallikak)* was printed in Germany in 1914.[3] Thus, the development of the German eugenic movement in some ways paralleled the U.S. experience, while in others it led the way.

There was early and regular contact between leaders of the U.S. eugenics movement and likeminded colleagues in Germany. For example, German psychiatrist Ernst Rudin contacted Eugenics Record Office Director Charles Davenport in 1910 for assistance on a race hygiene exhibit in Dresden.[4] As secretary of the American Breeders Association, Davenport helped organize the First International Congress of Eugenics at the University of London in 1912, where he met members of the German contingent, such as Alfred Ploetz.[5] Davenport shared his contacts in the international eugenics community with Harry Laughlin.

In late 1920, Erwin Baur wrote to the ERO for information on U.S. sterilization practices that he could distribute to "his committee of eugenic advisors for the German government."[6] Baur was an early proponent of the eugenics movement in Germany and coauthor of *Grundriss der Menschlichen Erblichkeitslehre* (translated in 1927 as *Outline of Human Genetics and Racial Hygiene*), a leading German text on genetics and eugenics.[7] Davenport referred the inquiry to Harry Laughlin, a recognized expert on sterilization who had just completed an article on national eugenics in Germany that would appear in the London-based *Eugenics Review*. There Laughlin praised the "Teutonic stock" and its instincts for self-preservation memorialized in a new constitution.[8] He believed the time was "ripe for the development of a [German] national eugenical policy" and told Davenport that he would be "especially interested in the success that Dr. Baur's committee has in developing eugenical interest in Germany."[9]

Laughlin's correspondents eventually included a virtual who's who of German eugenics. In addition to Baur, Ploetz, and Rudin, he knew Fritz Lenz, a coauthor of the influential eugenics text. Lenz wrote to Laughlin in 1928, asking permission to reprint a Laughlin paper on sterilization in a German eugenics journal.[10] Eugen Fischer was the third coauthor of that volume, and he developed a particularly collegial relationship with Laughlin. More than once, Fischer and Laughlin had each other's articles translated for publication in both Germany and the United States.[11]

The ruling passions of Laughlin's career were immigration restriction, eugenic sterilization, and prohibition of interracial mating; Adolf Hitler shared these interests. Even before his rise to power in Germany, Hitler praised U.S. immigration restrictions and condemned the automatic grant of citizenship extended to every child born in Germany as "thoughtless [and] hare-brained." America, "by simply excluding certain races from naturalization," was making "slow beginnings" toward a vision Hitler could support. Hitler's plans for sterilization were also outlined in his book *Mein Kampf.* Like many of his admirers in the United States, he was a moralist in matters of sexuality. Hitler condemned those who contracted syphilis to "terrible justice" and described the result of profligacy in oft-repeated biblical terms: "The sins of the fathers are avenged down to the tenth generation." Hitler also raged against prostitution, condemning popular culture as a "hothouse of sexual ideas and stimulations." To counter depraved and degenerate behavior, Hitler recommended the "gravest and most ruthless decisions," including support for sterilization. "The demand that defective people be prevented from propagating equally defective offspring is a demand of the clearest reason and if systematically executed represents the most humane act of mankind." He decried a culture in which preventing births among people with "syphilis, tuberculosis, hereditary diseases, cretins and cripples is a crime" but where the birth of "monstrosities halfway between man and ape" was allowed. His vision of good public policy required that "only the healthy beget children," and he believed the state should employ the "most modern medical means" toward that goal. The government must declare "unfit for propagation all who are in any way visibly sick or who have inherited a disease and can therefore pass it on."[12]

The potential for German legislation that would mandate sexual sterilization surfaced in 1914, the same year that Harry Laughlin's Model Law was first published in the United States. Like Americans, Germans were cautious about sterilization, and some commentators judged that "the time has not yet come for such a measure in Germany."[13]

As in America, some doctors in Germany performed operations for eugenic purposes. One even boasted about the sixty-three surgeries he had performed as a self-appointed guardian of the national health, but—like Hoyt Pilcher of Kansas and Indiana's Harry Sharp—without the benefit of legal sanction.[14] And like their counterparts in the United States, sterilization "pioneers" in Germany pointed to the financial benefits of eugenic surgery.

Worldwide depression was particularly devastating to the German economy, which had never recovered following World War I. In 1929 there were calls for reform of the welfare laws that supported asylums, sanitaria, and other institutions. By 1932 the Prussian Health Council considered a sterilization proposal that was designed to reduce costs generated by dependent populations.[15] That draft never became law, but when Hitler was named chancellor in January 1933, it provided a framework for the first sterilization legislation ever enacted at the national level. The "Law for the Prevention of Hereditarily Diseased Offspring" was passed in July 1933, with an effective date of January 1934.

At the time, Harry Laughlin was the editor of the *Eugenical News,* and during Hitler's rise to prominence he reported regularly on the Nazi eugenics program and the eugenicists whose work eventually informed Hitler's domestic policies. Laughlin reacted immediately to news of the new German legislation. Even before it went into effect, he secured a copy of the law from the German consul general, had it translated, and printed it as part of the lead article in the *Eugenical News.* Laughlin declared that the pathbreaking national policy put Germany in the vanguard of "the great nations of the world" that recognized the "biological foundations of the national character."[16]

Laughlin then obtained a copy of a speech delivered by Wilhelm Frick to the First Meeting of the Expert Council for Population and Race Politics. Frick was a member of Hitler's cabinet, and the speech identified Hitler's regime as a leader among governments "putting Eugenics and Race Culture (Race-hygiene) in the service of the state." Laughlin told his colleague Madison Grant, famous author of *The Passing of the Great Race* and a prominent U.S. eugenicist, of his plans for the Frick speech: "We sent to Germany for Dr. Frick's paper, . . . [and] propose devoting an early number of the *Eugenical News* entirely to Germany, and to make Dr. Frick's paper the leading article. Dr. Frick's address sounds exactly as though spoken by a perfectly good American eugenicist in reference to what 'ought to be done,' with this difference, that Dr. Frick, instead of being a mere scientist, is a powerful Reichsminister in a dictatorial government which is getting things done in a nation of sixty million people." Laughlin cheered the Frick speech as "a milepost in statesmanship" and predicted that future leaders would be "compelled to look primarily to eugenics" for the solution to national problems.[17]

The speech, titled "German Population and Race Politics," appeared in the *Eugenical News* in April 1934, an edition packed with German news. Frick focused attention on lowering governmental expenditures by reducing payments for "the antisocial, inferior and those suffering from hope-lessly hereditary diseases." Such people, said Frick, should also be prevented from reproducing. The public health system must be reorganized in order to "banish the dangers of hereditary defects" just as sanitary regulations were used to eradicate other diseases.[18]

In later editions Laughlin coyly referred to his own work, describing German legislation to allow sterilization of criminals drafted "after the pattern of foreign model laws."[19] Again emphasizing his contribution to German eugenics, Laughlin had an article written by Reich Education Council film specialist Kurt Thomalla translated for publication in *Eugenical News*. The article noted that "Germany learned from the United States" in enacting sterilization laws.[20]

Americans also read in the popular press of the German plan to steril-ize nearly half a million of the "unfit." "Germany continues to astound the world," announced *Newsweek*. Some of the first sterilizations were reported in the *Washington Post*. The *New York Times* ran a lengthy essay by Cam-bridge University geneticist C. C. Hurst, cautiously endorsing the German plan.[21] A German physician, working with his country's embassy in Wash-ington, prepared an officially sanctioned commentary on the promise of the new legislation, calling "interference with unfit life . . . a sound applica-tion of the true Christian love of one's fellow man."[22]

Professional publications also noticed the German law. In the *Journal of Heredity*, California eugenicist Paul Popenoe quoted Hitler's comments from *Mein Kampf* as examples of how Germany was "proceeding toward a policy that will accord with the best thought of eugenists in all civilized countries." Popenoe also helped organize an exhibit featuring sterilization in Germany for the 1934 meeting of the American Public Health Association.[23]

Editorialists at the *New England Journal of Medicine* thought Germany "the most progressive nation in restricting fecundity among the unfit." Looking forward to a shift in the "sentiment of the people" that would al-low Americans to adopt a "scientific spirit" and accommodate a plan like Germany's, the medical journal urged the use of "moral suasion" by doctors to get consent for surgery.[24]

Not everyone shared that view. At least one author felt that eugenics had become almost completely discredited by 1933. He pointed to the impend-ing period of "questionable publicity as Herr Hitler attempts to compel his followers to adopt his standard of the German breed and to close the national stud to all who depart from the set type."[25] Another critic described the sterilization law along with other Nazi mandates as a series of "hammer blows" used by the government to drive German citizens "into a swastika shaped hole."[26]

Legal scholar and sterilization expert J. H. Landman, whose own book about sterilization was published in 1932, disputed the value of a broad sterilization policy, saying that too little was known of the workings of heredity to justify such a program.[27] The Nazi program led Landman and others to speak out and sparked disagreement within the scientific community. According to one commentary, never had such an "extreme" program been initiated, and no plan had "opened so wide an avenue down which the outcast and despised could be driven."[28]

Writing in the *American Journal of Public Health,* physician and public health expert W. W. Peter declared that the German law "merits the attention of all public health workers in other countries." Peter was secretary of the American Public Health Association and, in addition to a medical degree, held degrees in public health from Harvard, MIT, and Yale. He boasted of having the "privilege to meet some of the leaders in the present political [Nazi] regime who are responsible for . . . reconstruction of the social order." He chronicled the costs associated with a "load of socially irresponsibles," judging such people "liabilities which represent a great deal of waste." His perspective was honed by the year (1933–34) he had spent in Germany during the Nazi ascendancy studying public health facilities. "To one who lives here for some time," he concluded, "such a sterilization program is a logical thing."[29] After a year's observation, another commentator said that the "most extensive experiment in sterilization for human betterment" was being administered with a "commendable conservatism."[30]

With the promulgation of the Nazi sterilization law, Laughlin's international standing among eugenicists increased. At home in Cold Spring Harbor, the situation was much less hospitable. From the time of his first declarations on the need to sterilize millions of Americans, Laughlin was a lightening rod for criticism. He had also been the subject of pejorative comments from prominent geneticists for his congressional testimony and other public statements during the debates on immigration restriction in the early 1920s. For example, Johns Hopkins University geneticist H. S. Jennings had challenged Laughlin's methods and arguments for immigration quotas and scornfully described some of Laughlin's "deduction[s]" as "illegitimate and incorrect."[31]

Laughlin's posture as spokesman for the most radical brand of eugenics became even more strident following the *Buck* decision. In response to the growing academic and public criticisms of eugenics, the president of the Carnegie Institution of Washington decided to audit the ERO in 1929 to determine if he could justify ongoing funding for its activities. Though the review unearthed serious questions about the value of past eugenic field research, funding nevertheless remained in place for both the ERO and Laughlin.[32]

Prudence would have dictated that Laughlin turn down the volume of eugenic propaganda and keep a low profile. But success at home and momentous events abroad apparently emboldened him further. The pro-Nazi tone of *Eugenical News* only added to the perception of Laughlin as eugenic provocateur. His good fortune and seeming immunity began to ebb when Charles Davenport decided to retire early in 1934. Davenport had been mentor and protector of Laughlin since his arrival from Missouri in 1910. He regularly shielded Laughlin's work from external attacks, at times dissembling to divert attention from his own positions on controversial topics like sterilization. With Davenport no longer in a position to provide insulation for Laughlin's activities, Carnegie scheduled a second audit in 1935.

The consultants' report was devastating. It described the records collected at the ERO as a "vast and inert accumulation . . . unsatisfactory for the scientific study of human genetics." The auditors recommended a new direction for Laughlin's program to ensure that it was "divorced from all forms of propaganda and the urging or sponsoring of programs for social reform or race betterment such as sterilization, birth-control, inculcation of race or national consciousness, restriction of immigration, etc." They recommended a new name for the *Eugenical News* to avoid the negative connotations of the word "eugenics," which was "not a science."[33]

Archaeologist A. V. Kidder, who chaired the Carnegie evaluation committee, said Laughlin had "a messiah attitude toward eugenics" and that he "would be happier as the organizing and propaganda agent of a group devoted to the promotion of race betterment. In a scientific institution he is really out of place." John C. Merriam, who headed the Carnegie Institution, accepted the committee's report and agreed with the plan to limit Laughlin's

ERO work to "matters which can clearly be handled on a scientific basis" and to "distinguish between fundamental research and propaganda."[34]

Another Carnegie advisor, geneticist L. C. Dunn, later ran the Cold Spring Harbor Department of Genetics. He offered comments as a supplement to the evaluation committee report, giving voice to concerns among geneticists that "eugenical research was not always activated by purely disinterested scientific motives. . . . 'Eugenics' has come to mean an effort to foster a program of social improvement rather than an effort to discover facts," he wrote. Dunn made a direct connection between Laughlin's habits of propaganda and events he had seen firsthand in Germany: "I have just observed in Germany some of the consequences of reversing the order as between [social] program and discovery. The incomplete knowledge of today, much of it based on a theory of the state which has been influenced by the racial, class and religious prejudices of the group in power, has been embalmed in law. . . . The geneological [sic] record offices have become powerful agencies of the state, and medical judgments, even when possible, appear to be subservient to political purposes." Dunn blamed much of the problem in Germany on Hitler's dictatorship but made it clear that his concern about Laughlin and the ERO was informed by the Nazi example, which showed the "dangers which all programs run which are not continually responsive to new knowledge."[35]

With his funding in jeopardy and his work under even more scrutiny, attendance at the 1935 Berlin International Congress for the Scientific Investigation of Population Problems was out of the question for Harry Laughlin. Unwilling to be muzzled, he wrote Eugen Fischer with a plan to have a "distinguished colleague" present a paper to the Berlin meeting on his behalf. Laughlin's paper, "Further Studies on the Historical and Legal Development of Eugenical Sterilization in the United States," would update the conferees on developments since his 1922 book. Laughlin even made slides to illustrate his paper, including a pedigree chart that connected "The Near Blood-Kin of a Feebleminded Woman Sterilized by the State of California."[36]

Fischer was also an expert on sterilization, and he held a seat as a judge on Berlin's appellate-level genetic health court to review cases determining who would face surgery under the sterilization law. Fischer had also been instrumental in the secret Gestapo sterilization of the hundreds of mixed-race children of French women and Algerian troops—the *Rhein-*

landbastarde—born in the occupied territory along the Rhine River following World War I.[37]

Laughlin's stand-in at the Berlin meeting was Dr. Clarence G. Campbell, who had retired from the practice of medicine to make eugenics his full-time vocation. A longtime Laughlin ally, he had served as president of the Eugenics Research Association and on the editorial advisory board of the *Eugenical News*. Laughlin also asked Fischer to extend a favorable reception to Wickliffe Draper, whom Laughlin described as "one of the staunchest supporters of eugenical research and policy in the United States."[38]

Wilhelm Frick was honorary chairman of the conference.[39] His cabinet post as Reichsminister of the Interior gave him jurisdiction over German domestic law, and promulgating the compulsory sterilization legislation had been among his first official acts in 1933.[40] Frick's keynote address reviewed laws the Nazis had designed to reinvigorate Germany; laws to reduce unemployment would focus on "economic security for the hereditarily sound family," and "hereditary degenerates" would be eugenically sterilized.[41]

German attorney Falk Ruttke, who was instrumental in drafting the 1933 German law and later became a member of the advisory board to the *Eugenical News,* delivered an essay discussing the German and even more recent Scandinavian sterilization laws.[42] Dr. Arthur Gutt, the "architect of Nazi public health" and likely coauthor of the German sterilization law, was also in attendance.[43] Gutt called forced surgery "the most important public-health measure since the discovery of bacteria."[44]

Time magazine quoted the Gutt speech on the "practical application of population science." Said Gutt: "Our penal code will shortly make compulsory a health examination for all marrying persons. The purpose of this is first to dissuade bodily or mental inferiors from marrying and especially from procreation."[45]

After reading Laughlin's paper, Clarence Campbell delivered his own lecture entitled "Biologic Postulates of Population Study." His remarks drew the attention of the *New York Times*, which described the physician as a "champion of Nazi racial principles" and someone recognized by Germans as "enthusiastic for the Nordic race."[46] *Time* magazine summarized Campbell's Berlin performance as "Praise for Nazis." Campbell, whom Laughlin had described earlier in "perfect agreement [with Laughlin] as to plans and policies," closed the meeting with a toast: "To that great leader, Adolf Hitler!"[47]

Many doctors in the United States, like their German counterparts, thought of eugenic surgery as a tool of enlightened preventive health practice. Following *Buck,* Dr. John Bell took to the lecture circuit, telling a meeting of the American Psychiatric Association that sterilization was an important spoke in the "wheel of social progress." He summarized his role in the *Buck* case and described trends among the more than one hundred operations—90 percent of them on women—that had taken place at the Colony by early 1929. Each case had been selected "with infinite precaution" and attention to their potential to spread inherited defects.[48]

Bell traveled throughout Virginia appearing as a regular speaker at meetings of the Medical Society of Virginia. Describing hereditary illnesses as "the protoplasmic blight," he lectured colleagues on the potential for a dystopian future, a time when the world would be populated by "a race of degenerates and defectives . . . a people sunk in moral and social obliquity." He called for "the united efforts of all socially minded people" to produce "a citizenry purged of mental and physical handicaps." He declared eugenic controls "a biological and economical necessity." Bell predicted for his medical colleagues that "the physician of the future will be an expert in preventative medicine and public health" as much concerned with a patient's "eugenic problems" as with medical maladies. Bell later announced that sterilization was progressing "as rapidly as is consistent with known facts in biology and eugenics" and boasted that among the fourteen hundred members of "the lower orders of mankind" who had been "rendered procreatively harmless" via surgery, only one fatality had occurred.[49]

Bell's reputation in eugenics was so well established that he was chosen a fellow and then a vice president of the American Psychiatric Association.

On December 9, 1934, after eighteen years' service at the Colony, Bell died of heart disease at the age of fifty-one.[50] In his last report for the Virginia Colony, Bell praised the "principle of genetic control" enshrined in Virginia law and heralded the new German law for sterilization. He predicted that the German law, which applied not merely to institutions but also to the country's entire population, would provide "a vast advantage in the elimination of the unfit." He called for a federal sterilization law in the United States, urging the nation to "apply the pruning knife with vigor."[51] Bell's death preceded by barely three weeks the effective date of the Nazi sterilization law. By then sterilization was also a regular feature at several

Virginia hospitals, each of which reported accelerating progress in the use of the eugenic measure.[52]

Few doctors in favor of sterilization were as vocal as Joseph DeJarnette, whose 1908 comments placed him among the trailblazers in eugenics. He was also quick to praise the use of surgery in the *Buck* case, remarking on the "blessing of sterilization" for families with "mentally defective" daughters who were often "over-sexed" and could only contribute burdens of "care and disgrace."[53] Quoting from his own poetry, DeJarnette said sterilization was the "kindest and best method to render the unfit fit." He boasted that in 1931 he and his medical colleagues were "just getting into good working trim, and expect to sterilize 500 per year in the five hospitals."[54]

Calling sterilization "one of the most progressive procedures," DeJarnette promised "to save the State millions of dollars" by extending its use. "The Department of Public Welfare is cooperating in this work," he noted, and most welfare workers were "heartily in accord" with the surgical program. By 1933 Virginia could point to nearly fifteen hundred sterilizations, surpassed in the states only by California. DeJarnette had increased his pace of operations to more than one hundred a year. "Sterilization . . . is our greatest work," he declared, and "the greatest economical measure we have." As proof that the idea of sterilization was "more or less universal," DeJarnette cited "Chancellor Adolf Hitler [who] has recommended a national law for Germany."[55]

Quoting again from the Holmes opinion in *Buck,* DeJarnette described a family in the mountains of Virginia that would "dethrone the Juke and Kallikak families" in generating "millions of dollars in relief work." These "clans who have no right to be born" lay "incalculable burdens" on the state budget. The "welfare workers" cooperated with DeJarnette, who was attempting to sterilize an entire family.[56]

From 1935 to 1939, DeJarnette tallied the numbers sterilized in Germany and in the U.S., chiding his countrymen for falling behind the pace set in Europe. At the end of the first year of the Nazi program, DeJarnette reported that Germany had already sterilized 56,224 people. "The Germans are beating us at our own game," said DeJarnette.[57] Another doctor at the Virginia Colony joined DeJarnette in praise for Hitler's sterilization program, describing it admiringly as the first of its type decreed by a national government.[58] Others applauding the value of the German law linked it to the sentiments contained in Holmes' *Buck* opinion, which "expressed the guiding spirit of a truly constructive social policy for any country."[59] By 1937 one thousand sterilizations had been done at the Colony.[60] Two years later, the state total passed three thousand, more than 11 percent of the operations done in the whole country. Through the application of "preventive eugenics," DeJarnette calculated a savings of more than $370 million in the century to come.[61]

Laughlin knew that officials like DeJarnette provided strong support for eugenics in Carrie Buck's home state. As his situation at Cold Spring Harbor worsened, he contacted friends in Virginia, searching for alternative employment.[62] He planned to solicit funds for an endowment from eugenical philanthropist Wickliffe Draper, whose largess he had cultivated over the years and whom he had only recently introduced to colleagues in the Nazi government. Laughlin proposed to bring like-minded eugenicists together in a national Institute of Eugenics based at the University of Virginia. Virginia was a hotbed of eugenical study where he could build on his many years of collaboration with local eugenics enthusiasts. Not incidentally, the institute would provide an academic home for him and a soft landing after the contentious years he had endured at the Eugenics Record Office.

Edwin Alderman, president of the University of Virginia, encouraged research in eugenics and attracted a faculty that was well versed in the field. Some had been mentored by Charles Davenport during summer sessions at Cold Spring Harbor. Davenport had even introduced Virginia medical professor Harvey Jordan to Francis Galton himself.[63]

Eugenics first appeared in the university's curriculum in 1911, but by then Jordan had already aroused some "hostile comment" because of his advocacy of sterilization.[64] While Laughlin's committee looked at the big picture, Jordan told the Virginia State Conference on Charities and Corrections that a full 5 percent of the nation's people were "seriously defective in some form or another," and the evidence was "overwhelming" that such conditions were linked by heredity "from ancestor to offspring." Looking forward to a day when waste of public resources on institutions to care for these social discards would no longer be needed, Jordan declared the time "fully ripe for . . . negative eugenics," which he clarified as "the prohibiting of parenthood to the unfit." Though he was not entirely persuaded that it was "the best alternative," he nevertheless saw its use in the "most extreme cases of hereditary degeneracy" and quoted others on the likelihood that sterilization laws "of increasing severity" would become common. Soon, sterilization could be used for "preventing the reproduction of ordinary defectives."[65]

The next year, Jordan, who would eventually become Dean of the Virginia medical faculty, contributed a paper titled "The Place of Eugenics in the Medical Curriculum" to the First International Eugenics Congress in London. He saw the doctors of the future as eugenic advisors who could help prevent "physical, mental and moral sickness and weakness" and set the stage for a "perfect society constituted of perfect individuals."[66] Jordan's vision for the future of medicine saw eugenics "embracing genetics."[67] In addition to writing articles with titles like "Eugenics: The Rearing of the Human Thoroughbred," Jordan was a member of the American Eugenics Society and the Eugenics Research Association, and he chaired the Eugenics Section of the Society for the Study and Prevention of Infant Mortality.[68]

Several Virginia faculty members, like physician-anthropologist Robert Bennett Bean and biologist Ivey Lewis, were even more ardent in their support for eugenics than Jordan. Bean would come to be known as one of the fathers of scientific racism, and Lewis, who would become Dean of Men at Virginia, was a close colleague and confidant of both Charles Davenport and Harry Laughlin.[69] Lewis served as a member of the Eugenics Research Association's committee on education for Virginia, along with Jordan. His graduate students at Virginia wrote theses on eugenics, parroting their teacher's attitudes and reprinting DeJarnette's eugenic poetry.[70]

But the affinity of faculty and students at Virginia with eugenicists was apparently insufficient reason for the Virginia administration to establish an academic center to nurture the controversial science, and once again Laughlin was left to work out his destiny at Cold Spring Harbor.[71] As Laughlin's fortunes declined at home, another invitation arrived from Germany.

The University of Heidelberg was planning a celebration commemorating its 550th anniversary, and Laughlin was chosen by the Nazi administrators there to receive the honorary degree of Doctor of Medicine for his work in the "science of racial cleansing." The letter of invitation made specific reference to Laughlin's achievements in immigration restriction, anti-miscegenation law, and, of course, eugenical sterilization. Laughlin expressed his

Die Universität Heidelberg
die älteste Hochschule des Deutschen Reiches, begeht in den
Tagen vom 27. bis 30. Juni 1936 die Feier ihres

550jährigen Bestehens.
Ich würde es mir zur Ehre anrechnen

Herrn Prof. Harry H. Laughlin

in diesen Tagen als Gast der Universität begrüßen zu dür-
fen. Ihre Antwort erbitte ich möglichst bis 31. Mai 1936,
damit Ihnen rechtzeitig die näheren Mitteilungen zugehen
können.

Heidelberg, im Mai 1936

Der Rektor der Ruprecht-Karls-Universität

Invitation to Harry Laughlin to attend honorary degree ceremony at the University of Heidelberg, Germany, 1936. Courtesy Truman State University.

"deep gratitude" for the honor, "because it will come from a nation which for many centuries nurtured the human seed-stock which later founded my own country."[72]

The date chosen for Heidelberg's celebration had already generated controversy because it was the anniversary of Hitler's 1934 purge of the Jews from the Heidelberg faculty. The press suggested that visitors who honored the Nazi invitation would provide a "foreign endorsement" to Hitler's regime.[73] A dwindling budget prevented Laughlin from making the voyage, but he did accept the degree in absentia and proudly announced it to his colleagues. The honorary degree, said Laughlin, was evidence of "a common understanding of German and American scientists of the nature of eugenics."[74]

After his success linking German supporters of eugenics and his colleagues from the United States at the Berlin Population Congress, Laughlin was in a good position to take advantage of his relationship with Wickliffe Draper. When Draper returned to the United States still excited about having seen Nazi eugenics in action, Laughlin put the finishing touches on another project that would advance his ambitions toward a national eugenic policy. Together, Draper and Laughlin completed plans for establishing a U.S. foundation called the Pioneer Fund that would subsidize "research into the problems of heredity and eugenics."[75]

Laughlin accepted Draper's invitation to be the first president of the Pioneer Fund and began his tenure by distributing a Nazi propaganda film entitled *Erbkrank* ("The Hereditary Defective").[76] The film portrayed horrible living conditions in the slums of a German city, contrasting those images with lavish pictures of expensive institutions housing disabled residents born to "the socially inadequate and degenerate family-stocks" of the country. It also catalogued the varieties of "human degeneracy" that could be traced to hereditary afflictions.[77] The message was clear: eugenic solutions like sterilization were needed to rid the country of the "socially inadequate." Laughlin distributed the German film for three years.

Laughlin pressed for more "eugenical education." He wanted to dramatize the social costs of institutions for the "mental deficient" with a U.S. film.[78] But Laughlin began to miss meetings of the Pioneer Fund's board due to epileptic seizures of increasing frequency. On one occasion, he had a seizure while he was behind the wheel of his car, causing it to crash into a retaining wall.[79]

When Vannevar Bush took over as the Carnegie executive in 1939, he reviewed the earlier audit reports on the Eugenics Record Office. Additional investigation convinced Bush that Laughlin was not accepted among geneticists.[80] In January 1940, Laughlin was forced into premature retirement.[81] He returned to his boyhood home of Kirksville, Missouri, and spent his last three years studying hereditary qualities of thoroughbred horses. There is no record of how he reacted to his own diagnosis of epilepsy, which under his Model Law would have marked him for sterilization. Laughlin died without children on January 23, 1943.[82]

Charles Davenport

Davenport survived his protégé Laughlin by just over a year, dying in February 1944. A lengthy biographical sketch described Davenport as a lonely, exceedingly competitive, and insecure man, a person who could not endure criticism.[83] By this account, Davenport was a man consumed by feelings of inferiority, whose enthusiasm for his own ideas led to exaggeration and even misrepresentation. He was persistent to a fault, and despite the tide of research that eventually left his early ideas on eugenics far behind, Davenport never adopted a more measured attitude toward the science he had fostered.

His feelings about "the criminal and degenerate" were clear. He declared to a European colleague, "Alas . . . it seems impracticable to put them all on ships" roaming the oceans forever. "I see no way out but to prevent breeding them."[84] Toward that end, Davenport appreciated the value of the *Buck* decision to the movement he had led. "There is little to add to the apt remarks of Justice Holmes," he mused. "Society has had to protect itself against the increasing menace of the hereditarily socially inadequate, for whose care millions of dollars are being annually paid out and whose bad behavior, while at liberty, is costing many millions more."[85]

During his ten years of retirement, Davenport maintained an extraordinary output of scientific publications and also became involved in civic affairs. Some of his attention focused on organizing an anti-tax association and protesting the role of government in subsidizing social welfare programs.[86] Another member of his generation who benefited from those social welfare programs would die less than two months later.

Emma Buck Dies

After she was married, Carrie Buck attempted to maintain the contact with her mother that she had renewed during several years of life at the Colony.

She wrote to Bell from her home in the southwest Virginia town of Bland:

Aug. 19, 1933

Dear Dr. Bell:

Will take the greatest of pleasure in writing you just a few lines to let you hear from me. This leaves me verry well and getting along just fine. I am still keeping house.

Dr Bell, I would just love to take my mother out for this winter and if nothing happens so I can't and if you will let her come I will make preparation for her to come and will meet her. I do not have but two rooms but still she is welcome to come and stay with me. We live out in the country. We have a pig and a nice garden and are putting up a lot of things this summer. I have planned on sending her a nice box but don't have the money to send it with. My husband works regularly but can't get any money for what he does, it sure is hard to get hold of in Bland, but he is going away to work for a week or two and I guess he will have some then. The people with who he works for furnished us things to live on. We will send her the money to come on and I will fix for her if you think she can make the trip alright. I am planning on sending her the money some time in September or October. I don't know for sure when, but as soon as I can get it. . . . I will see that my mother is well taken care of and plenty to eat if you will let her come. I am real anxious for her to come. I had a letter from Doris some time ago and said that her husband had left her. I am sorry about that—was hoping she would do well. He said she was going to get a divorce for her husband for not taking care of her—she wrote for me to send her a pair of shoes and a dress but I didn't have any money to buy them with.

Well, I guess I will close for now as it is bed time for me. Answer this real soon same as ever.

Yours truely, Mrs. W.D. Eagle[87]

Bell did not oppose Carrie's plan to take her mother in, but he urged her to give serious consideration to whether she and her husband could afford to support another dependent.[88] Apparently Carrie was unable to save enough cash to pay the train fare for her mother to travel. Emma Buck, then sixty years old, remained at the Colony. Her health had been reasonably good in her first ten years there, but as she aged conditions had worsened. At age fifty-eight, she had become severely ill with the mumps, and her health had began a steady decline. She had pulmonary problems including colds, pleurisy, bronchitis, influenza, and asthma. She had an enlarged, often inflamed heart, and her chronically high blood pressure led to fainting, headaches, dizziness, and shortness of breath. Her gums and jawbone were frequently infected, and she had had a half-dozen teeth pulled. She also had endured surgery for hemorrhoids.

In 1940 Carrie Buck Eagle wrote to G. B. Arnold, the new Colony superintendent, inquiring about her mother. "I would like to hear from her, and to find out how she is, and if she is well . . . to send her a box." Arnold described Emma as "well and hearty." Though she had lost weight and was "not quite as stout" as once, she was "happy most of the time," he wrote.[89] The next year Emma suffered a heart attack, and for the next three years she slowly declined, finally succumbing to congestive heart failure.

Emma Buck died on April 15, 1944, of bronchial pneumonia. She was seventy-one years old. The final notation in her medical record was made nearly two weeks after her death, when Carrie and her brother Roy traveled to the Colony for a visit. When the superintendent failed to receive a response to the telegram he had sent to notify Carrie, he went ahead with the burial at the Colony cemetery. Only afterward did Emma's children arrive. As her case file noted, "The son and daughter came to the hospital yesterday inquiring about their mother, Emma Buck. They did not know until their arrival to the hospital that she was dead. . . . The son and daughter were a bit upset. However, they were most considerate and accepted the explanation."[90]

Emma Buck's death left only one of Holmes' "three generations" alive. Carrie would survive for nearly forty more years, but the men whose lawsuit led to her sterilization were not as hardy.

Irving Whitehead and Aubrey Strode

After his appearance as Carrie Buck's lawyer, Whitehead's career as a banking attorney flourished. He was general counsel to the Federal Land Bank of Baltimore, became counsel to the Federal Intermediate Credit Bank, and then general counsel to the Farm Credit Administration, specializing in National Farm Loan policy during the Depression.

When he died in 1938, the Virginia Bar Association *Journal* asked Whitehead's wife for a sketch of his life. She asked Aubrey Strode to draft the obituary.[91] Strode and Whitehead had first met as boys on neighboring farms in Amherst, Virginia, and Strode's memorial to his friend recited the details of his long and distinguished life. He described Whitehead's role in politics, his legal career, and his prominence as an expert in finance but tastefully omitted any reference to Whitehead's role as lawyer for the losing party in the famous case of *Buck v. Bell*.

In the years following *Buck v. Bell*, Aubrey Strode maintained regular contact with Whitehead, as well as Dr. DeJarnette and Dr. Bell, who kept him apprised of progress with the sterilization program. Strode's wife also maintained contact with Arthur Estabrook.[92]

In 1934, after thirty years of legal practice, Strode was named Judge of the Corporation Court of Lynchburg, Virginia. When Amherst County Judge Bennett Gordon died, Strode was sometimes asked to preside over the court where Carrie Buck's sterilization trial had taken place. In that role he was responsible for approving guardians for sterilization hearings at the Colony.

Legislators and officials from the Virginia Colony approached him for advice when they amended the sterilization law in 1936. Strode also provided assistance to officials in other states who were involved in defending their own sterilization laws in the courts.[93]

In the 1940s, Strode suffered a series of strokes, but he delayed retirement until a friend in the General Assembly could get state law amended to pay full retirement benefits to judges forced to leave the bench because of physical disability.[94]

Strode spent his last four years at his family estate, and on May 17, 1946, he died in the house where he had been born almost three quarters of a century earlier. He was eulogized in the hometown newspaper as "A Humanist on the Bench." He was "like Jefferson, an aristocrat." His "great concern" throughout a long career was "the advancement of the welfare and happiness of the common man, and the improvement of society." His place in history was assured, according to the newspaper, by his role as author of Virginia's sterilization act and his victory in the case of *Buck v. Bell*.[95] The *New York Times* noted that Strode was known for "social legislation" and that the Virginia Sterilization Law he drafted became "a model for other States."[96]

Strode was buried in the Amherst village cemetery, only a few steps from the grave of his friend Irving Whitehead. While their roles in the *Buck* case have been all but forgotten, the impact of that case had worldwide repercussions. More than thirty U.S. sterilization laws would be enacted, and an additional twelve statutes made it into the law books of other countries. They included Alberta and British Columbia in Canada, Vera Cruz in Mexico,

Vaud in Switzerland, as well as Germany, Iceland, Norway, Sweden, Denmark, Estonia, Finland, and Japan. Harry Laughlin's dream of exporting eugenic sterilization around the world seemed to have become a reality. But even before Laughlin and the lawyers in the *Buck* case died, a promising opportunity to confront the *Buck* precedent appeared in a case from Oklahoma.

16

Skinner v. Oklahoma

The Supreme Court decision in *Buck* provided a firm precedent and legal foundation for state sterilization laws. Yet arguments against surgical eugenics persisted, and when public debate over the need for a sterilization law in Oklahoma began in 1929, the superintendent of one of the state's largest asylums spoke up against the measure. Dr. D. W. Griffin of Central Oklahoma State Hospital contended that a sterilization law would easily be open to abuse and that some "defectives" came from the best families in the state, making it impossible to predict who would have problem children.[1] When the law passed in 1931, doctors like Griffin were reluctant to operate without patient and family consent, and no surgeries were done for a year.

By 1932, the pressure to put the law into use was growing, and a test case was initiated at Griffin's Central Oklahoma State Hospital. The subject of the case was Samuel W. Main, a resident there. In some ways Sam Main's case resembled Carrie Buck's. Main was found to be "a probably potential parent of socially inadequate offspring" because his pregnant wife and four children lived on public relief. But unlike Buck, Main had a serious mental illness; he had been diagnosed as manic-depressive. Facing the prospect of five children and no source of income, Main and his wife agreed that he should be sterilized.[2]

By the time *In re Main* was heard in the Oklahoma Supreme Court, courts in Kansas, Utah, Nebraska, Idaho, and North Carolina had relied on the U.S. Supreme Court's reasoning in *Buck,* quoting the Holmes opinion in support of their own state laws. The Oklahoma court also followed the *Buck* precedent, and Main's surgery was first in a program that was soon extended to other state institutions, where the operation became a condition of release.[3]

Dissenting voices persisted. The fear of surgery moved one patient to pen an anonymous memoir entitled *Behind the Door of Illusion.* "Inmate Ward 8" declared that "the spectre of sex sterilization has been thrust over us." The patients at the Oklahoma asylum where he lived were "frightened, wrought up, angry and muttering." Newspapers, not wishing to offend a "narrow or prudish reader" avoided the subject, leaving patients in ignorance about the true scope of the sterilization program. The nameless author claimed that the sterilization law was designed for two types of patient: those with "social diseases, or abnormally sexed." The "legal safeguards" in the law were of no value, he said, because none of the patients could afford a lawyer and many were mentally unprepared for legal involvement. The law remained in place despite open disagreement among physicians about what forms of insanity might be inherited. "Inmate Ward 8" also revealed that before the law was enacted, secret castrations had been done, adding to the confusion among patients who did not understand that vasectomy, not castration, was the operation they might face.[4]

The Oklahoma accounts of abuse were not unique. As people were reading *Behind the Door of Illusion,* an Oregon newspaper reporter was nominated for a Pulitzer Prize for her series on the commitment and sterilization of children. Her investigation of over one hundred cases revealed "incompetent testing methods" for choosing surgical cases as well as a generally unfair application of the law at the Oregon State Institution for the Feebleminded.[5] When some linked Oklahoma sterilizations with events in Germany, a local columnist and eugenics booster did not dispute the characterization but simply noted that "Hitler is right for once."[6]

Francis Galton himself listed the eradication of crime on his eugenic agenda, and he saw sterilization as one method that advanced the crime-fighting plan. Similarly, U.S. criminologists saw a role for eugenics in sentencing. They argued that judges should be allowed to take hereditary propensities into account to set longer terms of imprisonment.[7] Prison, like custody in a colony for the feebleminded, segregated criminals from the rest of the population and effectively prevented parenthood. But many saw sterilization as a more permanent and much less expensive eugenic tool. Some of the earliest surgical experiments, such as Lloyd Pilcher's reform school operations and Harry Sharp's jailhouse "therapy," took place in a correctional setting. Crime prevention was a theme that persisted from the time of the earliest experimental surgeries through the date of the *Buck* decision, and by then most of the laws already enacted allowed states to sterilize criminals.

Harry Laughlin also believed that antisocial behavior sprang from the "germ plasm." As eugenics associate of the Psychopathic Laboratory of the Chicago Municipal Courts for more than ten years, he and his colleague Harry Olson argued constantly that sterilization would wipe out crime.[8] In his early attempts to publicize the Model Law, Laughlin claimed sterilization had met with "great success" in prisons.[9] But after years of debate with lawyers who opposed eugenic surgery for convicts, he knew that proving criminality itself to be hereditary was nearly impossible. Laughlin had also learned that the more sterilization looked like punishment, the more likely that it would be found constitutionally suspect. Even though the public clamored for a solution to the crime problem, the eradication of crime as a primary focus of sterilization policy seemed to have fallen by the wayside.

When Laughlin's book was published in 1922, he was careful to qualify his comments on sterilization of the convicted, concentrating instead on the value of surgery to clear the streets of women like Carrie Buck. Yet after the *Buck* case, the impetus to enact sterilization laws applying specifically to criminals resurfaced to generate a second Supreme Court decision on eugenics. It turned on an Oklahoma law passed after the *Main* case that extended the reach of sterilization, naming specific penal institutions where mandated involuntary sexual sterilization for recidivist prisoners could take place.

The new law was extraordinarily broad. It applied to all patients of child-bearing age (up to age forty-seven for women, sixty-five for men) who were ready for discharge. It applied to every variety of eugenic "defective," covering "patients likely to be a public charge, or to be supported by any form of charity," anyone suffering from hereditary forms of insanity, and "habitual criminals," defined as three-time felons.[10] The focus on the economic rationale for surgery was commonplace, and in the same year that the new law went into effect, a University of Oklahoma scientist gave a speech titled "Democracy and the Genes." He insisted that the "desirable members of society are being ever more heavily taxed to care for the undesirables," leading to lower birth rates among "healthy, substantial" citizens.[11] Fiscal stringency was popular, and before long, a proposal to require sterilization as a condition of receiving any kind of relief payment was on the legislative agenda.[12]

No person of intelligence would criticize the law, said Dr. Louis Ritzhaupt, the state senator who sponsored the legislation. The annual growth of jails, penitentiaries, and asylums "adds yearly to the tax burden," he said. Ritzhaupt's background as a surgeon who had served as president of the State Medical Association gave him a unique perspective on sterilization laws.[13] A man of many interests, Ritzhaupt also ran a restaurant and a chicken ranch in addition to his duties as a doctor and lawmaker. He raised prize Rhode Island Reds and Polish bantams that competed in shows as far away as New York. He was eventually elected President of the Oklahoma State Poultry Association.[14]

In contrast to mental patients like Samuel Main, Oklahoma's convicts did not agree to surgery. They tried to recruit Clarence Darrow as their attorney, but the aging lawyer was not available.[15] In time, other lawyers stepped in to represent the prison population; neither they nor their clients were willing to submit to the scalpel without a fight.

When McAlester State Penitentiary inmate and three-time burglar George W. Winkler volunteered to test the 1933 law, the case received extensive publicity. Three hundred prisoners who were subject to the law as "habitual criminals" pooled their canteen funds—money earned at the prison's snack bar—to pay two seasoned lawyers to argue Winkler's case. Claude Briggs, a state senator who would soon become president pro tempore of the Oklahoma Senate, led the defense. An experienced litigator, Briggs later helped found an organization that would become the American Trial Lawyers Association. Also arguing on behalf of the prisoners was Briggs's law partner E. F. Lester, former Chief Justice of the Oklahoma Supreme Court.[16]

The job of identifying convicts who would be liable for surgery fell to Sam Brown, the prison warden at McAlester. Brown was outspoken in his opposition to sterilization, saying that "serious prison unrest" had resulted from the news that the law would soon be put into effect. His records proved that few children of prisoners ended up in the penitentiary. But Brown felt that he had no legal option but to begin the hearing process.[17] Convicts posted a sign in the prison yard, soliciting funds to support their suit against sterilization. Their plea: "Keep your MANHOOD, Do your part, CONTRIBUTE."[18]

In July 1934, Winkler appeared before the Board of Affairs, which the law had empowered to order sterilization. He was surrounded by a committee of his fellow convicts at the hearing, a "brain trust" including a former attorney serving thirty-five years for bank robbery, another man serving life for murder, and a third colleague doing twenty years for auto theft. Back at the "big house," the warden openly opposed sterilization and convicts were "astir with mutterings and rumors."[19] At the end of the proceedings, Briggs announced that the defense was out of money, because the warden had prohibited the prison "defense committee" from soliciting funds.[20]

The board voted in favor of surgery, but Briggs took advantage of conflicts among the board members and opposition by prison administrators to delay the necessary court appeal of Winkler's case. He launched a series of legal maneuvers designed to stall the sterilization process and buy time.[21] The Board of Affairs finally relented, abandoning the *Winkler* case and asking the legislature for a revised statute that could not so easily be manipulated by defense lawyers.

Sterilization supporters returned to the Oklahoma legislature, again led by Ritzhaupt, who promised to repair the remaining shortcomings of the legislation. Ritzhaupt received death threats as he pressed for further sterilization authority, but he emerged from the legislative chambers in 1935 with the Oklahoma Habitual Criminal Sterilization Act.[22]

By then it was clear that the battle over sterilization in Oklahoma would be neither simple nor short. Briggs had begun to quote the work of Dr. Morris Fishbein, editor of the prestigious *Journal of the American Medical Association.* Fishbein had written that the success of eugenical sterilization was "far too slight" to command a medical endorsement.[23] He recommended that "political experiments in sterilization" be left to "Germany and the Fascist nations." Science had simply not progressed to the point where anyone could predict with precision who would be parents of unfit children.[24] Even in small-town Oklahoma, people knew of the German program, and it was the object of jokes in local newspapers. "Thousands of Germans are undergoing sterilization to blot out feeblemindedness," quipped the *Ada Evening News.* "If this keeps on, who will be left to support Hitler?"[25]

Ritzhaupt's new law contained a definition of "habitual criminal" that included any person "who, having been twice convicted . . . of crimes amounting to felonies involving moral turpitude, is thereafter convicted . . . of a crime involving moral turpitude." Lengthy procedures allowed a full hearing, with legal representation for the alleged "habitual criminal" and the right to several levels of appeal. Finally, several exceptions were listed. Acts that did not make a criminal liable for sterilization included "offenses arising out of the violation of the prohibitory laws, revenue acts, embezzlement, or political offenses."[26] These exclusions would later pose a problem for Oklahoma's eugenic aspirations.

Hoping to reactivate the sterilization program, prison officials looked for another volunteer but were unable to identify a willing inmate. However, the first person reported by a county attorney to be an "habitual criminal" under the new provisions was five-time felon Hubert L. Moore, who seemed likely, volunteer or not, to become the subject of the next test case.[27]

The tension in prison was palpable, and officials felt besieged. Before a hearing on Moore's case could be set, a number of convicts escaped.[28] Defense attorneys were also facing pressure. The press repeatedly noted that the prisoners put up one thousand dollars to pay their lawyers, and Briggs's decision to represent the convicts cost him dearly in the political realm. His colleagues in the state legislature "declared war" and vowed that they would "break him" for consorting with criminals.[29] Nevertheless, state officials knew that Briggs was a cunning adversary, and the Oklahoma attorney general expected a lengthy appeal process regardless of the outcome at trial.[30] To everyone's surprise, Moore decided not to contest the impending sterilization order. Instead, he bolted from the prison and fled for California.[31] Without a prisoner to object to the new law, there could be no test case.

Jack Skinner

Officials quickly identified another man who would more willingly represent the convict population. Jack Skinner worked as a prison clerk and believed that some people should be sterilized if they had "insanity in their family" or were unable to care for their children. "There is no point in government letting families on relief keep on having children. . . . It is nothing but pure ignorance that causes women to keep on having more children than they or society can afford," he said. But Skinner did not count himself among them. He had attended college, and despite his prison career, he had hopes of being a responsible parent with children of his own when he left the penitentiary. Volunteering to test the law did not bother Skinner, because if the law were upheld he would have to face surgery anyway. The Oklahoma attorney general agreed that Jack Skinner was a good candidate for surgery, and the case that would be known as *Skinner v. Oklahoma* began.[32]

Skinner's life had not been easy. For a time, he may not have even been

"Jack Skinner." His criminal records showed a first conviction as "Jaspar Ingram," which may have been his real name. For a second crime he was punished as "Joe Smith." At his sterilization trial, observers learned that Skinner's father died when Jack was "quite young," and his mother remarried when he was ten. Jack left home at the age of fifteen in search of work. He lost his foot in an accident, and, at age nineteen, while enrolled in business college, was convicted of stealing six chickens and sentenced to eleven months of hard labor.[33] Skinner was incarcerated for armed robbery in 1929 and again in 1934, just prior to the passage of the sterilization law. He chose not to fight the charges, pleading guilty to each crime and testifying that he had stolen because of his inability to work or support himself and his wife.[34]

Apparently under some pressure, D. W. Griffin set aside his earlier opposition to involuntary surgery and served as an expert for the prosecution at Skinner's trial. Griffin did not examine Skinner but testified that his courtroom observations had led him to conclude that Skinner was "psychopathic." Though he had never met Skinner's family, Griffin volunteered that if he knew Skinner's mother and father, perhaps he could identify something in the man's childhood or even earlier that would support the diagnosis. Griffin was joined by Ritzhaupt, one of the authors of the 1935 law. Dr. T. H. McCarley, a surgeon at the penitentiary and, like Ritzhaupt, a former president of the Oklahoma Medical Association, also testified in favor of sterilization.[35]

The Oklahoma law implied that anyone who committed three felonies involving "moral turpitude" must be eugenically unfit. Relying on this presumption, the prosecutor produced no evidence to show that Skinner's criminal disposition was inherited. Adopting a different strategy, the defense attacked the presumption directly. Briggs tried to disprove the theory of hereditary criminality by submitting a survey of Oklahoma convicts. It showed that relatively few prisoners had family members with a criminal record. But that evidence was ruled inadmissible, and the jury was allowed to consider only Skinner's past crimes.[36] Because Skinner had committed three felonies (at the time, any theft valued at twenty dollars or more constituted a felony in Oklahoma), he met the definition of a habitual criminal, so the judge ordered sterilization.

Considering Skinner's case on appeal, the majority of the Oklahoma Supreme Court upheld the sterilization law. But by 1941 enough questions had been raised about fairness in the use of sterilization that four of the nine judges on the Oklahoma Supreme Court dissented. Reminding their colleagues that "the right to beget children is one of the highest natural and inherent rights," the dissenters declared that without at least a chance to disprove the presumption that his children would be tainted, Skinner had been denied due process of law.[37]

When Oklahoma implemented its first sterilization law in the early 1930s, the recent *Buck* case seemed to have settled any constitutional questions about eugenic sterilization.[38] But by the time the *Skinner* case was appealed to the U.S. Supreme Court ten years later, several developments had brought sterilization back into the headlines and reopened the national debate over the surgical use of eugenics.

On the last day of 1930, Roman Catholic Pope Pius XI clarified what up until then had been a matter of speculation about Church doctrine regarding eugenics. In a Papal encyclical entitled *Casti Canubii* ("Of Chastity in Marriage"), he specifically denounced laws imposing eugenical sterilization.[39] This was perhaps the most authoritative religious proclamation against the practice. Another challenge to sterilization would come from the mental health world.

In a book edited by psychiatrist Abraham Myerson, the American Neurological Association went on record in opposition to most sterilization laws. The report condemned the "family studies" popularized in *The Jukes* and *The Kallikak Family,* judging such work, as well as the efforts of eugenicists at Cold Spring Harbor, "entirely invalid."[40] Commenting on the Myerson report, the *New York Times* said "considering how little is known about the mechanism of heredity . . . sterilization is scientifically justified in only a few rare, selected cases. . . . Before we can legislate the hereditarily unfit out of society we need facts."[41]

About that same time, the country was shocked by the news of the sterilization of heiress Ann Cooper Hewitt, who sued her mother and two doctors for a half-million dollars.[42] The trio, said Miss Hewitt, had conspired to render her barren so that she would not inherit her father's fortune. His will had left two thirds of his ten-million-dollar estate to her but had added that, if she died childless, the money would revert to her mother.[43] Criminal arrest warrants were issued for the accused mother, a five-time divorcee who promptly (but unsuccessfully) attempted suicide. The two doctors who performed the sterilization surgery, masking their intentions under the usual diagnostic dodge—they claimed that Ann had appendicitis—also faced trial. Eventually both Ann's mother and the physicians escaped criminal conviction.

Even though twenty-one-year-old Hewitt read Shakespeare and Dickens and spoke both French and Italian fluently, a San Francisco civil court heard evidence that she was "feebleminded" and "dangerously oversexed." Like Carrie Buck, she was characterized as a "high grade moron" who had written love notes to the chauffeur and flirted with both a bellhop and a "negro porter." Hewitt had a team of lawyers and ample resources to back her charges, but in the face of allegations that she was a "moral degenerate," her lawsuit ultimately failed.[44]

The Cooper Hewitt saga and stories of sterilization abuse in a Kansas home for girls may have heightened attention to the possibility that eugenic surgery could be abused, but neither these scandals nor reports by 1935 that at least 180,000 had already been sterilized in Germany stopped U.S. states from passing new sterilization laws.[45] Announcing a legislative campaign in favor of such a law, a Georgia Medical Association spokesman declared that the "sterilization project of Hitler in Germany is a step in the right direction." While the recent Nazi government enactment "might seem a bit drastic on the surface," he said, "it is being used wisely"—actually in a less expansive form than sterilization laws in some states.[46] Bills advanced in the Georgia legislature at the same time that Oklahoma was testing its new laws. In 1937, at the height of the Great Depression, Georgia became the last state to pass a sterilization statute. Those who championed the law explicitly described their goal as a step toward eradicating poverty and lowering taxes. In the face of significant protest, the Governor of Puerto Rico also signed a sterilization law that same year.[47]

Gallup polls showed ongoing widespread approval for eugenic sterilization. To the question "Do you favor sterilization of habitual criminals and the hopelessly insane?" 84 percent answered "yes." More people favored sterilization than those endorsing legal distribution of birth control information (70%), imposition of the death penalty for murder (65%), or bringing back the whipping post as a method of criminal punishment (61%). A majority (54%) also favored the most radical eugenic measure, government supervision of "mercy deaths" for "hopeless invalids."[48]

International events also added to the ongoing debate about sterilization. By the time Jack Skinner's request to be heard by the Supreme Court was granted, events in Europe had taken an ominous turn. In 1941, just before America entered World War II, Theodore Kaufman published a book entitled *Germany Must Perish*. The book was promoted to potential reviewers with a miniature casket that arrived in the mail. It contained a card printed in bold lettering: "Read GERMANY MUST PERISH! Tomorrow you will receive your copy." In the book, Kaufman called for the sterilization of all German citizens, starting with the Army and the entire male population within three months.[49] Describing the plan as "grisly," *Time* magazine reported that as part of an earlier antiwar scheme, Kaufman had advanced the bizarre suggestion that all Americans be sterilized to prevent them from having children to fight in future wars.[50]

U.S. commentary on the book dismissed it out of hand, with one critic calling it nothing but an "indulgence in dire vituperation."[51] The Nazis were quick to seize on Kaufman's proposal, however, incorporating it into their propaganda efforts. In a radio address from Berlin, master publicist Joseph Goebbels stirred his listeners' fears with supposed plans "for sterilization of our entire population under 60 years" of age. Such were the horrors that Germans would face, said Goebbels, if "democracies" won the war.[52] After America entered the conflict, Hitler himself said that "the sterilization of our male youth" was a primary U.S. goal.[53] Allied broadcasts attempted to counter the Nazi line, and newspapers highlighted how often Germans "harp on the suggestion made by one person" in the United States that Germans be subjected to surgery. The *New York Times* mocked Hitler's charge and, apparently oblivious of thirty-five years of U.S. eugenic policy on the state level, declared that sterilization was "a device first used by the Nazis against Germans."[54] Just a year earlier, American Eugenics Society President Frederick Osborn invoked the *Buck* case while arguing for the reduction of "defectives" in the name of national defense. Pointedly distancing U.S. practices from those of the Nazis, Osborn declared that "arbitrary control of births as in a dictatorship, is not eugenics; it may be only a distorted use of science for evil ends."[55]

Given the *Buck* precedent in favor of eugenics, it would not have been surprising if the Supreme Court had declined to review the *Skinner* case, allowing the Oklahoma decision to stand. But the Court that read the *Skinner* briefs was very different than the one that sat in judgment of Carrie Buck's case. Chief Justice Taft died in 1930, and Oliver Wendell Holmes died five years later. Only Harlan Stone, now Chief Justice, remained on the Court from the *Buck* era. He and his new colleagues voted unanimously to hear the *Skinner* appeal.[56]

Oral arguments revealed many defects in the Oklahoma case. When Chief Justice Stone asked whether any scientific evidence existed to prove that criminal traits were hereditary, Oklahoma Attorney General Mac Q. Williamson conceded, "I know of none." Pressed further about the role of heredity in crime, Williamson acknowledged that environment might be a more powerful factor. But he defended the state's prerogative to pass eugenic laws as a means of protecting "the purity and the virility of the race." Asked by the Justices why crimes like stealing chickens made a convict liable for sterilization while embezzlement did not, Williamson admitted frankly that the law's "exceptions are very difficult to reconcile."[57]

The Douglas Opinion

So many constitutional questions were raised about the Oklahoma law that there was no question among the Justices that it would be struck down. The job of writing the formal opinion for the Court was assigned to Justice William O. Douglas. Skinner's attorney provided Douglas with a multitude of reasons to reverse Skinner's conviction. Three of them stood out. First, there was the question of the ex post facto clause. Skinner's last conviction had come a year before the law passed, making this retrospective legislation. Second, the Fifth Amendment prohibited "double jeopardy." Even though the law was purportedly a civil enactment, prosecution testimony at trial had described the measure as a deterrent to future crime, indicating a penal motive. And finally, procedural due process was lacking. As a civil measure, the convict had no right to an attorney unless he could afford one, nor could he compel a witness to testify in his behalf.[58]

Setting these points aside, Douglas instead used the Equal Protection clause of the Fourteenth Amendment to fashion an entirely new constitutional standard. He began with a statement echoing Oklahoma's dissenting Justices: "This case touches a sensitive and important area of human rights. Oklahoma deprives certain individuals of a right which is basic to the perpetuation of a race—the right to have offspring." Later Douglas reiterated: "We are dealing here with legislation which involves one of the basic civil rights of man. Marriage and procreation are fundamental to the very existence and survival of the race." This was the first step along the way to later Supreme Court decisions describing reproduction as a "fundamental right." Douglas then analyzed the legal difference between embezzlement and larceny to point out the inequity of the Oklahoma law, using Jack Skinner's first offense as his example. "A person who enters a chicken coop and steals chickens commits a felony; and he may be sterilized if he is thrice convicted. If, however, he is the bailee of the property [that is, if he is entrusted with keeping someone else's chicken] and fraudulently appropriates it, he is an embezzler. Hence no matter how . . . often his conviction, he may not be sterilized."[59]

Douglas continued, coining a famous phrase: "strict scrutiny of the classification which a state makes in a sterilization law is essential," he wrote, "lest unwittingly or otherwise invidious discriminations are made against groups or types of individuals in violation of the . . . equal protection of the laws." The Douglas formula for "strict scrutiny" still guides constitutional inquiry into laws that threaten fundamental rights.

Emphasizing the illogic and the unfairness of the law, Douglas said that "sterilization of those who have thrice committed grand larceny with immunity for those who are embezzlers is a clear, pointed, unmistakable discrimination." He then undercut the science behind the Oklahoma law, noting that "we have not the slightest basis for inferring that that line [between larceny and embezzlement] has any significance in eugenics nor that the inheritability of criminal traits follows the neat legal distinctions which the law has marked between those two offenses."[60]

Interestingly, there was never the slightest suggestion in the *Skinner* opinion that the Court would use this occasion to overrule its earlier decision in *Buck.* Douglas's private notes show that he made a clear distinction between the two cases. He wrote, "Whether there are any scientific authorities in support [of sterilization] is not clear. Moronic minds [as in *Buck*] are different. [We have] no statistics as to criminals—if criminals do not produce their kind, then we have [a] serious question."[61] Douglas was not questioning the validity of all eugenic theory; he just could not find the eugenic sense in Oklahoma's law.

Douglas consulted several sources during his research on the *Skinner* case. A footnote in his opinion listed books that led him to repeat an argument made by Skinner's legal team. The sterilization law was faulty "in view of the state of scientific authorities respecting inheritability of criminal traits."[62] Those authorities included critics who had attacked Harry

Laughlin's Model Law years earlier and Abraham Myerson's American Neurological Association committee, which had found no justification to operate on criminals for eugenic purposes.[63]

Douglas also consulted work by leaders in the eugenics movement, like Leon F. Whitney. In 1926, as field secretary of the Eugenics Society of America, Whitney published a plan in *Christian Work Magazine* for eradicating crime. He called for sterilization of all children "who seemed to be genuinely bad," as recommended by school teachers reporting to a "city psychiatrist who really knew his business."[64]

Whitney was the executive secretary of the American Eugenics Society. In 1934 he wrote *The Case for Sterilization,* which praised "poor unfortunate" Carrie Buck for "how greatly, if unconsciously, she has served the world." Whitney boasted that Hitler himself had requested the book, though reviewers thought the *Buck* opinion provided "much more effective propaganda for sterilization."[65] Despite his passion for sterilization of the feebleminded, even Whitney was unable to articulate a good argument for sterilizing criminals for eugenic purposes. In fact, none of the "scientific authorities" Douglas had consulted condemned all sterilization. Even the most conservative, such as Abraham Myerson, thought that there should be "no hesitation" to operate on the feebleminded.[66]

All nine Justices voted to strike down the Oklahoma law, but two of Douglas's colleagues on the Court wrote separate opinions to clarify their reasoning. Chief Justice Harlan Stone wrote the first concurrence. He did not object to the Oklahoma law's eugenic foundations and left no doubt that "a state may . . . constitutionally interfere with the personal liberty of the individual to prevent the transmission by inheritance of his socially injurious tendencies." Echoing the *Buck* court, on which he had sat fifteen years earlier, Stone wrote, "Science has found and the law has recognized that there are certain types of mental deficiency associated with delinquency which are inheritable." But like the dissenting Oklahoma judges, Stone felt that Skinner should have been allowed, under the Fourteenth Amendment's due process clause, to show that his criminal acts were not based on an inherited defect. Not permitting Skinner that opportunity was a fatal flaw. "A law," wrote Stone, "which condemns, without hearing, all the individuals of a class to so harsh a measure as the present because some or even many merit condemnation, is lacking in the first principles of due process."[67]

Justice Robert Jackson offered a second concurrence, agreeing with both Douglas and Stone, endorsing both equal protection and due process attacks on the sterilization law. Of its scientific basis he wrote, "The present plan to sterilize the individual in pursuit of a eugenic plan to eliminate from the race characteristics that are only vaguely identified and which in our present state of knowledge are uncertain as to transmissibility presents other constitutional questions of gravity." Despite the Court's condemnation of Oklahoma's eugenic theory, Jackson pointed to the *Buck* case as a more precise use of eugenic criteria, writing, "This Court has sustained such an experiment with respect to an imbecile, a person with definite and observable characteristics, where the condition had persisted through three generations and afforded grounds for the belief that it was transmissible and would continue to manifest itself in generations to come." By directly referencing the alleged facts in *Buck,* Jackson, like Douglas, endorsed that decision's eugenic logic.

But to Jackson's eye, *Skinner* was not *Buck.* "There are limits," he concluded, "to the extent to which a legislatively represented majority may conduct biological experiments at the expense of the dignity and personality and natural powers of a minority—even those who have been guilty of what the majority defines as crimes."[68] Jackson's comment would echo when the world learned of sterilizations the Nazi doctors termed "biological experiments."

There is a common misconception that the Supreme Court decision in *Skinner* all but overruled *Buck* and that the postwar revelation of Nazi practices led to a general rejection of eugenics. But eugenically based assumptions about heredity as a basis for law survived well beyond *Skinner,* just as they outlived the Third Reich. Years after the case, Justice Douglas himself reiterated that there was no desire by the *Skinner* court to overrule *Buck.* Douglas was "very clear" on the case's constitutional validity. The eugenic foundations of *Buck* were safe because of the procedural protections that the Virginia law included.[69]

Jack Skinner was discharged from prison in 1939, his sentence served. His whereabouts were unknown as the case made its way to the Supreme Court. But when chicken breeder Ritzhaupt lost to chicken thief Skinner, the convicts at the McAlester penitentiary celebrated in "an air of jubilation" at news of the victory in their seven-year test of Oklahoma's law. Both Oklahoma's governor, Leon Phillips, and the attorney general, Mac Williamson, were unsurprised by the decision. Though he heard some encouragement to revise the law to meet the Court's constitutional critique, Ritzhaupt did not draft a new statute. Others suggested that the state would wait until war was over "before trying another experiment in biology."[70]

In September 1942, the *Eugenical News* analyzed the *Skinner* case. Said the editor, "In the case before the Court a man had been convicted twice for robbing with fire arms and once for stealing chickens. This is certainly not eugenic sterilization." He was equally certain that the decision had "no bearing" on the validity of the *Buck* decision. A report on the 38,087 sterilizations that had occurred in the United States was included in the same edition.[71] Several years later, the man who defended the sterilization law confirmed that sterilization was still the rule in Oklahoma, but only for the "insane," not for convicts.[72]

The *Skinner* case brought sterilization back into the headlines briefly, but it did nothing to lessen the impact of sterilization laws that remained in effect, authorizing surgery for people in state custody. The U.S. Public Health Service updated its listing of sterilization laws, showing twenty-seven states with compulsory statutes in 1940.[73] Nor did the decision muffle popular support for adding to the existing eugenic legal scheme. One doctor urged his colleagues at a meeting of the American Medical Association to consider a proposal for banning those who carried conditions as unremarkable as nearsightedness from marrying.[74] Harvard University anthropologist Earnest Hooton told a child welfare conference that America was "pampering the unfit." Rather than supporting more social programs, physicians could treat society with the tools of eugenics, including euthanasia, for the "hopelessly diseased and the congenitally deformed and deficient."[75] He repeated his message to the American College of Physicians, declaring that Germany had become a nation of "neo-simians" who were "the natural henchmen of paranoiacs and rascals." The "greatest human catastrophe" of the nineteenth century, Hooton said, was that medical science had failed to sterilize Adolf Hitler's mother.[76] Despite Justice Douglas's cautionary claim that sterilization could be abused "in reckless hands," support for increasing the use of eugenic surgery also remained popular.

When the United States entered World War II, critics of sterilization identified the procedure as a tool of the Nazis. Others distinguished the practice in the United States from the German experience. For example, in a treatise called *Civilization and Disease*, medical historian Henry Sigerist reminded his readers that all countries carry a "heavy burden" of people "disabled by hereditary diseases." He felt that the German "socio-biological experiment" with sterilization merited attention despite the unscientific ideology that drove it. Parroting social Darwinist logic, Sigerist highlighted the "preservation of hundreds of thousands of dysgenic individuals" who in the past would have failed in "the struggle for life." Outlining America's long history of sterilization law and quoting the "very important" Holmes opinion in *Buck,* Sigerist argued that it was a mistake to identify eugenics only with the Nazi program and dismiss it out of hand because of its recent association with an enemy regime. The problem sterilization addressed, he wrote, was "serious and acute" and would demand a response everywhere "sooner or later."[77]

While a shortage of physicians had lessened the number of operations performed in Virginia during the war years, the potential for a backlash in the face of Nazi excesses did not prevent physicians like Joseph DeJarnette from maintaining unrelenting support for the sterilization program. He published articles on sterilization, gave clinics to demonstrate the procedure, and made a standing offer to train "anyone who is interested" in the operation. A little training would, he noted, "save a great deal of trouble in working out the technique as we had to do."[78]

Many prominent Virginians gathered to celebrate DeJarnette's fiftieth anniversary as the superintendent of Western State Hospital. State legislators, judges, and two Virginia governors were in attendance that day as the poet who had written "Mendel's Law" explained how he had earned the name "Sterilization" DeJarnette. He regaled the crowd with his recollections of how Dr. Priddy had joined him in the cause and "won immortality" for his contribution. Of the others who played a part in the landmark case of *Buck v. Bell,* no one was more prominent in DeJarnette's praise that day than Aubrey Strode. He was, the doctor recalled, "a wonderful lawyer." "I knew him as a boy and I thought he was the handsomest young man I had ever seen," said DeJarnette. "He . . . has written the only law for sterilization of the unfit that has stood the tests of the Courts."[79]

In 1943 DeJarnette requested printed material from the Human Betterment Foundation in California in an attempt to "renew the interest of eugenic sterilization in Virginia." His belief, that curbing births among the "unfit" would save millions in the future, had not flagged.[80] When DeJarnette finally retired from state service after his eightieth birthday in 1947, he urged an extension of the sterilization law. Not satisfied that the state had operated on five thousand patients by then, DeJarnette argued that the number "ought to have been 20,000." He claimed that he had personally sterilized six hundred men. At age eighty-one, DeJarnette still welcomed attention as the "leading disciple of eugenic sterilization" in Virginia.[81] He was joined by professors at many Virginia schools who distributed propaganda on eugenic sterilization through the 1940s, as did Sarah Roller at the Juvenile Court in Richmond, one-time nemesis of the Mallory family.[82]

17
Buck, at Nuremberg and After

The many motives for government sterilization policies—ranging from concern for a deteriorating standard of national heredity to growing health and welfare expenditures to higher crime rates to ethnic or racial bigotry—would be dramatized as World War II came to an end. After the Allied victory in Europe was assured, President Harry S. Truman took the unprecedented step of appointing sitting U.S. Supreme Court Justice Robert Jackson to manage the prosecution of atrocities committed by the Nazis. Jackson became known as the architect of the Nuremberg Trials for his administration of the war crimes tribunals. The Nuremberg prosecutors would unwittingly provide a forum to highlight parallels between Nazi sterilizations and the policies endorsed in the *Buck* case.

The first trials involved more than twenty of the most notorious Nazi officials. They included Wilhelm Frick, Hitler's minister of the interior, whom Harry Laughlin had praised so lavishly for his application of eugenic principles through German law. Joining Frick in the dock were Hermann Göring, chief of the *Luftwaffe* (air force); Rudolf Hess, Hitler's deputy and leader of the Nazi Party; and Julius Streicher, publisher of virulently anti-Semitic newspapers and magazines. Streicher's numerous publications, issued in a calculated effort to fan public hatred for Jews, were critical in setting the stage for the Holocaust.

Streicher made one of the first references to sterilization at Nuremberg, but his comments had nothing to do with government programs. He spoke out in his own defense, arguing that his anger at Jews was a result of reading Theodore Kaufman's book *Germany Must Perish,* which advocated sterilization of all Germans.[1] Other accounts in the early war trials came from concentration camp victims who witnessed experimental sterilization and forced abortions.[2] Sterilization emerged in more detail during the later trials of the Nazi doctors.[3]

When the war ended, Leo Alexander was named medical consultant

236

for the Allied prosecutors at Nuremberg, and he made the most important contribution to understanding the scope of concentration camp sterilization. As a young doctor in Germany, Alexander did research on hereditary mental illness, collecting data on families much as eugenic field workers had.[4] Following several months of study in China just before Hitler came to power, the Jewish researcher Alexander could not return to Germany; his work in eugenics was abandoned when he emigrated to America.

Alexander met Abraham Myerson and became an important member of the American Neurological Association committee that Myerson chaired. Alexander was a coauthor of the committee's 1936 book criticizing sterilization policies.[5] He was well prepared to investigate the scope of sterilization abuse in Germany. But because Jackson and the prosecution team limited the criminal charges against the Nazi doctors to illegal acts taken in furtherance of the war effort, the trials largely ignored the hundreds of thousands of involuntary sterilizations carried out on German citizens under German law.[6]

At one level the unwillingness to address German domestic sterilization was consistent with the policy not to prosecute Nazis for matters that had occurred in Germany before the war. According to the prevailing legal opinion, such matters did not fall under the jurisdiction of an international tribunal. At another level, while the domestic sterilization program in Germany was significantly more expansive than U.S. state programs, it was promoted by very similar motives and fueled by similar propaganda. In both countries the cost of supporting hereditarily diseased asylum inmates was seen as a public social burden. Justice Robert Jackson believed that sterilizing the hereditarily defective could be appropriate social policy as long as it was based on science rather than mere bigotry. He had said as much in his *Skinner* opinion.

Jackson was not alone in trying to disentangle eugenics from prejudice. Early in 1936 Yale University political scientist Karl Loewenstein published a critique of the excesses of the Nazi regime in the *Yale Law Journal.* Loewenstein had been trained as a lawyer in Germany and was intimately aware of legal developments during Hitler's ascendancy. He believed German legislation during the Nazi period owed its origins to the "race myth" of Aryan supremacy that had made the Jews "a new class of untouchables" within Europe. At the same time, other legislation traceable to the "race myth" was focused on eliminating the physically unfit and providing for "healthy progeny in the future." Apart from its unscientific foundation in Nazi ideology, Loewenstein was not troubled by the application of law to the "unfit" in laws like the German sterilization statute. "Stripped of its racial exaggerations," said Loewenstein, "the public health legislation [marriage restriction and sterilization laws] seems soundly conceived." Loewenstein reported that in the first year of the sterilization law, during which approximately fifty thousand sterilizations were done, no "political abuse" had been recorded.[7]

A similar assessment led to the exclusion of Nazi scientists like Ernst Rudin from war crime prosecution. Rudin was a psychiatric geneticist whose work had bolstered "the legitimization and the popularization of the National Socialist Government."[8] He played an important role in the writing and passage of the 1933 German sterilization law. As a colleague of Harry Laughlin and Charles Davenport, he was also a member of the advisory board to the *Eugenical News*.[9] Targeted as a potential defendant at Nuremberg by Leo Alexander, Rudin was arrested in 1945. But he cited the widespread sterilization program in the United States and his collaboration with Americans as evidence that his involvement in the German domestic program was entirely legal.[10]

The extreme limitations of time and resources that faced prosecutors and their researchers led to inaccurate conclusions about the scope and duration of domestic sterilization in both countries. Alexander claimed that sterilization had no longer been used either in the United States or in Germany before the war. But this assessment missed the mark: it was publicly reported that more than a thousand sterilizations had occurred in the United States alone the year before Alexander's report.[11] Nevertheless, Rudin was released from custody in 1946 when it became clear that there would be no charges for domestic sterilization abuse at the trials of Nazi doctors.

The U.S. legal team eventually decided to focus on the sterilizations that took place in concentration camps, prosecuting them as "Crimes Committed in the Guise of Scientific Research." Alexander collected evidence of Nazi discussions before the war about using x-rays as a more efficient sterilization technique than surgery.[12] He assembled other evidence about herbal methods of sterilization and the plan to sterilize Jews, homosexuals, "gypsies" (Sinti and Roma), and non-German concentration camp victims.[13]

By the beginning of World War II, scientists understood how to use x-rays for sterilization. One researcher had even speculated about the use of radiation to sterilize people responsible for "bringing to life a generation of imbeciles."[14] This knowledge was put to use by the Germans. Among the many physicians implicated in the sterilization experiments, none was more prominent than Karl Brandt. As the general commissioner for sanitation and health, Brandt was the chief Nazi medical officer. He was also Adolf Hitler's personal physician.

Brandt's attorney introduced documents quoting extensively from the eugenics literature. He cited Harry Laughlin's 1914 proposal calling for the sterilization of fifteen million Americans and also quoted a translation of the *Buck* opinion from a German text on eugenics.[15] Brandt's defense of Nazi experiments resulting in the death of concentration camp prisoners seemed to echo the Holmes opinion: "In the same way as the state demands the death of its best men as soldiers, it is entitled to order the death of the condemned in its battle against epidemics and disease."[16] Other Nuremberg defendants also cited *Buck,* and a translation of the Holmes opinion appeared again as a defense example in the exhibit "Race Protection Laws of Other Countries."[17]

Even though "experimental" sterilization played a part in the conviction of Brandt and several of his Nazi colleagues at the Nuremberg trials, the *Buck* opinion remained the best defense of legally sanctioned sterilization, even in the immediate aftermath of the Holocaust.

Testimony at Nuremberg did not reveal that U.S. eugenicists had their own "experimental" sterilization programs years before the war. Harry Laughlin alerted his mentor Charles Davenport as early as 1921 that experts in x-ray technology at the American Roentgen Ray Society had developed a formula for producing sterility using x-rays. The formula included "dosage, strength, distance, target" and described the machine to be used. Laughlin noted that "there has been considerable demand for this sort of thing by institutional authorities."[18]

Laughlin later confirmed that x-rays were being "successfully employed experimentally for sexual sterilization." Of all the methods for effecting permanent sexual sterilization, he wrote, "X-ray treatment holds out the greatest promise."[19] Physicians like Joseph DeJarnette followed Laughlin's lead and used x-rays to sterilize women. DeJarnette also experimented with the use of electricity to sterilize patients at the hospital he directed.[20]

We should not be so surprised that residents at U.S. hospitals and asylums were the subjects of biomedical experimentation. Over the years, the

Virginia Colony had been the site of research not only by civilian doctors but by the military as well.[21] Research was also a critical activity on the agenda of the Eugenics Record Office, where there was open advocacy for using institutional inmates as research subjects. E. E. Southard, Harvard Medical School neuropathologist, director of the psychopathic department of the Boston State Hospital, and member of the Special Board of Directors of the ERO, said public institutions were like mines, waiting "to be explored for the ore of progress." Katherine Bement Davis, chair of the New York Parole Commission and executive secretary of the Bureau of Social Hygiene, expressed a common sentiment when she called "scientific investigation" of "the defective and delinquent classes" within institutions "the only possible hope for the future." Davis urged the use of "our great state institutions as human laboratories."[22]

Though it was never made public, one experiment conducted by Charles Davenport put the sentiments of Southard and Davis into practice, using a state institution as the venue for experimental science and the "defective" residents as research subjects. In 1929, Davenport planned and carried out the experimental castration of a dwarf in New York's Letchworth Village for the feebleminded. The surgery was not medically necessary but was done in an attempt to explore the chromosomal basis of what is now known as Down syndrome. Two men who would later become famous for their scientific accomplishments assisted Davenport. George Washington Corner was an anatomist who discovered the hormone progesterone; he performed the castrating surgery on the asylum inmate. T. H. Painter, renowned for his studies of chromosomes, analyzed the tissue Corner had removed. The experiment, which was designed by Davenport to provide insight into the "inherited defect" of "mongolism," addressed an important concern of the eugenicists.[23] It was completed fifteen years before similar experiments became the basis for prosecuting Nazis for war crimes.

Challenges to Buck

Despite the revelations of Nazi excesses, the *Buck* precedent went undisturbed. After World War II, a few scholarly challenges to the case appeared.[24] One legal commentator went so far as to say that the Holmes opinion was "a badly reasoned decision in which is wrapped up the theory of totalitarianism."[25]

J. E. Coogan, a Roman Catholic priest and professor of sociology at the University of Detroit, produced the most pointed challenge to the case's legitimacy. He was among the first to question Holmes' reference to "imbeciles." Coogan read Laughlin's booklet, *The Legal Status of Eugenical Sterilization*, published in 1930 after the *Buck* case and found "no reference whatever to imbeciles, properly so called; the three were morons at the worst." He also doubted the evidence used to label Emma Buck a "mental defective" based on her performance on a "much criticized intelligence test." He asked why "counsel for Carrie Buck did not oppose her sterilization, at the several stages of her case through the courts, on the grounds that none of the three generations were proven to suffer from 'mental defect,' not to say 'hereditary.'" He also asked for the subsequent history of Carrie Buck following the case as well as for the performance of her child, Vivian.[26]

Coogan's letter was forwarded to Laughlin in retirement in Missouri; he apparently chose not to respond. Coogan also queried staff at the Eugenics Record Office, who disavowed any official connection to the *Buck* case. Later Coogan wrote to Dr. John Bell at the Colony, seeking to learn Carrie's history as well as details on the mental condition and subsequent development of Carrie's child. But by then Bell had died. His successor replied to Coogan, indicating that Vivian was "reported to have been very bright" and noting that Colony records contained "no definite evidence" that the child was feebleminded.[27]

After the *Skinner* case, Coogan invited the Supreme Court to overturn *Buck,* calling sterilization a practice "smacking of Nazism" used to persecute "the defenseless poor."[28] A decade later, in a critique of the *Buck* case published to coincide with its twenty-fifth anniversary, Coogan made direct challenges to the factual accuracy of the Holmes opinion and the mythology that had grown up around the *Buck* case.[29] But Coogan's campaign had little impact on the public awareness of sterilizations that had occurred under the authority of the laws passed decades earlier.

Sterilization after World War II

At the end of World War II, eugenic sterilization drifted out of public consciousness for many years, but surgeries continued. In most states the number of operations declined drastically. California, which led all states in total operations, sterilized just over nineteen thousand from 1909 to 1949 but only 981 between 1949 and 1959. Similarly, from 1913 to 1949 Kansas did 3,001 operations—but only twenty-four in the next ten years. On the other hand, a few states increased their rate of surgery. North Carolina performed more than 3,500 operations, Georgia almost 2,500, and Virginia about 1,885. In all, almost twenty-three thousand sterilization operations were officially recorded in the United States between 1943 and 1959. After 1950, the rate of surgeries performed annually in Virginia surpassed every state but North Carolina.[30] Arguments continued to be made in medical journals and the popular press that sterilization would prevent the transmission of "hereditary diseases" and the birth of "mentally deficient children."[31]

By the middle of the century, some legislators wanted to extend the reach of compulsory sterilization laws in order to punish women who had had illegitimate children. They also repeated arguments linking sterilization with a decrease in welfare costs. The Virginia legislature considered a bill in 1956 that would empower public welfare officials to initiate sterilization proceedings against women who had given birth to more than one illegitimate child. Just as in earlier eugenic sterilization laws, the legislation required a court to find that "the welfare of the woman and of society" would be promoted by the operation. That bill died in a legislative committee.[32]

The perceived epidemic of illegitimate births moved the 1958 Virginia General Assembly to name a study commission to review the problem. The Commission to Study Problems Relating to Children Born out of Wedlock urged that voluntary sterilization be made available to anyone who wanted it. But it pointedly rejected the calls for a more expansive compulsory sterilization law, and it raised questions about the existing law, describing it as inconsistent with "advances made in medical science." Unable to investigate this topic thoroughly, the commission majority instead recommended further study of existing sterilization authority.[33]

However, three members of the commission filed a dissenting report. They attributed the problem of illegitimate births to an alarming trend "among the Negro race." Such a trend, they said, would persist "until illegitimacy becomes uneconomical or the Negro race develops a sufficient sense of pride or moral values" to reject illegitimacy. The dissenters also rejected the idea of voluntary sterilization as "a social experiment with dubious moral implications." One member added his support to many of Virginia's county welfare boards that had voted to recommend compulsory sterilization after the birth of a second illegitimate child.[34]

The report of the commission was taken up again by an Advisory Legislative Council appointed by the 1960 Virginia General Assembly. The council recommended a voluntary sterilization law that would protect physicians who performed elective, nontherapeutic surgery.[35] In 1962, other supporters of punitive sterilization proposed a bill in the Virginia Senate almost identical to the failed 1956 measure. That measure, applying only to women giving birth to illegitimate children while "receiving public welfare benefits," also failed.[36]

Virginia became the first state in the country to pass legislation specifically endorsing sterilization as a voluntary method of birth control. The law extended sterilization to anyone in Virginia, allowing an adult to have surgery after making a written application, receiving a "reasonable medical explanation" of the procedure, and returning after a thirty-day waiting period. Certain children could also have "voluntary" surgery if a parent or guardian made a request and if a court decided it was in the interest of the child and of "society." Like the 1924 legislation then still in effect for compulsory sterilization, the operation was specifically applicable to children "afflicted with any hereditary form of mental illness that is recurrent, mental deficiency or epilepsy."[37]

In 1960, the American Medical Association reviewed the operation of sterilization laws nationally, raising questions about the potential for their abuse and the viability of the *Buck* precedent.[38] But while the Virginia Advisory Legislative Council's study of compulsory sterilization also mentioned the 1924 sterilization law and the *Buck* precedent, it uncovered "no substantial complaint" about the law's operation. The council went on to say that there were "no medical or other scientific data" to justify revision of the law and that change was neither "imperative nor desirable." Virginia's unwillingness to revise its law was consistent with sentiment in the twenty-seven states with compulsory sterilization laws in 1962. That year, a national survey of physicians showed nearly 80 percent of respondents favoring sterilization when "mental deficiencies could be transmitted to children."[39]

Loving v. Virginia

Two social developments were responsible for bringing public attention back to the history of eugenics and the *Buck* case. The first was the civil rights movement, which effectively transformed U.S. ideas of race in the

VIRGINIA:

BEFORE THE STATE HOSPITAL BOARD

AT

_____
(Institution)

In re

_____, Register No._____ ⎞
⎟ Order for
⎟ Sexual Sterilization
Inmate ⎠

Upon the petition of_____,

Superintendent of _____,

and upon consideration of the evidence introduced at the hearing of this matter, the Board finds that the said inmate is

⎧ insane
⎪ idiotic
⎨ imbecile   and by the laws of heredity is the probable potential parent of socially
⎪ feeble-minded
⎩ epileptic

indequate offsprings likewise afflicted; that the said inmate may be sexually sterilized without deteriment to ⎰his ⎱ general health, and that the welfare of the inmate and of society will be promoted by such sterilization.
⎱her ⎰

Therefore, it appearing that all proper parties have been duly served with proper notice of these pro-

ceedings, and have been heard or given an opportunity to be heard, it is ordered that_____

_____ ⎰perform
(Superintendent)   ⎱have performed

by Dr._____, on the said inmate the operation of ⎰vasectomy⎱
⎱salpingectomy⎰

after not less than thirty (30) days from the date hereof.

_____
(Designated Member of Board)

Dated _____.

Note: Make two copies; one for guardian or committee and one for Record.

Sterilization form used in Virginia hospitals, circa 1960.

decade from the 1954 Supreme Court decision in *Brown v. Board of Education* to 1964, when the Civil Rights Act was passed. The second was the not unrelated campaign to gain an endorsement for privacy rights surrounding reproduction. In 1967 those two movements came together in the case of *Loving v. Virginia*.

Only three cases involving laws generated by the eugenics movement have been subjected to full Supreme Court review. The first was *Buck*; the second, *Skinner v. Oklahoma*. The third case, *Loving v. Virginia*, did not involve sterilization but was instead about another issue of burning importance to eugenicists: interracial marriage, or "miscegenation." *Loving* arose as a challenge to a law that owed its reasoning to the work of eugenic theorists and racial propagandists like Harry Laughlin. Laughlin owed some of his notoriety among the Nazis—as well as a portion of his honorary University of Heidelberg degree—to his advocacy of anti-miscegenation law. Laughlin had also consulted on the eugenic revision of an anti-miscegenation law passed by the Virginia General Assembly the same day as the sterilization statute challenged in *Buck*. He argued that interracial mixing was dysgenic, likely to pollute the white gene pool to the detriment of future generations of Americans. His efforts to reinvigorate existing anti-miscegenation laws using concepts and rhetoric borrowed from twentieth-century eugenics culminated in the revision of several state laws in the 1920s and 1930s and provided a boost to the "Jim Crow" legislation separating the races.[40]

Within Virginia, John Powell and Walter A. Plecker were the most important advocates of Virginia's racially restrictive marriage law.[41] Powell, a prominent musician and composer from Richmond, Virginia, lobbied for the anti-miscegenation law and founded the Anglo-Saxon Clubs of America. He led a branch of that organization while teaching at the University of Virginia.

Joining Powell as the most publicly strident supporter of marriage restriction law was Dr. Walter Plecker. From 1914 to 1942, Plecker served as registrar at the Bureau of Vital Statistics, a division of the State Board of Health. He lectured regularly as a public health expert and commented often in state publications about the role of racial interbreeding as the source of public health problems. Plecker used his office to lobby for laws that would classify all citizens by race and prohibit interracial mixing of any kind.[42]

The law that Powell and Plecker wrote included provisions redefining "white persons" and allowed them to marry only others who had "no trace whatever of any blood other than Caucasian." An exception was made for supposed descendants of Pocahontas who could show "one-sixteenth or less of the blood of the American Indian." Anyone who violated these prohibitions by marrying a "colored person" could be charged with a felony and imprisoned from one to five years.[43]

In the forty years following its 1924 passage, the Virginia law survived several challenges. One case made it all the way to the U.S. Supreme Court, where it was ultimately stricken from the Court docket—too controversial to decide.[44] But a case begun in 1958 eventually led to the law's demise. That year a Virginia grand jury indicted Mildred Jeter, a black woman, and Richard Loving, a white man, who had been married in the District of Columbia but later moved back to neighboring Virginia. After pleading guilty to violating the ban on interracial marriage, the couple was sentenced to one year in jail. The trial judge suspended their sentences on the condition that they accept banishment from the state for twenty-five years.

After living for five years in Washington, D.C., the Lovings returned to Virginia to have their convictions erased. Several more years of litigation in state and federal courts failed to overturn their official status as outlaws. But when the Supreme Court finally decided their case in 1966, it ruled that Virginia's law prohibiting interracial marriage was unconstitutional. Chief Justice Earl Warren's opinion for a unanimous Court cited Justice Douglas's earlier comment in the *Skinner* case: "Marriage is one of the 'basic civil rights of man' fundamental to our very existence and survival. . . . The Fourteenth Amendment requires that the freedom of choice to marry not be restricted by invidious racial discriminations."[45]

The *Skinner* case gave the Supreme Court an opportunity to expose some of the class bigotry that had made its way into law that had support from the eugenics movement. Twenty-five years later, the *Loving* case, as one of the final crucial decisions of the civil rights era, exposed the racial bias embedded in another kind of eugenic legislation. But the *Loving* opinion made only an oblique reference to the eugenic roots of the law it struck down, and the case brought no attention to its Virginia counterpart, upheld in the *Buck* litigation more than a generation earlier. The new sensitivity to racial discrimination born of the civil rights movement doomed anti-miscegenation laws in the mid-1960s, but the *Buck* case remained an obscure episode, unexamined and generally forgotten. It would take even more attention to sterilization abuse among minority populations to revive attention to *Buck*.

Sterilizing Minority Women

Roe v. Wade, the momentous 1973 decision establishing the constitutional right to abortion, forever changed the U.S. discussion of how the law could control reproduction. Supreme Court deliberations in *Roe* were still in process when sterilization once more appeared in the headlines. Again, the twin controversies of race and reproduction took center stage.

By then, the idea that bad heredity could be interrupted by sterilization had faded. Even at the high point of the eugenics movement, most eugenicists knew that sterilization was unlikely to have much impact on the gene pool. As early as 1935 Paul Popenoe noted that sterilization was being used more for purposes of "economics and social welfare" rather than eugenic considerations. He concluded that "the eugenic basis for sterilization has become more and more secondary."[46] Others completely discarded heredity as a motive for sterilization. They said that arguments about heredity were "endless and profitless," while saving a child from being born into a bad environment was the better justification for operating.[47] Though it was often obscured behind arguments about genetics or the power of heredity, the economic motive for sterilization was never hidden. By the 1970s money took center stage in discussions of sterilization, and the point became even simpler: impoverished or incompetent parents would have children who ended up on welfare. The fiscal logic of sterilization played out repeatedly among poor populations. The most important case came from Alabama.

Fourteen-year-old Minnie Lee Relf and her twelve-year-old sister, Mary Alice, lived in Montgomery, Alabama, with their disabled father and illiterate mother. Their income came from welfare payments, and the family lived in government-subsidized housing. Mary Alice was born with a speech impediment and had been diagnosed as mentally retarded; she had no right hand.

In the summer of 1973, Mrs. Relf placed her X on a form at a public birth control clinic, believing that she was authorizing medication for her girls. When she learned, too late, that the form indicated consent for sterilization, she filed a lawsuit on the girls' behalf.[48]

Several investigations followed the *Relf* lawsuit, revealing that federal funds had been used to sterilize between 100,000 and 150,000 low-income people in only a few years. Many were adults, and some were children; some were mentally incompetent. Some people had acquiesced to an operation after being threatened with the loss of welfare payments unless they consented. Many coerced sterilizations occurred at the time of childbirth, with some women reporting that delivery of their babies had been conditioned on agreeing to the surgery.

Within six months of the *Relf* lawsuit, the federal Department of Health, Education, and Welfare publicized new "Guidelines for Sterilization Procedures" that would apply to any surgery paid for with federal funds. The proposed regulations required clear disclosure to patients and written informed consent before an operation could take place. Roman Catholics fought the regulations, saying that no sterilizations should be federally funded. The changes were also opposed by some state legislators who judged the regulations too much of an impediment to surgery. They argued that all welfare recipients as well as all "mental incompetents" should be sterilized in order to relieve "the burden on taxpayers."[49]

In the midst of this controversy, another lawsuit was filed in California charging doctors with a pattern of coercion to sterilize Spanish-speaking women in Los Angeles. The women who initiated the lawsuit claimed that consent was solicited while they were in labor or anesthetized and that consent forms they were given were printed in English only.[50] Reports of extensive sterilization abuse in Puerto Rico and massive sterilization of Native Americans at facilities run by the Indian Health Service appeared at about the same time.[51]

Sterilization lawsuits began to crop up around the country. A case was filed in South Carolina in which all three obstetricians in Aiken County denied care to a woman on welfare unless she agreed to be sterilized. She won her case against one of the doctors at trial, but the decision was overturned. According to the appellate court, doctors—even those paid with federal money—were not required to discard their "personal economic philosophy" while they practiced medicine. That conclusion left them free to make sterilization a condition for receipt of other medical services.[52] The South Carolina Medical Association agreed, finding the doctors' behavior ethically appropriate.[53]

Another lawsuit made the headlines, this time from Indiana, where a woman had petitioned a local judge for authority to sterilize her fifteen-year-old daughter, whom she characterized as "somewhat retarded." The girl was going to school and had been promoted with her class each year, but her mother said that she was in the habit of associating with older boys and had stayed out all night more than once. The mother judged it best for the girl to be sterilized, in order "to prevent unfortunate circumstances." No Indiana law sanctioned this kind of operation, but without holding a hearing or even speaking to the daughter, a judge approved the surgery. Six days later, the girl was taken to a local hospital and sterilized under the pretense that she needed to have her appendix removed. Two years later the girl married and eventually learned that she would not have children.[54]

The subsequent lawsuit against the judge went all the way to the Supreme Court, which decided that judges were immune from civil suit, even when they acted without proper jurisdiction. Newspapers all over the country announced the controversial result, which coincided with the long-awaited adoption of federal regulations requiring doctors to explain the consequences of sterilizing operations to patients and a thirty-day waiting period between the time a person consented to sterilization and the date surgery actually occurred. The regulations also prohibited federal payments for operations on people under twenty-one years of age as well as anyone in a jail, prison, or mental institution.[55]

Though the regulations were first triggered by the extensive reports concerning sterilization abuse by private physicians in general hospitals or similar settings, this last provision effectively stopped any systematic program of sterilization under remaining eugenic statutes. The flood of negative publicity surrounding legal actions provided a catalyst for many states to review and rewrite sterilization laws.

During the 1970s, the civil rights revolution began to address the mistreatment of the disabled in state institutions, providing even more impetus to remove antique eugenic statutes from state codes. A federal court declared Alabama's rarely invoked law unconstitutional.[56] Between 1965 and 1979, at least fifteen other states repealed their eugenic sterilization laws. Yet despite changing laws and regulations, sterilization of the underage and incapacitated continued to occur.[57]

Rediscovering Buck

Virginia periodically amended its laws to satisfy developing definitions in the field of mental health, and small changes were also made to the sterilization law. The listing of the "insane, idiotic, imbecile, feeble-minded or epileptic" was shortened in 1950 to include the "mentally ill, mentally deficient or epileptic" among those subject to surgery. In 1968 the General Assembly removed epileptics from the list. Yet for fifty years, through two major recodifications of Virginia law, the eugenic criteria to prevent reproduction among the likely parents of "socially inadequate offspring" remained in place. In the wake of national publicity about sterilization abuse, the 1924 Virginia sterilization law was finally repealed in 1974.[1] Other references to sterilizing those with "hereditary forms of mental illness that are recurrent" were removed from Virginia law in 1979. The last documented sterilizations at the Virginia Colony occurred that year.[2]

Then, after more than fifty years of obscurity and in the midst of legal turmoil over sterilization in other states, America rediscovered the *Buck* case. In 1979, after several years of searching, K. Ray Nelson found Carrie Buck. Nelson was the chief administrator of the Lynchburg Training School, formerly known as the Virginia Colony, and he located Carrie and her sister Doris by searching Social Security records that crossed his desk. After he brought the sisters together, Nelson mentioned the *Buck* case to a Winchester, Virginia, reporter who was touring the Lynchburg facility. The series of articles that resulted triggered reporting all over the country that would bring Carrie Buck's story back into the American memory.[3] Accounts of her case appeared in the *Washington Post* and the *New York Times*, and she was interviewed on National Public Radio.[4]

Dr. John Bell had boasted about the benefits of sterilization to Carrie, saying that after surgery "she was immediately returned to society and made good."[5] But in 1980 she was living in a cinderblock shed in a muddy field just north of Charlottesville, Virginia. Her life had been defined by constant and sometimes abject poverty. As for her "moral delinquency," she recalled repeatedly that Clarence Garland, a nephew of her foster parents, had "forced himself upon me."

Her recollections of surgery at the Colony were limited, and she had no clear memory of being told that she would be sterilized. "They just told me I had to have an operation," she said.[6] Doris Buck's story was similarly bleak. After years of marriage, she was in her sixties before she realized that she has been sterilized during her supposed "appendectomy" at the Colony. "I wanted babies bad," she said.[7]

Reporters found other former patients and collected stories of their time at the Colony. One patient remembered the surgery "assembly line" at the Colony where patients gave up their fertility.[8] It became clear that the procedural protections of the 1924 law, while providing a pretense for legality, were rarely followed in any meaningful way. Many patients were sterilized as teenagers and remained uncertain about the nature of the surgery they had endured well into adulthood.

Following initial publicity about the Bucks, the American Civil Liberties Union urged Virginia officials to identify and notify people who had been sterilized.[9] The state resisted, fearing the appearance of more victims to make legal claims and more bad publicity. After months of negotiations, in late December of 1980 the ACLU filed a lawsuit against the state of Virginia on behalf of four unnamed patients and other victims of sterilization in Virginia. The suit was designed to overturn the precedent of *Buck v. Bell*. The case was named *Poe v. Lynchburg Training School and Hospital*.

Poe v. Lynchburg

Plaintiffs did not ask for money damages as part of the ACLU lawsuit but asked instead that the court declare their surgeries unconstitutional. The suit also demanded that the state notify all those who had been sterilized and provide them with free medical and mental health services to mitigate the damage of their involuntary sterilizations. Their lawyers charged that scores of sterilized people were never informed of the nature of the operation they had and that many were victims of the same vague and often inaccurate "diagnoses" that echoed the charges made against Carrie Buck. Like Carrie, they were provided with either inadequate legal counsel or none at all. By not informing the patients clearly of their intentions, state officials violated the procedural conditions contained in the sterilization law. The suit went on to charge that years later, many former Colony patients were unaware of the "facts, circumstances, effects and possible reversibility of their sterilizations." It demanded that the state locate all who had been sterilized and inform them of their condition.[10]

In the months that followed the announcement of litigation, the ACLU identified approximately thirty people who had been sterilized and collected information about their experiences at the Colony in interviews and depositions. Most of the plaintiffs were sterilized as teenagers. Many were told that the operation would "wear off" in five to seven years, and they could have children then. Often, multiple members of the same family were sterilized. Inappropriate sexual activity was still the most likely "diagnosis" for women, and staff members from the Colony testified that surgery was often used as punishment when sexual misbehavior was reported.[11]

While all residents were told that they must be sterilized before being released, a former employee verified that when surgery was scheduled, it was regular practice to tell girls they were having an appendectomy. This report was consistent with one physician's insistence in 1944 that the "matter of sterilization [should] be whipped up" so that all patients could be sterilized before they were paroled.[12] Documentation of legally required hearings was often missing, and some consent forms were forged. Some patients reported being raped at the Colony, and several described sexual abuse by staff, including doctors and nurses. One patient reported an operation on an eight-year-old girl. Despite all of these reports, patients who attempted to get their medical records for review were discouraged by staff members and told they could not sue the state.

The ACLU charges were difficult to dispute. Doctors knew that patients would be much more willing to endure an operation that was supposedly in their own medical interest, and as early as 1919 Dr. Joseph DeJarnette recommended that patients be sterilized without their knowledge.[13] Nor did claims of irregularities in the sterilization program seem exaggerated. In 1938 a lawyer from the Virginia Attorney General's Office warned that sterilization practices "ought to be corrected in order to prevent embarrassing, or even more serious consequences" such as a lawsuit challenging an already completed surgery.[14] In 1943, the Virginia State Hospital Board itself reported the need for "official action to bring the sterilization procedure at the Lynchburg Colony in line with the statute." Even then, patients often did not meet their legal guardians or discuss pending surgery until just before the official hearing, and representation for those facing sterilization was frequently absent.[15]

Though she was not listed under her real name in the lawsuit, the most prominent victim of sterilization was also the plaintiff who had been sterilized first, when she was thirteen years old. Doris Buck joined the suit to redress the fifty-year-old cloud over her family name. Other plaintiffs, while not famous, had lived extraordinary lives.

Raymond Hudlow was sterilized in 1942, when he was seventeen years old, only three months after the *Skinner* case was decided. He was sent to the Colony as punishment for running away from an abusive home. In interviews during the lawsuit, he recounted how he was taken to the operating room and strapped to a table to watch while the surgery was carried out without anesthesia. After he left the Colony, Hudlow was drafted into the Army. He served with distinction in World War II, landing on France's Omaha Beach and fighting in Belgium and Holland. Through several tours with the Army and the Air Force, he reached the rank of sergeant. At one point he supervised more than forty-five people. Hudlow gave more than twenty years of distinguished service to the military and earned both the Bronze Star and the Silver Star for valor as well as a Prisoner of War Medal and a Purple Heart. Stories like Hudlow's, however, had no impact on the court.

At a preliminary court hearing of the ACLU lawsuit, Virginia's lawyers attempted to have the suit dismissed. Reviewing their arguments, a federal court judge said that despite "current mores and social thought," the constitutionality of involuntary sterilization had been settled by the U.S. Supreme Court by the decision in *Buck v. Bell*. Additionally, most plaintiffs had been sterilized many years earlier, and any claims they might have against the state were blocked by the statute of limitations. The judge did allow the suit to continue to determine whether the state, by not informing former Colony inmates of their surgery, continued to cause them harm.[16] In its preparation for *Poe v. Lynchburg Training School and Hospital,* the state of Virginia estimated that 8,300 people had been subjected to eugenic surgery.[17] But absent centralized records, the number of sterilizations performed in Virginia—as in other states—evaded accurate calculation.

The litigation went on, but the sterilization law had been repealed, and the federal court's decision made it clear that no retrospective relief would be available for those sterilized at the Colony or other Virginia asylums. Attorneys searched for other former patients and researched possible grounds to expand the suit. But after four more years of litigation, the suit was settled in 1985. The plaintiffs initially sought individual notice, free medical and psychological counseling, and, where feasible, surgical reversal operations at state expense for each person sterilized in a state facility. The settlement the state of Virginia agreed to was much more modest.

It included a two-month media campaign, with radio and television announcements to inform the public that the sterilization program had been discontinued. A toll-free "sterilization hotline" would be maintained to handle inquires from former patients concerning their medical histories. Notices were posted in local mental health offices offering counseling—for a fee—to anyone who had been sterilized. The dismissal of the ACLU lawsuit provided an official legal conclusion to what had been called Virginia's "Sterilization Era"—but it left the precedent of *Buck v. Bell* intact.

Despite the disappointing ending of the *Poe* case, its impact on public understanding of sterilization history was inescapable. National media coverage gave rise to additional scrutiny by scholars. Stephen Jay Gould, Harvard paleontologist and celebrated science writer, completed his book *The Mismeasure of Man* while the case was still in court. The book, which provided a detailed history of the intelligence testing movement, described the part played by the eugenicists in popularizing IQ tests as a supposedly foolproof tool to identify the "feebleminded." Working in collaboration with historians like University of Maryland scholar Steven Selden, Gould challenged the mythology of the Kallikak family and other eugenic legends. Gould ended his book with a brief afterword, quoting from the Holmes opinion in *Buck* as an example of how scientific reductionism and biological determinism could twist the law into an instrument of tyranny. The book was widely read, once again focusing attention on the plight of Carrie Buck.

Carrie Buck Dies

In 1982 Doris Buck died. Within six months, her sister Carrie was also at death's door. I met Carrie Buck while she was living in the "District Home," a rural facility near Waynesboro, Virginia, that had been constructed in 1927 to replace county almshouses. It eventually was transformed into a retirement community for indigent seniors.[18] Because she was not well, our conversation was brief. By then she was clearly aware of her role in the infamous Supreme Court case and still somewhat embarrassed about it. She didn't want to go into detail about what had happened to her, but she reaffirmed a statement made earlier that she was pregnant in 1924 because she had been assaulted. She showed no anger, but she did convey her feeling that she had been treated unfairly.

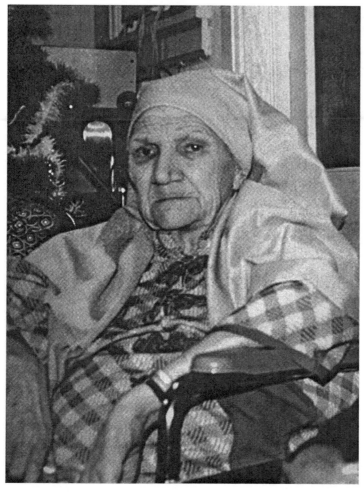

Last photo of Carrie Buck, December 1982. Courtesy District Home, Waynesboro, Virginia.

I talked in more detail with Carrie's friend, Elizabeth Pumphrey, a retired nurse. The two often spent mornings solving crossword puzzles together. She told me of Carrie Buck's efforts to tend to her infirm husband Charlie Detamore, whom she had married after her first husband died. Detamore also lived at the home.

A few weeks after our conversation, the *Charlottesville Daily Progress* ran a one-paragraph obituary for "Carrie Detamore." The notice said she was eighty years old (she was actually seventy-seven) and contained no information to identify her as the subject of the famous Supreme Court case. Twelve people witnessed her burial in Charlottesville on a cold, breezy day, January 31, 1983. Three people came from Front Royal, Virginia; Carrie had worked in their home. Mrs. Pumphrey and Carrie's husband, Charlie, came from the District Home. Accompanying the mourners were a minister, three employees of the funeral home, and a nurse and a social worker who attended to Carrie's husband and Mrs. Pumphrey. I also looked on.

Carrie's grave was placed near a maple tree beside the cemetery drive. The plot that later also accommodated her husband, Charlie, was marked that day only with a small tin nameplate. An inquiry with the cemetery staff identified the Dobbs family plot where Carrie's one-time foster parents John and Alice Dobbs rested. It was a short walk across the graveyard. Nearby on the same plot was a tiny marble pillar labeled simply V.A.E.D. It marked the grave where Vivian Alice Elaine Dobbs, Carrie's baby, was buried in 1932.

Carrie Buck's notoriety increased after her death. News of the *Poe* lawsuit settlement prompted additional commentary, but few publications had as much impact as Stephen Jay Gould's 1984 essay "Carrie Buck's Daughter." That article, appearing first in *Natural History Magazine* and reprinted a year later in Gould's anthology *The Flamingo's Smile*, is probably the most widely read description of the *Buck* case.[19]

In the decade after Gould's article, numerous studies of eugenics appeared. The story of sterilization in the United States received detailed attention from scholars, and the Holmes opinion in *Buck* always earned at least a quotation. But in 1993 Carrie Buck's story found a much larger audience when author and filmmaker Steven Trombley transformed parts of his earlier scholarship into a documentary film called *The Lynchburg Story*. This chronicle of Dr. Priddy's Virginia Colony and the *Buck* case was broadcast repeatedly for two years in the United States and Canada on the Discovery

Channel and in the United Kingdom on the BBC. It was also shown in more than a dozen other countries in Europe, Asia, and South America.[20] It introduced millions to the history of eugenics, clarifying America's pioneering role in enacting sterilization laws and exposing further the fraud of the *Buck* case. *The Lynchburg Story* also brought the Virginia sterilization story up to date with interviews of Mary Donald and Jesse Meadows, two former Colony inmates whose lives reflected the scars of sterilization. Broadcasts of *The Lynchburg Story* coincided with increasing media attention to the Human Genome Project.

As the Human Genome Project rekindled interest in the potential for using genetic science as a vehicle toward medical progress, it too heightened public awareness of earlier attempts to manipulate heredity during the eugenics era. The *Buck* saga came to stand for the potential for abuse of genetic technologies. In the 1990s, during the wave of memorials marking the fiftieth anniversary of the end of World War II, the liberation of the concentration camps and the prosecution of Nazi war criminals also added to the interest in the role of eugenics within Hitler's regime. By the turn of the millennium, these factors all provided a motive for governments to look back at the history of eugenic sterilization.

In the United States, attention to *Buck v. Bell* mushroomed in 2001 when eugenics met genomics in the popular press. For more than ten years, the media had lavished extraordinary attention on the scientific conquest known to the world as the Human Genome Project. In 2001, that project reached a milepost. In prearranged simultaneous publications, the prestigious journals *Science* and *Nature* presented special editions announcing that the sequencing of the human genome was complete. According to *Science*, sequencing of the genome provided a "powerful tool for unlocking our genetic heritage and finding our place among the participants in the adventure of life." The magazine focused on the efforts of J. Craig Venter and the private sector entrepreneurs at Celera Genomics, whose work provided a competitive tension for researchers from government-funded laboratories. *Science* reminded its readers that the public announcement of this achievement coincided with the anniversary week of the birth of Charles Darwin, setting genetics in historical context of evolutionary theory and emphasizing how the sequencing effort had "built on the scientific insights of centuries of investigators."[21]

Nature chose to focus on the publicly funded collaborative led by Francis

Collins of the National Human Genome Research Institute of the National Institutes of Health. Like its counterpart, *Nature* also recalled the history of genetics. It described the "scientific quest" that began with the "rediscovery of Mendel's laws of heredity" early in the twentieth century, launching the race "to understand the nature and content of genetic information that has propelled biology for the last hundred years."[22] *Science* and *Nature* led the coverage of the genome-sequencing story, and February saw an avalanche of headlines on the genome in the international press.

In their triumphant genome editions, neither *Nature* nor *Science* mentioned the dark term "eugenics." The same week of the media's genomania, the Virginia General Assembly passed a resolution that evoked memories of historical events also linked to genetic science but attracted significantly less attention.

Eugenic Apologies

Mitch Van Yahres, who represented the city of Charlottesville in the Virginia House of Delegates, introduced a resolution "Expressing Regret for Virginia's Experience with Eugenics."[23] Van Yahres argued that an examination of the past was critical at a time when we "face a future marked by great advances in understanding of genetics," and he emphasized that education was needed to avoid similar scientific disasters in the future. Press commentary accompanying the Van Yahres statement reminded readers that his warning "seemed especially topical amid news about the first analyses of the human genome" that appeared in scientific journals the same week.[24]

The Van Yahres proposal followed an extraordinary series of front-page articles in the *Richmond Times Dispatch* by journalist Peter Hardin. Hardin analyzed Virginia's eugenic experience, chronicling the careers of Albert Priddy and Joseph DeJarnette and explaining their roles in passing the state's eugenical sterilization law and in the *Buck* case. Hardin also explored how Dr. Walter Plecker engineered the "racial integrity" legislation that prohibited interracial marriage and was later used to erase whole tribes of the state's Native American population from demographic records.[25]

Following a Hardin article that highlighted DeJarnette's work, Van Yahres suggested to colleagues from the town of Staunton—home of DeJarnette's Western State Hospital—that it was wrong for a new building there to carry DeJarnette's name. At that point there was little interest in confronting the DeJarnette legacy, but before long, Virginia Delegate Ken Plum approached Van Yahres with a related idea. Plum had already drafted a resolution calling for a state apology, but it seemed more appropriate coming from Van Yahres, who represented Carrie Buck's hometown of Charlottesville. Van Yahres agreed to champion the DeJarnette resolution as well as the eugenic apology measure.

Plans to confront eugenic history in the legislature developed in private, but the public drama revealed through Hardin's stories increased when another reporter located sterilization victim and war hero Raymond Hudlow. Hudlow, one of the anonymous litigants in the case of *Poe v. Lynchburg* twenty years earlier, put a living face on the documentary history.[26]

Press attention in Carrie Buck's hometown echoed the debate on eugenics in the halls of the Virginia legislature.[27] Several stories detailed the controversy that arose when the original Van Yahres bill calling for an apology was introduced. Some citizens wanted the resolution to include more detail about how eugenic legislation was used to persecute Native Americans. From a contrasting perspective, a representative of the National Organization for European American Rights rejected any negative references to eugenics, particularly any condemnation of the "racial integrity" laws that had prohibited interracial marriage.[28]

Legislators also voiced opposition. Repeating a common objection, one lawmaker rejected the critique of past eugenic policies, since sterilization was "at the time, legal."[29] Others saw no benefit in revisiting past injustices and objected to "stirring up some history that none of us are proud of."[30] According to the *Washington Post*, Virginia leaders preferred to celebrate the state's role as the birthplace of presidents and rarely found time to recall the state's "prominent role in such historic evils as slavery, segregation and forced sterilizations." That Virginia was addressing its eugenic history at all was a subject worthy of comment to the *Post*, which saw the legislative resolution as "a remarkable moment."[31]

Predictably, the compromise emerging from the legislative debate did not satisfy everyone. The General Assembly eventually deleted the word "apology" in favor of a diluted declaration of "profound regret." When the Virginia Senate finally adopted the resolution on February 14, 2001, it was criticized as an inadequate response to living victims of eugenic laws. Highlighting the links that legislators made between old and new renditions of genetic science, *The Independent* newspaper in Britain quoted a Virginia lawmaker who described eugenics as "genetic engineering at its very worst." That paper described the legislative resolution as Virginia's attempt at "saying sorry, sort of."[32]

The connections between the historical misuse of science and the later rush toward new technologies were drawn by news analysts and public commentators. Local stories of Virginia's eugenic history shared front-page space with the Francis Collins and J. Craig Venter news conference on the sequencing of the human genome.[33] Editorial writers, echoing *Heart of Darkness* author Joseph Conrad, spoke of "the horror" of eugenics, characterizing it as a "past manipulation of the human gene pool." They cheered potentially "wondrous and welcome" developments such as gene therapy while warning against the "far more troubling" prospects of amniocentesis and genetic screening "to prevent the birth of children with serious physical defects—eugenics by pre-emption."[34]

Virginia Governor James Gilmore refused to go further than the General Assembly. There was no need for an apology, he said. Regret was enough. But the implications of a forgotten scandal with newly found relevance were not lost on other politicians. Looking forward to the November 2001 elections, three gubernatorial candidates pledged to issue a formal apology for Virginia's eugenic past. Lieutenant Governor John Hager emphasized the potential for both the positive and negative impact of science: "While the advocates of eugenics felt they were on the cutting edge of science, it was a terrible example of how science can be misused."[35]

In the months following Virginia's resolution "of profound regret," eugenics took center stage in public debates. A state agency decided to follow Van Yahres's lead and rename a building at a state mental hospital. Removing the name of self-proclaimed "Sterilization" DeJarnette from a building led to protests that the state was "sterilizing history."[36]

A newspaper in Virginia's neighboring state of North Carolina considered the need for apologizing for its eugenic history; another in Maryland termed the Virginia saga a "lesson in ethics for our brave new world."[37] The legal press also weighed in, placing Virginia's *Buck* decision alongside *Dred Scott v. Stanford*, *Plessy v. Ferguson*, and *Korematsu v. U.S.* in a "dubious pantheon" as one of the Supreme Court's "biggest blunders." According to *Legal Times*, the movement for a Virginia apology raised "uncomfortable reminders of the Supreme Court's role."[38]

As 2001 drew to an end, popular attention to the history of eugenics continued. One Florida newspaper pursued the eugenics story, recounting the debate in Virginia and finding others who had been sterilized.[39] Disability rights groups pressed the new governor, Mark Warner, for the apology he promised during his gubernatorial campaign, and the seventy-fifth anniversary of the *Buck v. Bell* decision gave rise to more legislative activity. The 2002 Virginia General Assembly adopted a resolution honoring the name of Raymond Hudlow, sterilization survivor and decorated veteran. A second resolution, calling the *Buck* decision the "embodiment of bigotry against the disabled," was drafted to honor "the memory of Carrie Buck on the occasion of the 75th anniversary of the *Buck v. Bell* Supreme Court decision." As the Virginia legislature debated memorial resolutions, other concerns about eugenics filled the legislative chambers. Some lawmakers highlighted "eugenic formulations" already used to screen stem cells.[40]

The media also monitored the impending date of May 2, 2002, which provided an occasion for the dedication of a state historical marker recalling the Holmes opinion in *Buck* exactly 75 years earlier.[41] As the anniversary date approached, the media again recalled the *Buck* case as a reference point for reflecting on uses of the new genetic technologies.[42] Journalists in other states focused on homegrown stalwarts of the eugenics movement, such as Harry Laughlin of Missouri, author of the Model Eugenical Sterilization Law, as they explored explicit parallels between the old eugenics and the new genetics.[43] The day before the *Buck* memorial event, people gathered in Lynchburg, Virginia, not far from the site of the institution formally known as the Virginia Colony for Epileptics and Feeble-minded, to present Raymond Hudlow with a copy of the legislative resolution passed in his honor.[44]

In Carrie Buck's hometown of Charlottesville, a short drive from the cemetery where she was buried, the Virginia Department of Historic Resources erected a marker just around the corner from the school Buck's daughter, Vivian, attended. The text of the Virginia Historic marker commemorating *Buck v. Bell* carried this inscription:

Buck v. Bell

In 1924, Virginia, like a majority of states then, enacted eugenic sterilization laws. Virginia's law allowed state institutions to operate on individuals to prevent the conception of what were believed to be "genetically inferior" children.

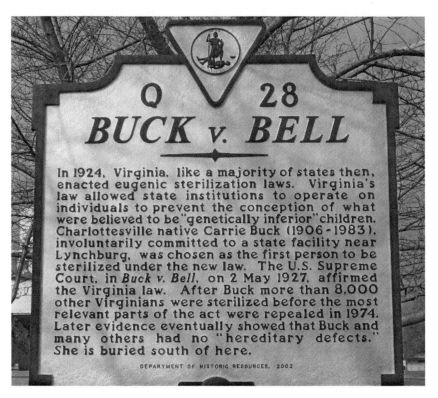

Virginia historical marker commemorating *Buck v. Bell*, erected May 2, 2002.

Charlottesville native Carrie Buck (1906–1983), involuntarily committed to a state facility near Lynchburg, was chosen as the first person to be sterilized under the new law. The U.S. Supreme Court, in *Buck v. Bell,* on 2 May 1927, affirmed the Virginia law. After Buck more than 8,000 other Virginians were sterilized before the most relevant parts of the act were repealed in 1974. Later evidence eventually showed that Buck and many others had no "hereditary defects." She is buried south of here.[45]

I invited Governor Mark Warner to fulfill his campaign promise on the *Buck* anniversary and make Virginia the first state to apologize officially for surgically sterilizing citizens using laws validated by the *Buck* decision. His statement of apology was read at the dedication ceremony.[46]

The history of eugenics and its contemporary genetic links reverberated through the articles commenting on the *Buck* marker.[47] The *Charlottesville*

Daily Progress reported Delegate Mitch Van Yahres's efforts to expose the state's schoolchildren to the history of eugenics.[48] This coverage was particularly poignant, since in 1927 the same paper had applauded the *Buck* decision and praised the Holmes opinion as "a genuine classic" while judging the sterilization law "eminently sane and beneficial."[49] A similar turnaround was evident in the *Richmond Times-Dispatch,* which had provided vigorous support for eugenics legislation in the 1920s. In addition to the prominent placement of articles on the history of eugenics, that newspaper ran an editorial entitled simply "Eugenics." It condemned "great crimes . . . committed in the name of progress" and "dubious theories" that provided justification for "state sanctioned butchery" as a part of recent history.[50]

Just how recent was brought home by the presence of two people at the *Buck* marker ceremony. As the guests of honor at the event, Mr. Jesse Meadows and Mrs. Rose Brooks helped to unveil the *Buck v. Bell* marker. Both had endured sterilization at the Virginia Colony. Their photos and comments to reporters were distributed worldwide via news service reports and feature articles in papers such as the *Washington Post* and the *Los Angeles Times* as well as news reports on National Public Radio and the British Broadcasting Corporation.[51]

Virginia's apology quickly gave rise to similar events in Oregon, where Governor John Kitzhaber, also a physician, said it was time "to apologize for misdeeds that resulted from widespread misconceptions, ignorance and bigotry." Approximately 2,600 people in Oregon had been sterilized, and Kitzhaber established an annual state Human Rights Day to memorialize that history.[52]

Governor Jim Hodges of South Carolina offered an apology for the 1935 eugenics law of his state. Hodges noted that some 250 operations occurred in that state between 1935 and 1985.

Winston-Salem Journal reporter Kevin Begos launched a series called "Against Their Will: North Carolina's Sterilization Program" that led to two different apologies.[53] One came from the dean at Wake Forest University's Medical School, where faculty members in its pioneering Department of Medical Genetics had participated in the state eugenics program.[54] North Carolina Governor Mike Easley made another apology, noting that the state had sterilized some 7,600 people between 1929 and 1974.[55] The North Carolina legislature then erased the last vestiges of eugenics from the state legal code by repealing the state sterilization law.[56]

As much as a third of all surgery done under eugenics laws in the United States occurred in California, whose 1909 sterilization act was among the first in the nation. California's Senate Select Committee on Genetics, Genetic Technology, and Public Policy took the fiftieth anniversary of the discovery of the DNA molecule as an opportunity to examine scientific history. A part of that history had broken into public consciousness in California as a result of the apologies made in eastern states. As the Virginia eugenics apology took place, Pulitzer Prize–winning science journalist Deborah Blum published an essay in the *Los Angeles Times* proposing a California version of the sterilization memorial and calling for "a trend in publicly rejecting *Buck v. Bell*."[57] *San Francisco Chronicle* columnist Tom Abate previewed the presentation on eugenics I was scheduled to make the next day to the Senate Select Committee. Emphasizing "California's role in the Nazis' goal of 'purification,'" as part of the history lesson his column provided on California eugenics, Abate printed the same pedigree chart Harry Laughlin had used to illustrate his 1935 lecture—delivered in absentia—to a Nazi population conference in Berlin. The slide showed "The Near Blood Kin of a Feeble-minded Woman Sterilized by the State of California," and it mimicked the pedigree Laughlin had developed to describe Carrie Buck's family.[58]

Following my presentation to the Senate Committee Hearing, both Governor Gray Davis and Attorney General Bill Lockyer made apologies to the victims of California's eugenic policies.[59] The California General Assembly later passed a resolution repudiating the state's eugenic laws.[60]

The seventieth anniversary of the last eugenic sterilization law passed in the United States, adopted in 1937 in Georgia, was in 2007. Thirty-three hundred operations occurred in the state, ranking Georgia fifth in the United States in eugenic surgery totals. I drafted a resolution decrying Georgia's eugenic history, which state Representative Mary Margaret Oliver then introduced in the Georgia House of Representatives. The *Atlanta Journal-Constitution* then printed a lengthy feature article including interviews with a doctor who had performed eugenic surgeries at one of Georgia's largest institutions and comments from families whose relatives had been sterilized.[61] A few days later, former patients appeared, reciting their own stories of institutional abuse. The editorial page of that paper advocated an apology for eugenic laws.[62]

Although Oliver, a Democrat, was unsuccessful in getting her proposal heard in the Republican-dominated House, Republican State Senator David Shafer drafted a resolution of "profound regret" for eugenics which was eventually adopted by the Georgia Senate.[63] Throughout the legislative session there was a strong reaction to the stories about sterilization in Georgia, and at least some of that reaction demonstrated one point with great clarity: many people still believe that eugenic laws and the bigotry they represented were a good idea. Some even argued that we should bring eugenic sterilization back.[64]

The year 2007 also marked the centennial of the 1907 Indiana sterilization law, the first of its kind in the world. Capping three days of events commemorating the state's eugenic history, Indiana's health commissioner, Dr. Judy Monroe, expressed regret for the state's role as the birthplace of eugenic sterilization law.[65] A state historic marker describing the now-defunct eugenics laws was unveiled by Linda Sparkman, plaintiff in the 1978 Supreme Court case challenging her own extra-legal sterilization.[66] Following memorial events, the Indiana legislature adopted a resolution repudiating eugenics, making Indiana the first state where representatives of the governor's office, the legislature, and the state supreme court were all involved in marking the history of state eugenics laws.

In several other countries, governments have decided or been forced by litigation to make payments to people who have been sterilized.

Reports emerging in the early 1990s from Sweden described how more than sixty thousand of its citizens had been sterilized under a law that borrowed both method and motive from U.S. and German sterilization precedents. A Swedish governmental commission authorized compensation of approximately $21,000 to each living victim.[67] The German compensation scheme for victims of Nazi sterilization provided five thousand German marks for each applicant.[68]

In Canada, the story of thirty-five-year-old Leilani Muir represented the most dramatic example of a government's response to its own sterilization history. In the early 1990s, Ms. Muir discovered that she was unable to have children. She had lived as an adolescent at the Red Deer Provincial Training School in Alberta. Investigation revealed that part of her "medical" experience there was an operation about which she had never been informed and whose nature she did not understand. Her experience was, unfortunately, not unique.

Alberta passed a Sexual Sterilization Act in 1928, following the *Buck* case. The law prescribed sterilization for inmates preparing for discharge from provincial mental hospitals but required consent of the patient or a family member before surgery could be performed.[69] But hospital officials and members of the provincial sterilization board complained that the consent requirement slowed the pace of sterilizations and opened the door for opposition to the procedure by patients and their families. Sterilization proponents successfully lobbied for a change to the law in 1937. In the case of surgery on the "mentally defective," the consent requirement was simply removed.[70] Under that version of the Alberta sterilization law, Leilani Muir and thousands of others had operations that were never explained to them.

Shocked at the news that she had been sterilized, Ms. Muir filed suit in provincial court. When she eventually prevailed several years later, a jury awarded her approximately $1 million (in Canadian currency) as compensation for her loss. The *Muir* case became the first of some seven hundred claims made against the Alberta government for sterilization of youths in provincial institutions between 1928 and 1972. The Alberta litigation produced an extraordinary result. Representatives for some five hundred victims of sterilization who still live within protective institutions agreed to accept approximately $150,000 each for the operations they had endured. More than two hundred other plaintiffs maintained a suit that concluded in a 1999 settlement paying them over $80 million. From the time of Leilani Muir's case to the final settlement of the Alberta litigation, more than 120 million Canadian dollars was distributed to victims. Before the ultimate settlement, the Alberta government attempted to set aside human rights protections available under the Canadian Charter of Rights and Freedom in order to avoid legal liability to sterilization victims. In the end, the government of Alberta apologized, but the story of official obfuscation, denial, and deceit in a campaign to avoid the blame for Alberta's sterilization history rivaled Virginia's efforts in the *Poe* case.

British Columbia, the only other Canadian province that passed a sterilization law (1933), faced a similar lawsuit in 2001. Claims in that case were eventually settled for approximately $450,000.[71] Despite extensive recent media coverage of eugenics history and seven different state apologies or proclamations, no U.S. state has ever paid reparations to the victims of sterilization laws.

Reconsidering Buck

How would Carrie Buck fare under the law if she actually were a person with diminished mental capacity related to an inherited disease, living in Virginia today? Virginia adopted a new law on involuntary sterilization in the early 1980s. Among the several states having such laws, Virginia is one of the most protective. It incorporates extensive due process steps before a court may order surgery for someone unable to give informed consent. The person who will be sterilized must have a need for contraception—that is, they must be sexually active now or likely to be in the near future. The court must determine that no reasonable alternative method of contraception is available. The proposed sterilization must not pose an unreasonable risk to the life or health of the patient, and the patient must be permanently incapable of raising a child. Specific provisions are included to determine, as completely as possible, the wishes of the patient.[1] Laws in other states have similar provisions that focus the court inquiry on "the best interests" of the person subject to surgery rather than the "convenience" of a guardian, family member, or the public more generally.[2] If we apply all of these standards, which are part of Virginia law today, to the facts of the *Buck* case, Carrie Buck would probably not be sterilized.

On the other end of the spectrum, the least protective involuntary sterilization statute is that of Arkansas, which permits sterilization by court order of "incompetents" who are unable to care for themselves "by reason of mental retardation, mental illness, imbecility, idiocy or other mental incapacity." Males may be sterilized by x-ray or vasectomy. In cases of "obvious hardship," three physicians can certify, pursuant to the wishes of a guardian, that "a sterilizing procedure is justified" without going to court at all. No standards are provided to determine the factors that might justify the operation.[3] Buck's prospects in Arkansas would be less promising.

Courts in several other states allow "voluntary" sterilization at the request of parents or guardians of people with mental retardation, with widely varying standards for determining when surgery would be appropriate. If a family member wishes to have someone sterilized in those states, surgery can occur, as long as genuine evidence of mental retardation is available. Sterilization operations involving people with mental retardation happen regularly in some states; but there are also operations done in private at the request of families or guardians to consider. The extent to which this practice may reflect the same attitudes played out in the *Buck* case—that the disabled are worthy of contempt and that the social costs such people generate justify court orders for unwanted surgery—is troubling.

Private, family-directed, sexual surgery raises many separate ethical and policy issues that go beyond state-controlled surgery. Some would argue that elective sterilizing surgery for those who cannot consent for themselves should never occur, while others believe that depriving people with disabilities of the most effective means of birth control is a type of discrimination. Although there is ample evidence that abuse occurs routinely under the mantle of familial care, it is also true that some families request sterilization of a disabled relative with the most noble of intentions. The complex and serious question of whether surgery is ever the most appropriate means to bring about involuntary sterility as well as the related and equally momentous question of how to measure the burdens on families who act as caretakers for the mentally infirm are significant ones and should not be dismissed lightly. As important as those questions are, they do not proceed based on legal mandates exercised by state agencies or institutions. The *Buck* case raised a different, more fundamental question about the relationship of individual citizens and the government. When and for what purposes should we ever use the most intrusive medical interventions as tools of state policy? Today, the answer to that question depends on a legal evaluation of *Buck* and the place it holds not as an historical artifact but as U.S. Supreme Court precedent.

Buck and the other eugenics cases heard by the Supreme Court fill an important niche that helps define the contours of the constitutionally protected right to reproduce. *Buck* allowed sterilization of the "socially inadequate," seeming to recognize no limit to the government's power over reproduction. *Skinner* provided some restriction on that power, prohibiting states from sterilizing criminals arbitrarily and establishing the principle that marriage and reproduction are fundamental rights. Later cases on reproduction built on those decisions.

The 1965 decision in *Griswold v. Connecticut* introduced a right to privacy that protected the distribution of birth control information and devices.[4] In *Griswold*, the Court referred to the premise from *Skinner* that marriage and procreation are fundamental rights. *Loving v. Virginia* followed *Griswold* and, in striking down the prohibition of interracial marriage, echoed the "fundamental rights" language of *Skinner* as well as the use of heightened "strict scrutiny" to invalidate governmental encroachments on those rights.[5] In the most important of all the reproductive rights cases, *Roe v. Wade*, the right to an abortion was explained and qualified by all three of the eugenics decisions.[6]

In *Roe*, the Court said that past decisions have made "it clear that only personal rights that can be deemed 'fundamental' or 'implicit in the concept of ordered liberty' are included in this guarantee of personal privacy." Relying on *Griswold* and referring to both *Loving* and *Skinner*, the Court noted that the right to privacy had been extended to activities related to marriage and procreation, including the right for women to decide if they wished to terminate a pregnancy. But the Court refused to declare the right to privacy absolute, recalling from past decisions that there is not "an unlimited right to do with one's body as one pleases."[7] The only two cases the Court referenced for that proposition were *Jacobson* and *Buck*. *Jacobson* provides the precedent for state-ordered vaccination; *Buck* endorsed state powers to order sterilization. The *Roe* formula gives women control over the earliest stages of pregnancy but allows states to regulate abortions in later stages of fetal development. In the most important reproductive rights case in Supreme Court history, the only case involving a woman's potential to be a mother that could be cited in support to the governmental prerogative to limit reproductive decisions in the later stages of a pregnancy was *Buck*.

All three eugenics cases—*Skinner* and *Loving* in repudiating eugenic statutes and *Buck* in upholding one—have contributed principles that are critical to understanding the right to reproduce as a constitutional issue today. Language that first appeared in those cases provides grounding for the case law that now defines reproductive rights more generally.[8] *Buck* bolsters the power of government to limit reproductive choices and supplies a boundary to the personal autonomy that cases like *Griswold* and *Roe* established.

How sturdy is *Buck* as a precedent? To answer that question, we might consider a hypothetical case in a state where Carrie Buck would find a law almost exactly like the one she challenged in the 1920s. Mississippi is alone among all the states in having a 1928 eugenic sterilization statute that has survived intact, complete with phrases from the *Buck* era.[9] The law marks any person in a state mental institution as a potential subject of a sterilizing operation. The institutional board may order surgery if it concludes that an inmate is "insane, idiotic, imbecile or feebleminded, and by the laws of heredity is the probable potential parent of socially inadequate offspring likewise afflicted." There must be some evidence that "the welfare of the inmate and of society will be promoted" by surgery, exactly as the Virginia law tested in *Buck* prescribed.[10]

Suppose that an official in Mississippi decided to have someone in a state institution sterilized under existing law. The hospital director could petition the hospital's board of trustees for authority to proceed with surgery. If the board were to approve the petition, the case could be appealed through state courts and up to the U.S. Supreme Court.

How would the Supreme Court rule? It passed up the best opportunity for addressing *Buck* and striking down all sterilization laws when it distinguished between sterilizing criminals and sterilizing the "feebleminded" in the *Skinner* case. In the thirty years after *Skinner,* a time in which more than twenty thousand documented operations took place, no serious constitutional challenge to *Buck* reached the Court again.

Buck has been cited more than 150 times in judicial opinions since it was decided in 1927. As recently as 2001, a federal court in Missouri referred to *Buck* as an authority for a rule stating that "involuntary sterilization is not always unconstitutional if it is a narrowly tailored means to achieve a compelling government interest."[11] Although few judges applaud the Holmes opinion today, *Buck* has not been overturned as a matter of law. Of course, most states have repealed or severely amended their eugenic sterilization laws, and even the Mississippi law seems moribund. The consensus among legal scholars is that attitudes toward eugenics coupled with more recent endorsements of the right to reproduce leaves the case with almost no vitality as a precedent.

It is tempting to say that the revolution in reproductive rights culminating in *Roe* and the more general expansion of the ethic of autonomy in medical jurisprudence have changed the constitutional calculus. If so, it is possible that our hypothetical Mississippi case would provide an occasion for overturning *Buck.* If the principle of reproductive privacy extends to the use of birth control, as *Griswold* says, or protects the decision to terminate a pregnancy, as *Roe* says, one might confidently assert that the same principle insulates a person against state-mandated sterilization surgery. From that perspective, we would predict that the Court would simply take the belated but necessary steps of striking down the Mississippi statute as unconstitutional under the principles outlined in *Roe* and finally extending reproductive rights to some latter-day Carrie Buck. For those who find the prospect of coercive surgery troubling, that is a hopeful assumption.

It does not, however, reflect the direction the law has taken. Because even in the *Roe* case, where the right to reproductive privacy received its most expansive expression, the Court emphatically stated that such a right was not absolute. As we saw earlier, *Buck* was the precedent upon which that limitation was based. It is also likely that the current Court—with a much different group of justices than the *Roe* court—will act to minimize the principle of reproductive privacy, rather than strengthen it further. That would necessarily leave the core of *Buck* entirely intact, allowing states to order sterilization under their own laws. To some extent, that outcome depends on what happens to the key precedent of *Roe v. Wade.*

Roe was decided thirty-five years ago. While numerous cases since then have affirmed reproductive rights, the judicial tide on that question appears recently to have turned. Serious legal commentators now readily admit that *Roe* may soon be subject to reversal, and the current Supreme Court majority might no longer sustain its fundamental premise. If those commentators are correct and *Roe* falls, other precedent cases could fall as well.

Though it is impossible to predict the degree to which tomorrow's Supreme Court will reverse its past decisions that support reproductive autonomy, at least some current Justices would have no hesitation in overturning cases like *Skinner*. How farfetched is that prospect? Consider the responses of one Supreme Court nominee. During the 1987 confirmation hearings for Judge Robert Bork, a discussion of *Buck v. Bell* and *Skinner* occurred. Senator Alan Simpson of Wyoming asked Bork whether the equal protection clause in the Constitution protected people from "invidious discrimination"—a phrase taken directly from Justice Douglas's opinion in *Skinner*. Bork's answer suggested that "invidious discrimination" was only relevant to cases involving unequal treatment based on race.

Bork had criticized *Skinner* earlier, but during the nomination process he said that the Oklahoma statute it challenged might have been unconstitutional not for the reasons Justice Douglas cited but because it was based on "racial animus." Said Bork: "It struck at, in effect, crimes that at that time were more likely to be committed by poor blacks than by middle class white-collar whites." How he might have known that was anyone's guess, since nothing in the Oklahoma legislative record made any reference to race. Perhaps even more telling, all three men Oklahoma officials chose as subjects to test the law were white.

Bork went on to say that the Court had never called all sterilization unconstitutional. He recited the "three generations of imbeciles are enough" comment from *Buck* and said he thought the language was "terrible." The Holmes line caught Senator Simpson's attention:

Simpson: Justice Holmes said that?

Bork: Justice Holmes said that.

Simpson: I think we ought to get him back here.

People in the Senate gallery laughed. Perhaps Simpson was applauding the Holmes sentiment in *Buck;* perhaps he was just making a joke. How we interpret the comment makes no difference at this point, except as a reminder that in 1987 *Buck* was still an occasion for laughter. Neither Simpson nor Bork argued that it should be overturned.

During those same confirmation hearings, Bork was repeatedly asked if there was a "basic right not to be forcibly sterilized by the State."[12] Though he had been very critical of *Skinner* and the foundation it provided for later cases extending reproductive privacy, he deftly deflected the question a number of times. Bork finally admitted that he felt *Skinner* was wrongly decided. He said that at some future time a constitutional rationale for the *Skinner* result might be discovered, but he did not then nor has he yet offered such a rationale. In fact, later, with no Supreme Court nomination hanging in the balance, Bork described the *Skinner* case as the beginning of "judicial lawlessness." He declared that "the right to procreate is not guaranteed, explicitly or implicitly, by the Constitution."[13] Under Judge Bork's analysis, even the kind of sterilization struck down in *Skinner* would be permissible today.

Though Judge Bork was not confirmed and he is not on the Court today, other judges who agree with his reasoning certainly are. The most recent case involving reproductive rights, *Gonzales v. Carhart,* made that clear. While that decision allowed new restrictions on a woman's options in ter-

minating pregnancies, it also provided an occasion for two justices to reiter-
ate clearly their desire to overturn *Roe*. In *Carhart* Justice Clarence Thomas
reasserted his belief that *Roe* "has no basis in the Constitution." Judge An-
tonin Scalia joined him in that opinion.[14] Those Justices and others who
may join the Court in the future may seriously limit the reach of *Roe v.
Wade*, or even overturn it completely. But that would not change the outcome of our hypothetical Mississippi sterilization, which would continue
to be constitutional under *Buck*.

Would an anti-*Roe* majority on the Supreme Court also mean the complete demise of principles of reproductive privacy? The expectation of personal autonomy in medical decision-making would clearly diminish in the
absence of *Roe*, but it would take an even more radical revision in the jurisprudence of reproduction to erase the implications of *Skinner, Griswold*,
and related cases. Even though *Buck* would survive the overturning of *Roe*,
removing constitutional protections for abortion would not rule out continuing constitutional protection for birth control. More importantly, overturning *Roe* would not completely erase all rights to reproductive privacy,
nor would it abolish the fundamental distinction between government control of some decisions surrounding reproduction and individual control of
those same decisions.

Nevertheless, the choice between *Roe*, which affirms reproductive autonomy, and *Buck*, which affirms the state's power to control reproduction,
presents a quandary. Those who condemn *Buck* as strong evidence of the
worst tendencies of government to run roughshod over decisions normally
left to individuals are often the same people who are most opposed to abortion. If, in rejecting *Buck*, you assert that the government should not be
in the business of supervising reproduction, you run the risk of endorsing
principles crucial to decisions like *Skinner, Griswold*, and *Roe*. On the other
hand, those who emphasize reproductive liberty and want *Roe* to remain
intact should realize that in its present form, even *Roe* doesn't trump the
power of a state to issue a sterilization order of the kind affirmed in *Buck*.

But any prediction of judicial acquiescence in the face of a new regime of
sterilization is probably naive. It puts too much stock in logical consistency
and denies too quickly both the ability and the will of Supreme Court Justices to find a way around some state's future desire to return to the practice
of state-sponsored coercive surgery. It was Justice Holmes, after all, who
said that "the life of the law has not been logic, it has been experience."[15]

Perhaps we have learned from the U.S. experience of eugenics that reproduction is an area that governments are ill suited to police.

But if reproductive decisions are protected only by precedents like *Skinner, Griswold,* and *Roe,* the demise of that line of cases would destroy the most significant *legal* barrier that could protect us from the return of sterilization laws. As the Supreme Court said in *Planned Parenthood v. Casey,* "If indeed the woman's interest in deciding whether to bear and beget a child had not been recognized as in *Roe,* the State might as readily restrict a woman's right to choose to carry a pregnancy to term as to terminate it, to further asserted interests in population control, or eugenics, for example."[16] If government does not protect the liberties we exercise in reproduction, then restricting childbirth for "the good of society" and in the name of public health remains a possibility. If deciding for or against being a parent is a state rather than a personal decision, laws to force sterilization, or more, could change with "the prevailing political winds," and a return to the science-fiction nightmare of eugenics is, in principle, "as close as the next election."[17]

Our Mississippi hypothetical is imaginary. No state is actively sterilizing people for overtly "eugenic" purposes, and in the absence of new sterilization laws, *Buck* is unlikely ever to be challenged. But what if we set aside speculation about a future Supreme Court role in revisiting *Buck* and ask whether public opinion or political will prevent a return to eugenic sterilization. We might want to believe that the popularity of eugenic sterilization died out with Harry Laughlin and others like him, but public reactions to state apologies suggests that a number of people today would support a revival of the old eugenics laws. Recent stories of legislative apologies for eugenics in both Georgia and Indiana drew instant responses on newspaper websites. Those reactions demonstrated that some people still believe that eugenics laws and the bigotry against the poor, the disabled, and minorities were a good idea. It is worth remembering one of the important lessons of the *Buck* story: a small number of zealous advocates can have an impact on the law that defies both science and conventional wisdom.

There are still other unapologetic fans of sterilization and the *Buck* case. Most conspicuous among them is Richard Lynn, retired professor of psychology at the University of Ulster in the United Kingdom. His work is still subsidized by the Pioneer Fund, the eugenics foundation that named Harry Laughlin as its first president. Although the mythologies of families like the Jukes and the Kallikaks have long been discredited, Lynn insists on reaffirming their importance. He claims that an "underclass" is kept alive by the reproduction "down the generations in certain sociopathological families." Supporters of modern eugenics, says Lynn, should make "elimination" of that underclass a goal.[18] Lynn has sought to legitimize *Buck v. Bell* as a good example of successful eugenic politics, and he endorses the "IQ tests" performed by 1920s eugenic enthusiasts to prove Buck's "feeble-mindedness."[19]

Politicians also regularly signal their willingness to reintroduce a regime of governmental eugenics. In the 1990s, a bill designed "to assist in reducing the crime rate" was approved by a Colorado legislative committee. It would have required the Colorado Department of Corrections to institute "family planning incentives," such as reducing the sentences of inmates who agreed to sterilization.[20] After critical press scrutiny, the proposal died for lack of support.[21]

Another proposal appeared in Florida, where legislators offered to pay an incentive of up to four hundred dollars per year to women on public assistance if they would agree to surgical implantation of the birth control drug Norplant.[22]

In 2004, James Hart ran for a place in Tennessee's congressional delegation as the "eugenic candidate." He trumpeted his agenda—a "war on poverty genes" focusing on the costs of welfare and the need to protect the country from the "less favored [African] race"—as a reminder that the class-based, racist eugenics of previous generations is not yet dead. In his losing bid to represent Tennessee's 8th Congressional District, Hart received slightly under sixty thousand votes.

In the 2006 Virginia legislative session, a bill was introduced to offer "voluntary" castration to convicted sex offenders as a way to avoid an indefinite stay in state mental health facilities as "sexually violent predators." Virginia State Senator Emmett Hanger Jr., who introduced the measure—already law in three other states—said that surgery would interrupt "an unending cost to the taxpayer." As a Senate committee member, Hanger had voted against the Virginia eugenics "regret" resolution five years earlier.[23]

New Mexico State Senator Rod Adair introduced a bill in the state legislature in 2007 requiring mothers of any baby born with fetal alcohol syndrome to receive injections of Depo-Provera, a chemical contraceptive. If a second child with fetal alcohol syndrome is born to a mother already under a contraceptive order, the bill directs local courts to order surgical sterilization for both mother and father. Failure to comply with the court order would be a felony.[24]

Adair introduced other legislation in 2007 to criminalize parents of children born addicted to methamphetamine. Citing the "damage done to society at large" (and echoing Holmes), Adair reiterated his view that "two generations of drug-addicted babies are enough."[25]

Much like advocates of eugenics in the early twentieth century, the judiciary also regularly uses surgery as an incentive to motivate responsible behavior. Some judges offer men who fall seriously behind in their child support payments the option of going to jail or having a vasectomy.[26] Other judges, while not mandating surgery, have repeatedly threatened drug-addicted mothers that if they become pregnant, they will not be given custody of their existing children. An extensive catalogue of the costs of raising a foster child, including additional costs for raising children who are "handicapped" or have "special needs," has been clearly set out in judicial opinions.[27] The motive for such orders is explicit: saving tax money.

Drug abuse among women also led to a private campaign for sterilization. In 1997, Californian Barbara Harris founded a program called CRACK (Children Requiring a Caring Kommunity). The organization bought billboard space across the country, offering two hundred dollars to any cocaine-addicted woman who could provide medical proof of sterilization.[28] Little did Harris know that her plan mirrored a similar proposal made by journalist H. L. Mencken sixty years earlier. At the time of the *Buck* decision, Mencken had scoffed at eugenics as "mainly blather."[29] Ten years later, Mencken had changed his tune; he decided that allowing the unlimited increase in Southern sharecroppers would make the region "swarm like a nest of maggots." His solution was "Utopia by Sterilization," and he proposed, like Ms. Harris, a payment of one hundred dollars out of private funds to any man who would submit to a vasectomy. Mencken said that this would forestall the need for taxpayers to support a growing "herd of morons."[30]

Some sterilization advocates have gone even further than financial incentive programs. As his justification for "enforced sterilization," Virginia Judge Ralph B. Robertson cited the expense associated with premature birth and neonatal care for "crack babies." Noted Robertson, "I believe that it was Oliver Wendell Holmes who wrote 'Three generations of imbeciles are enough.'"[31]

In 2006, Charleston, South Carolina, City Councilman Larry Shirley echoed Harry Laughlin's prescription for fighting crime: "We pick up stray animals and spay them. These mothers [of delinquent children] need to be spayed if they can't take care of theirs. Once they have a child and it's running the street, to let them continue to have children is totally unacceptable."[32]

The regularity with which citizens, legislators, and judges call for and in some cases order involuntary sterilization suggests that the idea endorsed in the *Buck* case has never been fully rejected. If sterilization is ever revisited in America on a wider scale, it will probably rely on the same motives that recent proposals have invoked for castration of sex criminals or for incentives to sterilize drug addicts or welfare recipients. In the *Buck* era, alarmism about the potential for a genetic apocalypse was often only a pretext to expand state powers over reproduction. Experience has shown that, even when science failed to carry the day, arguments to lower taxes provided many legislatures an adequate policy basis for enacting sterilization laws.

Because *Buck v. Bell* was the most dramatic U.S. legal expression of eugenics, it has become a symbol for our national revulsion toward the "well-born science" itself. Many people cringe at the Holmes opinion because *Buck* violates fundamental principles of our legal and ethical traditions and represents a chilling statement of totalitarian sentiment. To those who see depriving anyone of the freedom to reproduce as a violation of human rights, it is an emblem of government at its most tyrannical. The patent injustice of a policy that targeted the poor and disabled represents another reason to reject *Buck*. And people who condemn any interference with the reproductive process as "unnatural" see coercive sterilization as even more profoundly immoral. These discomforting conclusions about *Buck* are intensified by the knowledge that the case was a sham. *Buck* is certainly among the most cruel Supreme Court opinions, and as we now know, it also stands among the most false. Carrie Buck was betrayed by her foster parents and her lawyer; her story was manipulated to protect the doctors who were charged with caring for her. Perhaps this makes the case all the more tragic; but perhaps it makes our moral judgments about it all too easy.

Because the Supreme Court got it wrong—to use Holmes' language, Carrie was no "imbecile," and no sound evidence of hereditary disease was demonstrated in her case—it is easy to generate scorn for the case and the movement it represented. But a moralistic backward judgment about eugenics is naively ahistorical. To impute only corrupt motives to supporters of the eugenic agenda because of our disgust at the worst of those who claimed the label is to miss the myriad ways other motives guided their efforts. It also obscures the many ways our current practices and motives find a parallel in them. It also may imply that had Holmes' commentary been accurate—if Carrie Buck actually were likely to pass on a genetically diagnosed disabling condition—we would endorse the Holmes conclusion and the type of law it affirmed as well.

We are startled today by Holmes' language in *Buck,* but our revulsion is driven to some extent by changes in attitudes toward mental illness and other cognitive disorders. While much that is done in the name of eugenics has rightly earned our contempt, we should not forget that the impulse for social improvement, the wish to eradicate suffering and the problems that plague society were motives that attracted many Americans to eugenics. Those motives fueled support for sterilization. Eugenics was a successful ideology because it was driven at least as much by hope as it was by hate.

In the shadow of the Holocaust and in the light of Carrie Buck's saga, *eugenics* is now almost universally considered a dirty word. But many of our motives today are no different from those of the *Buck* era; we continue to hope that science can be used to improve the human condition. We all want to eradicate disease; we all hope to have healthy children. We also all want lower taxes. Whether or not we use the word *eugenics* to describe those motives, we must recognize their power, both in historical context as well as today.

The flawed genetic guesswork that led to Carrie Buck's sterilization is, little by little, being replaced by more precise diagnosis of some genetic diseases. Yet *Buck v. Bell* and the history of sterilization remains to haunt our national conscience at a time when the search for "defective" genes is no longer the province of a flawed social movement but a priority on the agenda of national health research.

Today we can diagnose some forms of deafness, blindness, and cancer as well as numerous other diseases, where we know the genes that lead to disease and we can reliably predict its onset. The search for the cause of mental retardation has not abated since the time of *Buck,* and many genetic markers for cognitive impairments remain under study. How much does it matter if we use a technique—less troubling to some than coercive surgery—to "cleanse the germ plasm," as the eugenicists would have said? Does our embrace of techniques such as preimplantation selection of "normal" fetuses or prenatal genetic diagnosis and selective abortion make our motives in "eradicating defects" less suspect? Our modern emphasis on autonomy as a principle important to both law and ethics does not free us from the hard questions posed by our newest technologies. Even overturning *Buck* would not exorcise the demons of eugenics.

L. C. Dunn was president of the American Society for Human Genetics in 1961. In his presidential address that year, Dunn noted that in the early decades of the twentieth century, the excitement that surrounded scientific discoveries that seemed to have such a direct application to human development fed the eugenics movement. "Rapid translation of new knowledge into terms applicable to improvements of man's lot is at such times," Dunn warned, "likely to take precedence over objective and skeptical evaluation of the facts."[33] Dunn was concerned that this tendency, like other "defects in the adolescent period of human genetics," had not disappeared, even late in his own career. Much of what Dunn said more than forty-five years ago is pertinent today, in the headlong rush to apply the insights gleaned from genetic research.

The field of genetics has a history, often forgotten in today's excitement over its explosive growth as a subject of scientific study, cultural fascination, and commercial potential. The story of Carrie Buck is an inescapable part of that history. How we remember her, or whether we choose to remember her at all, will continue to be a matter of consequence in public policy debates about the uses of new genetic technologies as well as the ongoing debate about constitutional dimensions of reproductive freedom.

ACKNOWLEDGMENTS

I owe thanks to many people for their support while this book was in its long gestation, but I would be remiss if I did not thank James H. Jones first. Jim read my dissertation more than twenty-five years ago, and he responded to it with honest criticism that challenged me to pursue the *Buck* story further. Many years later, he repeatedly encouraged me to start thinking seriously about writing this book. It is not an exaggeration to say that it would not have been completed without Jim's help.

From 1998 to 2005, I made numerous trips to Cold Spring Harbor, New York, to work on the Image Archive on the American Eugenics Movement. That exhibit is now the most extensive online compilation of images and textual material related to the history of eugenics in the United States. Helping to edit and compile those materials was a wonderful opportunity for me to learn more about the history of eugenics, and it was also one of the richest professional experiences of my life. Many insights that made their way into this book resulted from conversations with my colleagues on that project as we—both literally and figuratively—retraced the steps of Harry Laughlin and the eugenicists through their Long Island haunts. I thank Garland Allen, Elof Carlson, and Steven Selden for their friendship, their good humor, and their generosity during our time on the historians' working group of the Image Archive of the American Eugenics Movement. Similarly, the hospitality of all the staff at the Cold Spring Harbor Laboratories, particularly Dave Micklos at the Dolan DNA Learning Center and Jan Witkowski at the Banbury Center, made my visits to Cold Spring Harbor a pleasure. My special thanks are also due to Sue Lauter at the Dolan DNA Learning Center, who delivered very timely help in retrieving many of the images for this volume.

For several years during his tenure as director of the Center for Biomedical Ethics at the University of Virginia, my friend and colleague Jonathan Moreno exhorted me to get this book into print. He was careful to navigate the delicate territory between nudge and nag, and the motivation he provided for me to complete my account of *Buck* was a great help.

Five people read this book in manuscript. Jim Jones provided important perspective, and Gregory Dorr's penetrating historical critique was invaluable. I turned to Wendy Parmet for her view as a lawyer with a keen sensitivity to history and to my wife, Conni, who reminded me that clarity was universally valued in prose, regardless of the topic. My editor, Jacqueline Wehmueller, has been unrelentingly positive from the moment of her initial invitation to submit this book to the Johns Hopkins University Press. All five were, as a valued mentor once put it, "kind enough to put kindness aside and tell me what they really thought." My thanks to all of them. I suspect that I did not always succeed in clearing the bar they set for me, so whatever missteps a reader may discover in the text should be attributed to my limitations alone.

Many people sent material over the years that helped me flesh out the *Buck* story. Four friends I have known my entire adult life will recognize in these pages the little gems they tossed my way. To John Migliaccio, David O'Leary, Maryvelma Smith O'Neil, and Paul Zurkuhlen: *Mille grazie.*

The current public awareness of *Buck v. Bell* is largely due to the efforts of the reporters, science writers, and filmmakers who rediscovered the case and told it to larger audiences in the last twenty years. Often they revived dated and generally unnoticed scholarship, but they also tracked down and gave voice to living survivors of sterilization.

Peter Hardin's work as the Washington correspondent for the *Richmond Times-Dispatch* deserves first notice in this group. Hardin's numerous page-one features on Virginia eugenics created momentum both for the first legislative denunciation of eugenics and the first gubernatorial apology on behalf of a state. Mary Bishop, formerly of the *Roanoke Times,* was also a stalwart in keeping the story of Virginia eugenics and the faces of its survivors in the public eye. Kevin Begos again called attention to the *Buck* story during his extraordinary series on North Carolina eugenics. And it would be hard to overestimate the international impact of writer-filmmaker Stephen Trombley in telling Carrie Buck's story to millions around the world.

My early interest in *Buck v. Bell* was fueled by the support of the late paleontologist and science writer Stephen Jay Gould. At Steve's urging, I pursued documents and records about the case I might otherwise never have discovered. I am thankful to him for giving me the opportunity to collaborate in telling an important part of the *Buck* saga and for his regular encouragement through the years.

Julius Paul is an emeritus professor at the State University of New York at Fredonia. Upon his retirement in the early 1990s, Professor Paul gave me the fruit of his more than thirty years' research on eugenics—books, articles, papers, and more. I have mined this material, as well as his lengthy unpublished manuscript on eugenic sterilization, for many insights on eugenics in America after *Buck*. I know no more generous scholar than Professor Paul.

Ralph Brave, a freelance writer and political activist whose science reportage appeared in *The Nation* and numerous newspapers, died as this book was going to press. Brave was almost singlehandedly responsible for working behind the scenes with public officials, scholars, and colleagues in the press to orchestrate the events that led to California's apology for eugenics. His contribution, for which he never even requested acknowledgment, should be celebrated.

I have been fortunate to benefit from the talents of many librarians and archivists who maintain the records that form the foundation for historical scholarship. For years, Judy May Sapko, of the Pickler Library at Truman State University, managed Harry Laughlin's papers. She provided me with more help than I had any right to expect in navigating that collection. At the University of Virginia, Joan E. Klein and Kent Olson often supplied me with material they discovered that enriched the *Buck* story. I thank them and all their colleagues who provided guidance through many collections of records and in archives all over the country.

Steve Kaminshine, dean of the College of Law at Georgia State University, provided leave time from teaching that allowed me to complete this book. I am grateful to Steve and also to my colleague Charity Scott, director of the Center for Law, Health, and Society at Georgia State, for their continuing support and encouragement.

Finally, thanks to my wife, Conni and my children, Christopher and Clare. Without your confidence in me, I would not have started; without your patience, I could not have finished.

The Supreme Court Opinion in *Buck v. Bell*,
by Justice Oliver Wendell Holmes Jr.

BUCK

v.

BELL,

Superintendent of State Colony for Epileptics and Feeble Minded.

Decided May 2, 1927

274 U.S. 200

Mr. Justice HOLMES delivered the opinion of the Court.

This is a writ of error to review a judgment of the Supreme Court of Appeals of the State of Virginia, affirming a judgment of the Circuit Court of Amherst County, by which the defendant in error, the superintendent of the State Colony for Epileptics and Feeble Minded, was ordered to perform the operation of salpingectomy upon Carrie Buck, the plaintiff in error, for the purpose of making her sterile. The case comes here upon the contention that the statute authorizing the judgment is void under the Fourteenth Amendment as denying to the plaintiff in error due process of law and the equal protection of the laws.

Carrie Buck is a feeble-minded white woman who was committed to the State Colony above mentioned in due form. She is the daughter of a feeble-minded mother in the same institution, and the mother of an illegitimate feeble-minded child. She was eighteen years old at the time of the trial of her case in the Circuit Court in the latter part of 1924.

An Act of Virginia approved March 20, 1924 recites that the health of the patient and the welfare of society may be promoted in certain cases by the sterilization of mental defectives, under careful safeguard, etc.; that the sterilization may be effected in males by vasectomy and in females by salpingectomy, without serious pain or substantial danger to life; that the Commonwealth is supporting in various institutions many defective persons who if now discharged would become a menace but if incapable of procreating might be discharged with safety and become self-supporting with benefit to themselves and to society; and that experience has shown that heredity plays an important part in the transmission of insanity, imbecility, etc.

The statute then enacts that whenever the superintendent of certain institutions including the above named State Colony shall be of opinion that it is for the best interest of the patients and of society that an inmate under his care should be sexually sterilized, he may have the operation performed upon any patient afflicted with hereditary forms of insanity, imbecility, etc., on complying with the very careful provisions by which the act protects the patients from possible abuse. The superintendent first presents a petition to the special board of directors of his hospital or colony, stating the facts and the grounds for his opinion, verified by affidavit. Notice of the petition and of the time and place of the hearing in the institution is to be served upon the inmate, and also upon his guardian, and if there is no guardian the superintendent is to apply to the Circuit Court of the County to appoint one. If the inmate is a minor notice also is to be given to his parents, if any, with a copy of the petition. The board is to see to it that the inmate may attend the hearings if desired by him or his guardian.

The evidence is all to be reduced to writing, and after the board has made its order for or against the operation, the superintendent, or the inmate, or his guardian, may appeal to the Circuit Court of the County. The Circuit Court may consider the record of the board and the evidence before it and such other admissible evidence as may be offered, and may affirm, revise, or reverse the order of the board and enter such order as it deems just. Finally any party may apply to the Supreme Court of Appeals, which, if it grants the appeal, is to hear the case upon the record of the trial in the Circuit Court and may enter such order as it thinks the Circuit Court should have entered.

There can be no doubt that so far as procedure is concerned the rights of the patient are most carefully considered, and as every step in this case was taken in scrupulous compliance with the statute and after months of observation, there is no doubt that in that respect the plaintiff in error has had due process at law. The attack is not upon the procedure but upon the substantive law. It seems to be contended that in no circumstances could such an order be justified. It certainly is contended that the order cannot be justified upon the existing grounds.

The judgment finds the facts that have been recited and that Carrie Buck "is the probable potential parent of socially inadequate offspring, likewise afflicted, that she may be sexually sterilized without detriment to her general health and that her welfare and that of society will be promoted by her sterilization," and thereupon makes the order. In view of the general declarations of the Legislature and the specific findings of the Court obviously we cannot say as matter of law that the grounds do not exist, and if they exist they justify the result.

We have seen more than once that the public welfare may call upon the best citizens for their lives. It would be strange if it could not call upon those who already sap the strength of the State for these lesser sacrifices, often not felt to be such by those concerned, in order to prevent our being swamped with incompetence. It is better for all the world, if instead of waiting to execute degenerate offspring for crime, or to let them starve for their imbecility, society can prevent those who are manifestly unfit from continuing their kind. The principle that sustains compulsory vaccination is broad enough to cover cutting the Fallopian tubes. *Jacobson v. Massachusetts,* 197 U.S. 11. Three generations of imbeciles are enough.

But, it is said, however it might be if this reasoning were applied generally, it fails when it is confined to the small number who are in the institutions named and is not applied to the multitudes outside. It is the usual last resort of constitutional arguments to point out shortcomings of this sort. But the answer is that the law does all that is needed when it does all that it can, indicates a policy, applies it to all within the lines, and seeks to bring within the lines all similarly situated so far and so fast as its means allow. Of course so far as the operations enable those who otherwise must be kept confined to be returned to the world, and thus open the asylum to others, the equality aimed at will be more nearly reached.

Judgment affirmed.

Mr. Justice Butler dissents.

Virginia Eugenical Sterilization Act, 1924

An ACT to provide for the sexual sterilization
of inmates of State institutions in certain cases.

(Virginia Acts of Assembly, Chapter 394, 1924)

Approved March 24, 1924

Whereas, both the health of the individual patient and the welfare of society may be promoted in certain cases by the sterilization of mental defectives under careful safeguard and by competent and conscientious authority, and

Whereas, such sterilization may be effected in males by the operation of vasectomy and in females by the operation of salpingectomy, both of which said operations may be performed without serious pain or substantial danger to the life of the patient, and

Whereas, the Commonwealth has in custodial care and is supporting in various State institutions many defective persons who if now discharged or paroled would likely become by the propagation of their kind a menace to society but who if incapable of procreating might properly and safely be discharged or paroled and become self-supporting with benefit both to themselves and to society, and

Whereas human experience has demonstrated that heredity plays an important part in the transmission of insanity, idiocy, imbecility, epilepsy and crime, now, therefore

1. Be it enacted by the general assembly of Virginia, That whenever the superintendent of the Western State Hospital, or of the Eastern State Hospital, or of the Southwestern State Hospital, or of the Central State Hospital, or of

the State Colony for Epileptics and Feeble-Minded, shall be of opinion that it is for the best interests of the patients and of society that any inmate of the institution under his care should be sexually sterilized, such superintendent is hereby authorized to perform, or cause to be performed by some capable physician or surgeon, the operation of sterilization on any such patient confined in such institution afflicted with hereditary forms of insanity that are recurrent, idiocy, imbecility, feeble-mindedness or epilepsy; provided that such superintendent shall have first complied with the requirements of this act.

2. Such superintendent shall first present to the special board of directors of his hospital or colony a petition stating the facts of the case and the grounds of his opinion, verified by his affidavit to the best of his knowledge and belief, and praying that an order may be entered by said board requiring him to perform or to have performed by some competent physician to be designated by him in his said petition or by said board in its order, upon the inmate of his institution named in such petition, the operation of vasectomy if upon a male and of salpingectomy if upon a female.

A copy of said petition must be served upon the inmate together with a notice in writing designating the time and place in the said institution, not less than thirty days before the presentation of such petition to said special board of directors when and where said board may hear and act upon such petition.

A copy of the said petition and notice shall also be so served upon the legal guardian or committee of the said inmate if such guardian or committee be known to the said superintendent, and if there be no such guardian or committee or none such be known to the said superintendent, then the said superintendent shall apply to the circuit court of the county or city in which his said institution is situated, or to the judge thereof in vacation, who by a proper order entered in the common law order book of the said court shall appoint some suitable person to act as guardian of the said inmate during and for the purposes of proceedings under this act, to defend the rights and interests of the said inmate, and the guardian so appointed shall be paid by the said institution a fee of not exceeding twenty-five dollars as may be determined by the judge of the said court for his services under said appointment and such guardian shall be served likewise with a copy of the aforesaid petition and notice. Such guardian may be removed or discharged at any time by the said court or the judge thereof in vacation and a new guardian appointed and substituted in his place.

If the said inmate be an infant having a living parent or parents whose names and addresses are known to the said superintendent, they or either of them as the case may be shall be served likewise with a copy of the said petition and notice.

After the notice required by this act shall have been so given, the said special board at the time and place named therein, with such reasonable continuances from time to time and from place to place as the said special board may determine, shall proceed to hear and consider the said petition and the evidence offered in support of and against the same, provided that the said special board shall see to it that the said inmate shall have opportunity and leave to attend the said hearings in person if desired by him or if requested by his committee, guardian or parent served with the notice and petition aforesaid.

The said special board may receive and consider as evidence at the said hearing the commitment papers and other records of the said inmate with or in any of the aforesaid named institutions as certified by the superintendent or superintendents thereof, together with such other legal evidence as may be offered by any party to the proceedings.

Any member of the said special board shall have power to administer oaths to any witnesses at such hearing.

Depositions may be taken by any party after due notice and read in evidence if otherwise pertinent.

The said special board shall preserve and keep all record evidence offered at such hearings and shall have reduced to writing in duplicate all oral evidence so heard to be kept with its records.

Any party to the said proceedings shall have the right to be represented by counsel at such hearings.

The said special board may deny the prayer of the said petition or if the said special board shall find that the said inmate is insane, idiotic, imbecile, feeble-minded or epileptic, and by the laws of heredity is the probably potential parent of socially inadequate offspring likewise afflicted, the said inmate may be sexually sterilized without detriment to his or her general health, and that the welfare of the inmate and of society will be promoted by such sterilization, the said special board may order the said superintendent to perform or to have performed by some competent physician to be named in such order upon the said inmate, after not less than thirty days from the date of such order, the operation of vasectomy if a male or of salpingectomy if a female; provided that nothing in this act shall be construed to authorize the operation of castration nor the removal of sound organs from the body.

3. From any order so entered by the said special board the said superintendent or the said inmate or his committee or guardian or parent or next friend shall within thirty days after the date of such order have an appeal of right to the circuit court of the county or city in which the said institution is situated, which appeal may be taken by giving notice thereof in writing to any member of the said special board and to the other parties to the said proceeding, whereupon the said superintendent shall forthwith cause a copy of the petition, notice, evidence and orders of the said special board certified by the chairman or in his absence by any other member thereof, to the clerk of the said circuit court, who shall file the same and docket the appeal to be heard and determined by the said court as soon thereafter as may be practicable.

The said circuit court in determining such appeal may consider the record of the proceedings before the said special board, including the evidence therein appearing together with such other legal evidence as the said court may consider pertinent and proper that may be offered to the said court by any party to the appeal.

Upon such appeal the said circuit court may affirm, revise or reverse the orders of the said special board appealed from and may enter such order as it deems just and right and which it shall certify to the said special board of directors.

The pendency of such appeal shall stay proceedings under the order of the special board until the appeal be determined.

4. Any party to such appeal in the circuit court may within ninety days after the date of the final order therein, apply for an appeal to the supreme court

of appeals, which may grant or refuse such appeal and shall have jurisdiction to hear and determine the same upon the record of trial in the circuit court and to enter such order as it may find that the circuit court should have entered.

The pendency of an appeal in the supreme court of appeals shall operate as a stay of proceedings under any orders of the special board or of the circuit court until the appeal be determined by the said supreme court of appeals.

5. Neither any of said superintendents nor any other person legally participating in the execution of the provisions of this act shall be liable either civilly or criminally on account of said participation.

6. Nothing in this act shall be construed so as to prevent the medical or surgical treatment for sound therapeutic reasons of any person in this State, by a physician or surgeon licensed by this State, which treatment may incidentally involve the nullification or destruction of the reproductive functions.

Laws and Sterilizations by State

By the time of U.S. entry into the Second World War, most states—thirty out of forty-eight—had sterilization laws. There was, however, disagreement among experts about which laws counted as properly "eugenic," and a few early, purely punitive statutes were sometimes omitted from listings. The chart below includes all statutes counted by Harry Laughlin (1922) and J. H. Landman (1932) as well as the last two laws passed in South Carolina and Georgia. The chart does not include Puerto Rico, which passed a law in 1937 and repealed it in 1960.

Many states passed, revised, and repealed laws; other enactments were struck down by courts. The chart indicates the first law passed in each state followed by the last action to remove "eugenic" language from state legal codes.

There is no reliable accounting of the number of sterilization operations performed in the United States. The chart is based on a survey done in the 1960s by Julius Paul, described in his unpublished manuscript "Three Generations of Imbeciles Are Enough: State Eugenic Sterilization Laws in American Thought and Practice" (1965). Paul's study is probably the most thorough and systematic state-by-state investigation of sterilization practices since the 1930s. In it, he assembled data on operations from existing state records, institutional reports, and surveys of officials in all the states. As he reported, accurate totals were extremely elusive. Sometimes surgeries occurred in states with no sterilization laws; in states with laws, sometimes surgeries occurred in hospitals, asylums, and other facilities not governed by those laws. None of these numbers are to be found in official records. All of these factors suggest that the totals appearing in the chart are at best conservative estimates of the total surgeries performed under the rubric of "eugenics."

Laws and Sterilizations by State

State	Adopted	Repealed	Number of Operations
Indiana	1907	1974	2,424
Washington	1909	intact	685
California	1909	1979	20,108
Connecticut	1909	1986	557
Nevada	1911	1961	0
Iowa	1911	1977	1,910
New Jersey	1911	1920	0
New York	1912	1920	42
Oregon	1917	1983	2,341
North Dakota	1913	1965	1,049
Kansas	1913	1965	3,032
Michigan	1913	1974	3,786
Wisconsin	1913	1977	1,823
Nebraska	1915	1969	902
South Dakota	1917	1974	789
New Hampshire	1917	1975	679
North Carolina	1919	2003	6,851
Alabama	1919	1974	224
Montana	1923	1981	256
Delaware	1923	1985	945
Virginia	1924	1979	8,300
Idaho	1925	1971	38
Utah	1925	1988	772
Minnesota	1925	1974	2,350
Maine	1925	1981	326
Mississippi	1928	intact	683
West Virginia	1929	1974	98
Arizona	1929	1974	30
Vermont	1931	1967	253
Oklahoma	1931	1983	556
South Carolina	1935	1985	277
Georgia	1937	1970	3,284

NOTES

PROLOGUE. The Expert Witness

1. Wilhelm to Priddy, October 15, 1924, Colony File no. 1392. Each inmate was assigned a number at admission to the Colony; case files containing all notes on admission and subsequent progress reports, medical notes, and correspondence are collected under that number.

2. Strode to Estabrook, November 6, 1924, Arthur H. Estabrook Papers, University at Albany, Albany, New York; hereafter AHE Papers.

3. Strode to Estabrook, October 10, 1924, AHE Papers.

4. Jessie Estabrook to A. H. Estabrook, October 23, 1924, AHE Papers.

5. Arthur Estabrook to Aubrey Strode, November 8, 1924; and Aubrey Strode to Arthur Estabrook, November 6, 1924, AHE Papers.

6. Trial transcript, *Buck v. Priddy,* 82–93.

7. Arthur H. Estabrook and Charles B. Davenport, *The Nam Family: A Study in Cacogenics,* Eugenics Record Office Memoir no. 2 (Cold Spring Harbor, NY: Eugenics Record Office, 1912) 1.

8. Charles L. Brace, "Pauperism," *North American Review,* vol. 120, April 1875, 315–34, quotation at 321.

9. Arthur H. Estabrook, *The Jukes in 1915* (Washington, D.C.: Carnegie Institution, 1916) iii, 78.

10. Estabrook, *Jukes,* 56–58, 85.

11. *Buck v. Bell,* 274 U.S. 200 (1927) at 207.

CHAPTER 1. Problem Families

1. Francis Galton, *Inquiries into Human Faculty and Its Development* (London: Macmillan, 1883) 24n11.

2. Nicholas Wright Gillham, *A Life of Sir Francis Galton: From African Exploration to the Birth of Eugenics* (New York: Oxford, 2001) 13.

3. Francis Galton, "Studies in Eugenics," *American Journal of Sociology,* vol. 11, July 1905, 11–25, quotation at 25.

4. Daniel J. Kevles, *In the Name of Eugenics: Genetics and the Uses of Human Heredity* (New York: Alfred A. Knopf, 1985) 4, 85.

5. Karl Pearson, *The Life, Letters, and Labours of Francis Galton,* vol. 3a (Cambridge, U.K.: Cambridge University Press, 1930) 411–25.

6. Francis Galton, *Memories of My Life* (New York: E. P. Dutton, 1909) 311, 323.

7. Charles Loring Brace, *The Dangerous Classes of New York, and Twenty Years' Work among Them* (New York: Wynkoop and Hallenbeck, 1872) 42–43; and Samuel Gridley Howe, quoted in Penny L. Richards, "Beside Her Sat Her Idiot Child: Families and Developmental Disabilities in Mid-Nineteenth-Century America," in *Mental Retardation in America: A Historical Reader,* ed. Steven Noll and James W. Trent Jr. (New York: New York University Press, 2004) 65–84, quotation at 70. On degeneracy generally, see Daniel Pick, *Faces of Degeneration: A European Disorder, c. 1848–c. 1918* (New York: Cambridge University Press, 1989).

8. Exodus 34:6–7; Deuteronomy 5:9.

9. Nicole Hahn Rafter, *Creating Born Criminals* (Urbana: University of Illinois, 1997) 36–37.

10. "Hereditary Crime," *Scientific American,* vol. 32, January 9, 1875, 18.

11. "The Generation of the Wicked," *Scientific American,* vol. 32, February 27, 1875, 128.

12. "Medical Notes," *Boston Medical and Surgical Journal,* vol. 42, January 28, 1875, 112–13, quoting the *American Medical Weekly.*

13. Oliver Wendell Holmes [Sr.], "Crime and Automatism," *Atlantic Monthly,* vol. 35, April 1875, 466–81, quotation at 475–76. Writers in other popular journals echoed Margaret's story; see, for example, Charles L. Brace, "Pauperism," *North American Review,* vol. 120, April 1875, 315–34.

14. Richard Dugdale, *The Jukes: A Study of Crime, Pauperism, Disease, and Heredity,* 4th ed. (New York: G. P. Putnam, 1884) 8, 13, 15, 27, 70. Both Dugdale and Holmes Sr. would later be consulted for books like T. W. Shannon, *Eugenics* (Marietta, OH: S. A. Mullikin Co., 1916) 5.

15. An extended account of the environmental emphasis of the Jukes study may be found in Elof Axel Carlson, *The Unfit: A History of a Bad Idea* (Cold Spring Harbor, NY: Cold Spring Harbor Laboratory Press, 2001).

16. Galton, *Inquiries into Human Faculty,* 43.

17. J. H. Kellogg, *Plain Facts for Old and Young* (Burlington, IA: Segner and Condit, 1884) 109–11.

18. Rev. Oscar C. McCulloch, "The Tribe of Ishmael: A Study in Social Degradation," in *Proceedings of the National Conference of Charities and Corrections,* ed. Isabel C. Barrows (Boston: George Ellis, 1888) 154–59.

19. Frank W. Blackmar, "The Smoky Pilgrims," *American Journal of Sociology,* vol. 2, January 1897, 485–500; and Florence H. Danielson and Charles B. Davenport, *The Hill Folk: Report on a Rural Community of Hereditary Defectives,* Eugenics Record Office Memoir no. 1 (Cold Spring Harbor, NY: Eugenics Record Office, 1912). Several of the earliest studies were collected in a single volume by Nicole Hahn Rafter, *White Trash:*

The Eugenic Family Studies, 1877–1919 (Boston: Northeastern University Press, 1988).

20. Walter E. Fernald, "The History of the Treatment of the Feeble-Minded," in *Proceedings of the National Conference of Charities and Corrections*, ed. Isabel C. Barrows (Boston: George Ellis, 1893) 203–21, quotations at 203, 211, 212, 221.

21. "Society Proceedings," *Medical News*, June 9, 1900, vol. 76, 922–25, quotation at 924.

22. Alexander Johnson, "The Segregation of Defectives," in *Proceedings of the National Conference of Charities and Corrections*, ed. Isabel C. Barrows (Boston: Fred J. Herr, 1903) 246, 250.

23. John Fiske, *Old Virginia and her Neighbors*, vol. 2 (Boston: Houghton Mifflin, 1897) 25–26.

24. Charles B. Davenport, *Heredity in Relation to Eugenics* (New York: Henry Holt, 1913) 207.

25. See Albert Deutsch, *Mentally Ill in America*, 2nd ed. (New York: Columbia University Press, 1949) 69–71, 112.

26. Report of the State Epileptic Colony (Richmond, VA: Superintendent of Public Printing, 1910) 8, hereafter, Colony Report.

27. Report of the Central State Hospital (Richmond, VA: Superintendent of Public Printing, 1903) 17; hereafter, Report, CSH.

28. Report, CSH (1903), 17.

29. See, for example, William F. Drewry to Strode, November 29, 1908, and December 10, 1908, Aubrey Ellis Strode Papers, Special Collections, University of Virginia Library, Charlottesville, Virginia, hereafter AES Papers.

30. *Virginia Acts of Assembly*, 1906, ch. 48.

31. *Virginia Acts of Assembly*, 1906, ch. 129; Colony Report, 1910, 9.

32. Colony Report, 1924–25, 9.

33. Colony Report, 1910, 7; 1911, 10.

34. Colony Record no. 1.

35. Colony Record nos. 3, 305.

36. Colony Report, 1915, 15–16.

37. *Proceedings of the National Conference of Charities and Corrections*, vol. 20, 1893, 213.

38. David J. Pivar, *Purity Crusade: Sexual Morality and Social Control, 1868–1900* (Westport, CT: Greenwood Press, 1973).

39. Mark Thomas Connelly, *The Response to Prostitution in the Progressive Era* (Chapel Hill: University of North Carolina Press, 1980) 41–43.

40. Pivar, *Purity Crusade*, 243–44.

41. On the social hygiene movement, see Allan M. Brandt, *No Magic Bullet: A Social History of Venereal Disease in the United States since 1880* (New York: Oxford University Press, 1987).

42. Diane B. Paul, *Controlling Human Heredity: 1865 to the Present* (Atlantic Highlands, NJ: Humanities Press, 1995) 77–78.

43. Mark H. Haller, *Eugenics: Hereditarian Attitudes in American Thought* (New Brunswick, NJ: Rutgers University Press, 1963) 106.

44. Martin W. Barr, "The How, the Why, and the Wherefore of the Training of Feeble-Minded Children," *Journal and Proceedings and Addresses of the Thirty-Seventh Annual Meeting of the National Educational Association* (Chicago: University of Chicago Press, 1898) 1045–51, quotation at 1049.

45. *Virginia Acts of Assembly, 1908*, ch. 276.

46. *Report of the State Board of Charities and Corrections* (Richmond, VA: Superintendent of Public Printing, 1909) 23, 33, 34, 217.

47. *Virginia Acts of Assembly*, 1910, ch. 83; 1912, ch. 143.

48. *Proceedings of the Virginia Medical Society*, October 24, 1911, 254.

49. *Virginia Acts of Assembly*, 1912, ch. 196. For example, in the late nineteenth century, the Rules and Regulations of the New York State Custodial Asylum for the Feebleminded in Newark, New York, specifically stated that those admitted should be "feebleminded women of childbearing age." Rafter, *Creating Born Criminals*, 44, 53n49.

50. *Report of the State Board of Charities and Corrections* (Richmond: Superintendent of Public Printing, 1913) 8–23.

51. *Report of the State Board of Charities and Corrections*, 8–23.

52. *Virginia Acts of Assembly*, 1914, ch. 147.

53. *Mental Defectives in Virginia: A Special Report of the State Board of Charities and Corrections to the General Assembly of 1916, on Weak Mindedness in the State of Virginia; together with a Plan for the Training, Segregation, and Prevention of the Procreation of the Feeble-Minded* (Richmond: Superintendent of Public Printing, 1916), 5: "Miss Ball . . . holds a certificate of graduation from the summer training school at Vineland, N.J."

54. *Eugenical News*, vol. 1, 1916, 4.

55. The Strode-Massie Building opened in April 1914 and received one hundred of the first epileptic women admitted to the Colony. See W. I. Prichard, "History-Lynchburg Training School and Hospital," *Mental Health in Virginia*, vol. 11, Summer 1960, 44.

56. *Virginia Acts of Assembly, 1916*, ch. 388: "The words 'feebleminded person' in this act shall be construed to mean any person with mental defectiveness from birth or an early age, . . . so pronounced that he is incapable of caring for himself or managing his affairs, or of being taught to do so, and is unsafe and dangerous to himself and to others . . . but who is not classible as an 'insane person' as usually interpreted."

57. *Virginia Acts of Assembly*, 1916, ch. 104.

58. *Virginia Acts of Assembly*, 1916, ch. 106.

CHAPTER 2. Sex and Surgery

1. J. H. Kellogg, *Plain Facts for Old and Young* (Burlington, IA: Segner and Condit, 1884) 370–71, 456.

2. "Removal of the Ovaries as a Therapeutic Measure in Public Institutions for the Insane," *Journal of the American Medical Association*, vol. 20, February 4, 1893, 135–37, quotation at 135, 136.

3. H. B. Young, "Removal of the Ovaries etc., in Public Institutions for the Insane," *Journal of the American Medical Association*, vol. 20, March 4, 1893, 258.

4. Mark H. Haller, *Eugenics: Hereditarian Attitudes in American Thought* (New Brunswick, NJ: Rutgers University Press, 1963) 48.

5. Philip R. Reilly, *The Surgical Solution: A History of Involuntary Sterilization in the United States* (Baltimore: Johns Hopkins University Press, 1991) 29.

6. Ro. J. Preston, "Sexual Vices—Their Relation to Insanity—Causative or Consequent," *Virginia Medical Monthly*, vol. 19, June 1892, 197, 200.

7. Martin W. Barr, *Mental Defectives: Their History, Treatment, and Training* (Philadelphia: Blakiston's Son, 1904) 196.

8. Issac N. Kerlin, "President's Annual Address," *Proceedings of the Association of Medical Officers of American Institutions for Idiotic and Feeble-Minded Persons*, 1892, 274–85, quotation at 278.

9. Barr, *Mental Defectives*, 192–93.

10. "Asexualization of Criminals and Degenerates," *Michigan Law Journal*, vol. 6, December 1897, 289–315, quotations at 291, 294, 315.

11. J. Ewing Mears, "Asexualization as a Remedial Measure in the Relief of Certain Forms of Mental, Moral, and Physical Degeneration," *Boston Medical and Surgical Journal*, vol. 161, October 21, 1909, 584–86, quotation at 584, 585.

12. "Whipping and Castration as Punishments for Crime," *Yale Law Journal*, vol. 8, June 1899, 382.

13. Mears, "Asexualization as a Remedial Measure," 585.

14. *Pennsylvania Bills*, Session of 1901, No. 511.

15. Barr, *Mental Defectives*, 193–97.

16. "To Check Idiocy by Surgeon's Knife," *Philadelphia Evening Bulletin*, March 21, 1905, 2.

17. "Surgery on Idiots," *Philadelphia Evening Bulletin*, March 22, 1905, 6.

18. Harry H. Laughlin, *Eugenical Sterilization in the United States* (Chicago: Psychopathic Laboratory, Municipal Court of Chicago, 1922) 35–36.

19. Edward Leroy Van Roden, "The Legal Trend of Sterilization in the United States," *Pennsylvania Bar Association Quarterly*, vol. 22, April 1951, 292.

20. J. Ewing Mears, "Ligature of the Spermatic Cord in the Treatment of Hypertrophy of the Prostate Gland," *Medical and Surgical Reporter*, December 15, 1894, vol. 71, 830–32.

21. A. J. Ochsner, "Surgical Treatment of Habitual Criminals," *Journal of the American Medical Association*, vol. 32, April 22, 1899, 867–68.

22. G. Frank Lydston, *The Diseases of Society* (Philadelphia: J. B. Lippincott, 1904) 562, 566.

23. "Race Suicide for Social Parasites," *Journal of the American Medical Association,* vol. 50, January 4, 1908, 55–56, quotation at 55.

24. C. H. Preston, "Vasotomy or Castration for Perverts and Defectives?" *Journal of the American Medical Association,* vol. 50, March 7, 1908, 785.

25. "Sterilization of Criminals," *Journal of the American Medical Association,* vol. 52, April 10, 1909, 1211.

26. 1907 Indiana Laws, ch. 215. All of the sterilization laws passed between 1907 and 1922 are analyzed in Laughlin, *Eugenical Sterilization.* Sharp's saga is related in detail in Elof Axel Carlson, *The Unfit: A History of a Bad Idea* (Cold Spring Harbor, NY: Cold Spring Harbor Laboratory Press, 2001) 207–18.

27. Harry C. Sharp, "Vasectomy as a Means of Preventing Procreation in Defectives," *Journal of the American Medical Association,* vol. 53, December 4, 1909, 1897–1902, quotations at 1898, 1899, 1900.

28. Thurman B. Rice, "A Chapter in the Early History of Eugenics in Indiana," *Eugenical News,* vol. 33, March-June 1948, 24–28, quotation at 28.

29. H. C. Sharp, "Rendering Sterile of Confirmed Criminals and Mental Defectives," *Proceedings of the Annual Congress of the National Prison Association of the United States, 1907* (Indianapolis: William Burford, 1908) 178, 179.

30. H. C. Sharp, "The Indiana Plan," *Proceedings of the Annual Congress of the American Prison Association of the United States, 1909* (Indianapolis: William Burford, 1910) 38.

31. Frank Wade Robertson, "Sterilization for the Criminal Unfit," *American Medicine,* vol. 5, July 1910, 354.

32. Warren W. Foster, "Hereditary Criminality and its Certain Cure," *Pearson's Magazine,* vol. 22, November 1909, 571.

33. *Washington Laws,* 1909, ch. 249.

34. Appellant Brief, *State v. Feilen,* quoted in Laughlin, *Eugenical Sterilization,* 150–51.

35. *State v. Feilen,* 126 Pac. 75, 77 (Wash. 1912).

36. *California Acts,* 1909, ch. 720.

37. *Connecticut Public Acts,* 1909, ch. 209.

38. "Gov. Wilson Signs the Sterilization Bill," *New York Tribune,* May 4, 1911, 1. Wilson also signed into law the bill to change determinate to indeterminate sentences for criminals, another pet project of the eugenicists. Woodrow Wilson, "A Statement on the Work of the New Jersey Legislature Session of 1911," *The Papers of Woodrow Wilson,* ed. Arthur Link (Princeton: Princeton University Press, 1976) 579. For an example of contemporary praise of Wilson for his eugenic views, see "New Jersey Enacts Social Legislation," *The Survey,* vol. 28, June 3, 1912, 346.

39. *New Jersey Statutes of 1911,* ch. 190.

40. Charles Virgil Tevis, "The Future American to be the Perfect World Type, Says Thos. A. Edison," *Washington Post,* September 10, 1911, p. SM4.

41. *Smith v. Board of Examiners of Feebleminded,* 88 A. 963 (N.J. 1913) at 965.

42. Washington's law was challenged earlier in *State v. Feilen*, 126 Pac. 75 (Wash. 1912). It allowed surgery solely on rapists and "habitual criminals."

43. See Maynard C. Harding, "The Solution of the Negro Problem," *Pacific Medical Journal*, vol. 53, 1910, 343–46, quotation at 344. A similar "remedy" was suggested in 1921 by Oregon physician Bethenia Owens-Adair. See Michael Anne Sullivan, *Healing Bodies and Saving the Race: Women, Public Health, Eugenics, and Sexuality, 1890–1950*, Ph.D. dissertation, University of New Mexico, 2001, 201.

44. *Smith v. Board of Examiners*, at 966, 967.

45. Laughlin, *Eugenical Sterilization*, 177; on later attempts at legislation, see James Leiby, *Charity and Correction in New Jersey* (New Brunswick, NJ: Rutgers University Press, 1967) 237.

46. Quoted in Robert D. Johnston, *The Radical Middle Class: Populist Democracy and the Question of Capitalism in Progressive Era: Portland, Oregon* (Princeton: Princeton University Press, 2003) 202–3.

47. Laughlin, *Eugenical Sterilization*, 21.

48. Exhibit A. Transcript of Sentencing, *State of Nevada v. Pearley C. Mickle*, August 14, 1915, 6, in *Briefs and Records, Mickle v. Hendrichs, 262 Fed. 687 (Nev. 1918)*.

49. *Mickle v. Hendrichs*, 262 Fed. 687 (Nev. 1918) at 691.

50. *Iowa Acts of Assembly*, 1911, ch. 129.

51. *Laws of the 35th General Assembly of Iowa*, 1913, ch. 187.

52. *Davis v. Berry*, 216 Fed. 413 (S.D. Iowa 1914) 417.

53. *Laws of the 36th General Assembly of Iowa*, 1915, ch. 202.

54. "Brief for Plaintiffs in Error" August 30, 1916, *Davis v. Berry*, quoted in Laughlin, *Eugenical Sterilization*, 191.

55. *Davis v. Berry*, 242 U.S. 468 (1917).

CHAPTER 3. The Pedigree Factory

1. Donald Pickens, *Eugenics and the Progressives* (Nashville: Vanderbilt University Press, 1968) 46–48.

2. C. B. Davenport, *Eugenics: The Science of Human Improvement by Better Breeding* (New York: Henry Holt, 1910) 6.

3. Garland E. Allen, "The Eugenics Record Office at Cold Spring Harbor, 1910–1940: An Essay in Institutional History," *Osiris*, vol. 2, 2nd series, 1986, 228.

4. Charles Davenport, "Animal Morphology in its Relation to other Sciences," *Science*, vol. 20, November 25, 1904, 697–706, quotation at 698.

5. Davenport, *Eugenics*, 31–32.

6. Davenport, *Eugenics*, 33–35.

7. Allen, "Eugenics Record Office," 235, 260–62. On Rockefeller's involvement, see "Extends Work in Eugenics," *New York Times*, March 30, 1913, 10.

8. See Harry H. Laughlin, *Eugenics Record Office Report*, no. 1 (Cold Spring Harbor, NY: Eugenics Record Office, 1913) 29.

9. Theodore Roosevelt to Davenport, January 3, 1913, Charles B. Davenport Papers, American Philosophical Society, Philadelphia, Pennsylvania, hereafter CBD Papers.

10. Theodore Roosevelt, "Twisted Eugenics," *The Outlook,* vol. 106, January 3, 1914, 32.

11. James A. Field, "The Progress of Eugenics," *Quarterly Journal of Economics,* vol. 26, November 1911, 37.

12. "Progress of Genetics in Thirty Years," *Annual Report of the Director of the Department of Genetics* (Washington, D.C.: Carnegie Institution, 1934) 32.

13. H. H. Laughlin, "Report on the Organization and the First Eight Months' Work of the Eugenics Record Office," *American Breeders Magazine,* vol. 2, April-June 1911, 111.

14. H. H. Laughlin, *Eugenics Record Office Report,* no. 1.

15. Henry H. Goddard, "Heredity of Feeble-Mindedness," *Eugenics Record Office Bulletin,* no. 1 (Cold Spring Harbor, NY: Eugenics Record Office, 1911).

16. Charles B. Davenport, H. H. Laughlin, David F. Weeks, E. R. Johnstone, Henry H. Goddard, "The Study of Human Heredity," *Eugenics Record Office Bulletin,* no. 2 (Cold Spring Harbor, NY: Eugenics Record Office, 1911) 4, 5, 12–16.

17. Charles B. Davenport, "The Trait Book," *Eugenics Record Office Bulletin,* no. 6 (Cold Spring Harbor, NY: Eugenics Record Office, 1912) 1.

18. Charles B. Davenport, "The Family History Book," *Eugenics Record Office Bulletin,* no. 7 (Cold Spring Harbor, NY: Eugenics Record Office, 1912).

19. H. H. Laughlin, "First Annual Conference of the Eugenics Field Workers," *American Breeders Magazine,* vol. 3, 1912, 265–68, quotation at 266.

20. Davenport, "Family History Book," 100.

21. "Memorandum for High School Teachers in Reference to Family History Study as Laboratory Work in Science," item 1736, and "Memorandum of Suggestions to Instructors Who Are Using the Record of Family Traits," item 1735; available at www.eugenicsarchive.org, accessed January 17, 2008.

22. Amy Sue Bix, "Experiences and Voices of Eugenics Field-Workers: 'Women's Work' in Biology," *Social Studies of Science,* vol. 27, 1997, 625–68, esp. 634–35.

23. Biographical note, AES Papers, available at http://library.albany.edu/speccoll/findaids/apap069.htm#history, accessed January 7, 2008.

24. Arthur H. Estabrook and Charles B. Davenport, *The Nam Family: A Study in Cacogenics* (Cold Spring Harbor, NY: Eugenics Record Office, 1912) 1, 45, 66, 84.

25. Laughlin, "Report on the Organization," 111.

26. "Degenerate Family Costly to State," *New York Times,* December 17, 1911, 12.

27. A. H. Estabrook, "A Two-Family Apartment," *The Survey,* vol. 29, March 15, 1913, 853–54.

28. Estabrook to Laughlin, April 20, 1914, Eugenics Record Office Collection, American Philosophical Society, Philadelphia, Pennsylvania.

29. "The Trend of the Science of Eugenics," *The Survey,* vol. 32, July 11, 1914, 388.

30. Arthur H. Estabrook, *The Jukes in 1915* (Washington, D.C.: Carnegie Institution, 1916) 1, 58, 85.

31. The term *cacogenic,* meaning "ill-born," was coined by Boston psychiatrist and ERO Board member E. E. Southard. See Estabrook and Davenport, *Nam Family,* 1.

32. See Goddard, "Who is a Moron?" *Scientific Monthly,* vol. 24, January 1927, 41–46.

33. Leila Zenderland, *Measuring Minds: Henry Herbert Goddard and the Origins of American Intelligence Testing* (Cambridge, U.K.: Cambridge University Press, 1998) 2.

34. Henry Herbert Goddard, *The Kallikak Family: A Study of the Heredity of Feeble-Mindedness* (New York: Macmillan, 1912) 51.

35. Goddard, *Kallikak Family,* 50, 19, 83.

36. Goddard, *Kallikak Family,* 50, 99.

37. Goddard, *Kallikak Family,* 18–19.

38. Goddard, *Kallikak Family,* 30.

39. The charting system had been developed in conjunction with Charles Davenport in 1909. See Zenderland, *Measuring Minds,* 158.

40. Goddard, *Kallikak Family,* 71, 101, 108, 115, 117.

41. Book reviews, *American Journal of Psychology,* vol. 24, 1913, 290–91.

42. George William Hunter, *A Civic Biology* (New York: American Book Company, 1914) 261–65; and Zenderland, *Measuring Minds,* 144.

43. Exodus 20:5, Deuteronomy 5:9.

44. Henry Herbert Goddard, *Feeble-Mindedness: Its Causes and Consequences* (New York: Macmillan, 1914) 4.

45. Goddard, *Feeble-Mindedness,* 18, 539.

46. Goddard, *Feeble-Mindedness,* 588.

47. W. E. Castle, "Review of *Feeblemindedness, its Causes and Consequences,*" *Journal of Abnormal Psychology,* vol. 10, August-September 1915, 213–17, quotation at 217.

48. H. C. Stevens, "Eugenics and Feeblemindedness," *Journal of Criminal Law and Criminology,* vol. 6, July 1915, 193, 196.

49. Charles Davenport, "Scientific Books," *Science,* vol. 42, December 10, 1915, 837–38.

CHAPTER 4. Studying Sterilization

1. "Report of the Committee on Eugenics," *Proceedings of the American Breeders Association,* vol. 6, 1908, 201–14. See Barbara A. Kimmelman, "The American Breeders' Association: Genetics and Eugenics in an Agricultural Context, 1903–1913," *Social Studies of Science,* vol. 13, May 1983, 163–204.

2. Chas. B. Davenport, "Eugenics, A Subject for Investigation Rather than Instruction," *American Breeders Magazine,* vol. 1, 1910, 68–69.

3. C. B. Davenport, "News and Notes," *American Breeders Magazine,* vol. 1, 1910, 304–7, quotation at 306.

4. C. B. Davenport, "Report on Conference of Research Committees of the Eugenics Section," *American Breeders Magazine*, vol. 2, 1911, 145–46, quotation at 145.

5. Bleeker Van Wagenen [*sic*], "Preliminary Report of the Committee of the Eugenic Section of the American Breeders' Association to Study and to Report on the Best Practical Means for Cutting Off the Defective Germ-Plasm in the Human Population," *Problems in Eugenics: Papers Communicated to the First International Eugenics Congress* (London: Eugenics Education Society, 1912) 460–79. Laughlin credited Van Wagenen for providing funds to subsidize the committee's activities; see Harry H. Laughlin, *Eugenics Record Office Report*, no. 1 (Cold Spring Harbor, NY: Eugenics Record Office, 1913) 25.

6. Van Wagenen, "Preliminary Report," 462, 464.

7. Van Wagenen, "Preliminary Report," 470.

8. Van Wagenen, "Preliminary Report," 477.

9. "Report of Proceedings," *Problems in Eugenics: Papers Communicated to the First International Eugenics Congress*, vol. 2 (London: Eugenics Education Society, 1913) 30–34, esp. 31.

10. Lawrence F. Flick, "Eugenics and Mental Diseases," *Ecclesiastical Review*, vol. 51, 1914, 151–58, quotation at 153.

11. C. B. Davenport, "Marriage Laws and Customs," *Problems in Eugenics: Papers Communicated to the First International Eugenics Congress* (London: Eugenics Education Society, 1912) 151–55, quotation at 155.

12. David Heron, "English Expert Attacks American Eugenic Work," *New York Times*, November 9, 1913, p. SM2.

13. Charles B. Davenport, "American Work Strongly Defended—English Attack on Our Eugenics," *New York Times*, November 9, 1913, p. SM2–3. Emphasis added.

14. See Charles Davenport, "A Reply to Dr. Heron's Strictures," *Science*, vol. 38, November 28, 1913, 773–74; David Heron, "A Rejoinder to Dr. Davenport," *Science*, vol. 39, January 2, 1914, 24–25; and C. B. Davenport and A. J. Rosanoff, "Reply to the Criticism of the Recent American Work by Dr. Heron of the Galton Laboratory," *Eugenics Record Office Bulletin*, no. 11 (Cold Spring Harbor, NY: Cold Spring Harbor Laboratory, 1914).

15. David Heron, "English Expert Again Attacks Davenport," *New York Times*, January 4, 1914, p. SM14.

16. "Doctor Ridicules Laws for Eugenics," *New York Times*, June 21, 1914, 14.

17. Charles B. Davenport, "State Laws Limiting Marriage Selection Examined in Light of Eugenics," *Eugenics Record Office Bulletin*, no. 9 (Cold Spring Harbor, NY: Eugenics Record Office, 1913) 37–40.

18. "Savants Aid Eugenics," *Washington Post*, September 4, 1915, 3.

19. H. H. Laughlin, "Account of the Work of the Eugenics Record Office," *American Breeders Magazine*, vol. 3, 1912, 119–23, esp. 120.

20. H. H. Laughlin, "First Annual Conference of the Eugenics Field Workers," *American Breeders Magazine*, vol. 3, 1912, 265–69, quotation at 267.

21. H. H. Laughlin, quoted in *American Breeders Magazine*, vol. 4, 1913, 52.

22. "Asexualization of Criminals and Degenerates," *Michigan Law Journal*, vol. 6, December 1897, 289–316, quotation at 315.

23. "A Bill to Establish a Department of Public Health and to Define its Duties," *American Journal of Public Health and the Nation's Health*, vol. 28, June 19, 1897, 1191–94, quotation at 1194.

24. Surgeons General Rupert Blue and Hugh Cumming were also active participants in the eugenics movement. See "Membership: The Eugenics Committee," *Journal of Heredity*, vol. 5, 1914, 340; and Paul A. Lombardo and Gregory M. Dorr, "Eugenics, Medical Education, and the Public Health Service: Another Perspective on the Tuskegee Syphilis Experiment," *Bulletin of the History of Medicine*, vol. 80, 2006, 291–316, esp. 306–10.

25. W. C. Rucker, "More Eugenic Laws: Four States Consider Sterilization Legislation and Nine Contemplate Restrictions on Marriage," *Journal of Heredity*, vol. 6, 1915, 219–26, quotation at 219.

26. A. Einer, "Medical Supervision of Matrimony," *Virginia Medical Semi-Monthly*, vol. 9, February 10, 1905, 489–92, quotation at 490.

27. "Gets Eugenic Certificate: Prospective Bridegroom Receives the First Issued by the Government," *New York Times*, October 22, 1913, 1.

28. "Evil Stalks Bare," *Washington Post*, June 3, 1913, 1. See also "Board to Promote Eugenics," *Washington Post*, June 4, 1913, 2.

29. Georgina D. Feldberg, *Disease and Class: Tuberculosis and the Shaping of Modern North American Society* (New Brunswick, NJ: Rutgers University Press, 1995) 86–87.

30. "The Restriction of Marriage," *The Outlook*, April 6, 1912, 760.

31. Rev. John Haynes Holmes, "Deny Marriage to the Unfit," *New York Times*, June 2, 1912, magazine, sec. 5, p. 4.

32. Rev. George C. Peck, "Mawkish Sentiment Must Yield," *New York Times*, June 2, 1912, magazine, sec. 5, p. 4.

33. "Should Minister Marry the Physically Unfit?" *New York Times*, June 2, 1912, magazine, sec. 5, p. 4; and "Pastors for Eugenics," *New York Times*, June 6, 1913, 10.

34. "Two 'Health Marriages,'" *New York Times*, April 11, 1912, 1.

35. "Burns Robe, Quits Pulpit," *New York Times*, July 27, 1913, 1.

36. "Defend Their Church," *Washington Post*, July 28, 1913, 10.

37. G. Frank Lydston, *The Diseases of Society* (Philadelphia: J. B. Lippincott, 1904) 562.

38. "Extends Work in Eugenics," *New York Times*, March 30, 1913, 10.

39. *Pennsylvania Laws*, Session of 1913, Act No. 458, "Regulating the issuance of licenses to marry," 1013–14. See also "Eugenic Marriage Law: Pennsylvania Adopts Drastic Measure in Interest of Health," *New York Times*, July 5, 1913, 10.

40. "Wisconsin's Eugenic Laws," *New York Times*, July 26, 1913, 1.

41. Charles B. Davenport, "The Duty of Society," *New York Times*, July 8, 1913, 6.

42. H. H. Laughlin, "Calculations on the Working out of a Proposed Program of Sterilization," in *Proceedings of the First National Conference on Race Betterment,* ed. E. F. Robbins (Battle Creek, MI: Race Betterment Foundation, 1914) 478–94, esp. 478, 480 (chart: "Rate of Efficiency of the Proposed Segregation and Sterilization Program").

43. Laughlin, "Calculations," 485–86.

44. Laughlin, "Calculations," 484.

45. Laughlin, "Calculations," 494.

46. "Slender Women are Criticized at Eugenics Session," *New York World,* January 9, 1914, 1; "Talk on Nicotine and Better Race," *Chicago Tribune,* January 11, 1914; "France Bows to Teuton Mothers: Dr. Kellogg Urges Eugenistry at Race Betterment Conference," *Chicago Tribune,* January 11, 1914, 7; "Curves and Steady Nerves Mean a Perfect Woman," *Cincinnati Enquirer,* January 9, 1914, 9; "Babyless Age Is Eugenist's Fear for 2014," *Detroit Free Press,* January 8, 1914; "Bankruptcy and World Famine in 50 Years Is Prophecy of Eugenist," *Detroit Free Press,* January 9, 1914, 1; and "Pet Projects of Eugenists Assailed," *Detroit Free Press,* January 13, 1914, 3.

47. Davenport to the Editor of the *New York American,* September 4, 1915, Charles B. Davenport Papers, American Philosophical Society, Philadelphia, Pennsylvania, hereafter, CBD Papers.

48. C. B. Davenport, *Eugenics: The Science of Human Improvement by Better Breeding* (New York: Henry Holt, 1910) 16, 33–34; and Charles B. Davenport, *Heredity in Relation to Eugenics* (New York: Henry Holt, 1913) 258.

49. Harry H. Laughlin, "Report of the Committee to Study and to Report on the Best Practical Means of Cutting Off the Defective Germ-Plasm in the American Population: The Scope of the Committee's Work," *Eugenics Record Office Bulletin,* no. 10a (Cold Spring Harbor, NY: Eugenics Record Office, 1914) 7, 11, 15–16, 47.

50. Laughlin, "Report: The Scope of the Committee's Work," 55 , 57, 47.

51. Laughlin, "Report: The Scope of the Committee's Work," 60.

52. Laughlin, "Report: The Scope of the Committee's Work," 59.

53. Harry H. Laughlin, "Report of the Committee to Study and to Report on the Best Practical Means of Cutting Off the Defective Germ-Plasm in the American Population: Legal, Legislative, and Administrative Aspects of Sterilization," *Eugenics Record Office Bulletin,* no. 10b (Cold Spring Harbor, NY: Eugenics Record Office, 1914) 7, 9.

54. Laughlin, "Report: Legal, Legislative, and Administrative Aspects," 117.

55. Laughlin, "Report: Legal, Legislative, and Administrative Aspects," 118, 125.

56. Edward G. Conklin, "Heredity and Responsibility," *Science,* vol. 37, January 10, 1913, 46–54, quotation at 52.

57. "The Limitations of Eugenics," *Journal of the American Medical Association,* vol. 60, February 8, 1913, 450–51.

58. This concept was well known by the time Laughlin's Model Law was published. See Elof Axel Carlson, *The Unfit: A History of a Bad Idea* (Cold Spring Harbor, NY: Cold Spring Harbor Laboratory Press, 2001) 342–44.

59. "Eugenic Legislation," *Journal of Heredity,* vol. 6, March 1915, 142.

60. "A Warning to Eugenists," *Journal of Heredity,* vol. 5, October 1914, 430.

61. "Experience with Eugenics in Philadelphia," *Boston Medical and Surgical Journal,* vol. 170, April 30, 1914, 706–7.

62. Charles B. Davenport, *The Feebly Inhibited: Nomadism, or the Wandering Impulse, with Special Reference to Heredity* (Washington, D.C.: Carnegie Institution, 1915) 1, 121, 141 (fig. 43, "a shiftless, worthless epileptic"); 126 (fig. 2, "always peculiar"); 142 (fig. 46, "set ways and peculiar ideas").

63. "Bare Secrets of Life," *Washington Post,* November 14, 1914, 4.

64. Augusta F. Bronner, "Reviews and Criticisms," *Journal of Criminal Law and Criminology,* vol. 7, 1916–1917, 311–13.

65. Samuel C. Kohs, "New Light on Eugenics," *Journal of Heredity,* vol. 6, October 1915, 446–52, quotation at 446.

66. "Heredity and Environment," *Journal of Criminal Law and Criminology,* vol. 2, 1911–12, 123.

67. Adolf Meyer, "Reviews and Criticisms," *Journal of Criminal Law and Criminology,* vol. 2, 1911–12, 956.

68. Laughlin, "Report: Legal, Legislative, and Administrative Aspects," 23–24.

69. Charles A. Boston, "A Protest against Laws Authorizing the Sterilization of Criminals and Imbeciles," *Journal of Criminal Law and Criminology,* vol. 4, 1913–14, 326–58, quotations at 327, 330, 332, 336, 340, 358.

70. See Arthur Todd, "Sterilization of Criminals and Defectives," *Journal of Criminal Law and Criminology,* vol. 4, 1913–14, 420; and Frederick A. Fenning, "Sterilization Laws from a Legal Standpoint," *Journal of Criminal Law and Criminology,* vol. 4, 1913–14, 804–14.

71. Edith R. Spaulding and William Healy, "Inheritance as a Factor in Criminality: A Study of a Thousand Cases of Young Repeated Offenders," *Journal of Criminal Law and Criminology,* vol. 4, 1913–14, 837–58, quotations at 837, 858.

72. Joel D. Hunter, "Sterilization of Criminals: Report of Committee H of the Institute," *Journal of Criminal Law and Criminology,* vol. 5, 1914–1915, 514–39.

73. Joel D. Hunter, "Sterilization of Criminals," *American Bar Association Journal,* vol. 2, 1916, 128–34, quotations at 132, 133.

74. Hunter, "Sterilization of Criminals," 134.

75. William A. White, "Sterilization of Criminals," *Journal of Criminal Law and Criminology,* vol. 7, 1917–1918, 499–501, quotation at 500.

76. "Jury of Scientists Hung on Sterilization," *The Survey,* vol. 39, November 24, 1917, 206.

77. Victor Robinson, ed., "A Symposium on Sterilization of the Unfit," *Medical Review of Reviews,* vol. 20, January 1914, 13–21, quotations at 13, 19.

78. Alexander Graham Bell, "How to Improve the Race," *Journal of Heredity,* vol. 5, January 1914, 6. In 1914 the American Breeders Association changed its name to the

American Genetics Association, and the *American Breeders Magazine* was renamed the *Journal of Heredity*.

79. "Primitive Brains Bigger than Ours?" *New York Times*, August 12, 1913, 3.

80. "Sees Peril in Freak Laws on Eugenics," *New York Times*, August 30, 1913, 3.

81. "Overdoing Eugenics Here," *New York Times*, August 21, 1914, 8.

82. Garland Allen, *Thomas Hunt Morgan: The Man and His Science* (Princeton: Princeton University Press, 1978) 228–29. Morgan's critique was reiterated in 1916 as part of the Vanuxem Lectures, Princeton University, February-March 1916, reprinted in Thomas Hunt Morgan, *Evolution and Genetics* (Princeton: Princeton University Press, 1925) 201–5.

83. "Doctors Oppose Eugenics," *New York Times*, May 28, 1914, 1.

84. "Sermon of the Rev. Wm. A. Sunday: Chickens Come Home to Roost," New York City, April 29, 1917, 4. Papers of William Ashley "Billy" Sunday and Helen Amelia (Thompson) Sunday, Archives of the Billy Graham Center, Wheaton College, Wheaton, Illinois.

85. "35,000 Hear Sunday Talk to Men Only," *New York Times*, April 30, 1917, 20.

CHAPTER 5. The Mallory Case

1. "No More Kissing in Virginia?" *Washington Post*, December 3, 1902, 6.

2. "Opposes Bible Kiss," *Washington Post*, December 20, 1902, 9; and "Committees at Work," *Washington Post*, January 22, 1903, 10. Former state senator James G. McCune, who proposed the bill, later made headlines after eloping with a sixteen-year-old girl; see "Anti-Kisser Elopes," *Washington Post*, September 22, 1905, 5.

3. "An Act prohibiting expectorating or spitting in public places," *Virginia Acts of Assembly*, ch. 302, March 17, 1906, 533.

4. Jesse Ewell, "Plea for Castration to Prevent Criminal Assault," *Virginia Medical Semi-Monthly*, vol. 11, January 11, 1907, 463–64, quotation at 464.

5. Charles Carrington, "Sterilization of Habitual Criminals with Report of Cases," *Virginia Medical Semi-Monthly*, vol. 13, December 12, 1908, 389–90; and Charles V. Carrington, "Hereditary Criminals: The One Sure Cure," *Virginia Medical Semi-Monthly*, vol. 15, April 8, 1910, 4–8.

6. Charles V. Carrington, "Sterilization of Habitual Criminals," *Virginia Medical Semi-Monthly*, vol. 14, December 24, 1909, 421–22.

7. H. B. Mahood, "Sterilization of Rape Fiends," *Virginia Medical Semi-Monthly*, vol. 14, December 24, 1909, 437–38.

8. "Asexualization of Hereditary Criminals," *Journal of Criminal Law and Criminology*, vol. 1, 1910–1911, 124–25.

9. "Sterilization of Habitual Criminals, etc.," *Virginia Medical Semi-Monthly*, vol. 14, February 1, 1910, 512.

10. *Virginia Bills*, 1910 (Senate), no. 298.

11. "To Prevent Procreation by Confirmed Criminals, Idiots, Imbeciles, and Rapists," *Virginia Medical Semi-Monthly*, vol. 15, April 8, 1910, 23–24.

12. L. S. Foster, "Feeble-minded Children," *Virginia Medical Semi-Monthly*, vol. 17, January 10, 1913, 469–74.

13. H. W. Dew, "Sterilization of the Feeble-minded, Insane, and Habitual Criminals," *Virginia Medical Semi-Monthly*, vol. 17, April 11, 1913, 4–8, quotation at 5.

14. C. P. Wertenbaker, "Should Virginia Have a Marriage Law Based on Eugenics?" *Virginia Medical Semi-Monthly*, vol. 18, December 12, 1913, 420–24, quotations at 420, 423.

15. Charles V. Carrington, "Keep the Race Pure," *Virginia Medical Semi-Monthly*, vol. 18, December 12, 1913, 434–38, quotations at 434, 435.

16. "Diocesan Convention: Southern Virginia," *The Churchman*, vol. 107, June 21, 1913, 809.

17. "Governor Mann Covers Many New Subjects in Second Message to Assembly," *Richmond (VA) News Leader*, January 16, 1914, 11.

18. "Governor Mann for Eugenic Marriages," *Richmond (VA) Times-Dispatch*, January 17, 1914, 2; and "Eugenic Marriage Law for Virginia," *New York Times*, January 17, 1914, 18.

19. J. Miller Kenyon, "Sterilization of the Unfit," *Virginia Law Review*, vol. 1, 1914, 458–66, quotations at 461–62, 469.

20. Colony Report, 1911, 10; 1913, 18; 1915, 17.

21. "Report of Visit to Virginia State Epileptic Colony, September 24, 1914," in *Report of the State Commission to Investigate Provision for the Mentally Deficient* (Albany, NY: J. B. Lyon, 1915) 269–71, quotation at 271.

22. Priddy traveled to the 1915 Pacific Exposition. He also traveled to Indiana along with Irving Whitehead to visit the Indiana State Epileptic Village. See letter to L. T. Royster, October 30, 1915, Colony File no. 537; and *Minutes of the Meeting of the Special Board of Directors of the State Colony for Epileptics and Feebleminded*, August 12, 1912.

23. *California Statutes*, 1909, ch. 720.

24. Colony Report, 1915, 12.

25. *Minutes of the Meeting of the Special Board of Directors of the State Colony for Epileptics and Feebleminded*, June 9, 1916.

26. Colony Report, 1916, 15; 1917, 13.

27. Colony Record nos. 248, 252.

28. Colony Record no. 520.

29. J. T. C. to Priddy, July 9, 1914, Colony Record no. 403. Patients and their families who corresponded with Priddy are not identified by name.

30. Priddy to J. T. C., July 9, 1914, Colony Record no. 403; *Minutes of the Meeting of the Special Board of Directors of the State Colony for Epileptics and Feebleminded*, August 11, 1916.

31. Colony Record no. 567.

32. Affidavit, January 15, 1917, Colony Record no. 713.

33. CEM to Priddy, February 20, 1916, Colony Record no 588.

34. Progress notes, February 1, 1920, Colony Record no. 588.

35. Priddy to B. T., October 15, 1918; Priddy to Dr. J. R. C., February 5, 1921; and J. R. C. to Priddy, February 6, 1921, Colony Record no. 537.

36. Priddy to A. H., March 28, 1917; and Priddy to A. H., May 17, 1917, Colony Record no. 765.

37. Priddy to M. J. P., February 6, 1918, Colony Record no. 748.

38. Priddy to Parke L. Poindexter, Justice of the Peace, December 22, 1917, Colony Record no. 561.

39. Priddy to D. C., April 9, 1919, Colony Record no. 537.

40. Colony Report, 1916, 10.

41. Colony Record no. 188.

42. *Minutes of the Meeting of the Special Board of Directors of the State Colony for Epileptics and Feebleminded,* August 11, 1916.

43. Colony Record no. 197.

44. Priddy to R. E. P., September 11, 1916, Colony Record no. 454; *Minutes of the Meeting of the Special Board of Directors of the State Colony for Epileptics and Feebleminded,* September 8, 1916; and Priddy to M. G. N. C., September 3, 1918, Colony Record no. 454.

45. *Minutes of the Meeting of the Special Board of Directors of the State Colony for Epileptics and Feebleminded,* August 11, 1916. See also History and Examination papers; G. B. to Dr. A. S. Priddy, October 17, 1921; and Priddy to W. B., July 7, 1924, Colony Record no. 538.

46. Commitment papers; I. W. to Priddy, June 19, 1915; Priddy to I. W., June 22, 1915; and Priddy to G. P., September 11, 1916, Colony Record no. 474.

47. Priddy to Roller, October 5, 1916, Willie Mallory File, Colony Record no. 691.

48. Willie Mallory File, Colony Record no. 691. See also Clinical Record, Willie Mallory, Eastern State Hospital, May 21, 1904. Available records vary significantly on Mrs. Mallory's age. Her husband testified that she was forty-two as of the arrest, even though her own account put her at thirty-seven years old.

49. Deposition, George Mallory, *Mallory v. Children's Home Society,* 23. Record, *Mallory v. Priddy* (Va. Cir. Ct. Richmond, February 16, 1918), and the records in the related cases, Record, *Ex Parte Mallory,* 122 Va. 298 (1918), and Record, *Mallory v. Va. Colony for the Feeble-Minded,* 123 Va. 205 (1918), File Drawer no. 383, item 2711, Virginia State Archives; hereafter, Record, *Mallory v. Priddy;* Record, *Ex Parte Mallory;* Record, *Mallory v. Colony,* respectively.

50. Deposition, George Mallory, *Mallory v. Children's Home Society,* 14, Record, *Mallory v. Priddy.*

51. The fiancée would later marry Jessie, see Deposition, Alonzo Bowles, *Mallory v. Children's Home Society*, 62.

52. See *Ex Parte Mallory*, 122 Va. 298, 299 (1918). Petition at 2, Record, *Ex Parte Mallory*. One of the policemen was later dismissed from the Richmond police force for his involvement in bootlegging and other corrupt practices. See *Sherry v. Lumpkin*, 127 Va. 116, 1920.

53. Deposition of Willie Mallory at 32, Record, *Ex Parte Mallory*.

54. For the typical commitment procedure before 1916, see Act of April 7, 1903, ch. 139, *Virginia Acts of Assembly*, 1903, 121–30.

55. Willie and Jessie were committed on September 28, 1916. After being held in jail and at the Children's Detention Home for almost a month, they arrived at the Colony on October 13, 1916. Papers of Commitment, Willie Mallory File, Colony Record no. 691; and Nannie Mallory File, Colony Record no. 692. See also Jessie Mallory File, Colony Record no. 693. Administrative procedures for commissions of feeblemindedness were described in the law drafted by Aubrey Strode, see Act of March 20, 1916, ch. 388, *Virginia Acts of Assembly*, 1916, 662–68.

56. See *Ex Parte Mallory*, 122 Va. 298, 299 (1918). Petition at 2, Record, *Ex Parte Mallory*.

57. See Arthur W. James, *Virginia's Social Awakening* (Richmond, VA: Garrett and Massie, 1939) 65.

58. Interrogatories and Papers of Commitment, Willie Mallory, September 28, 1916, Colony Record no. 691.

59. Physical exam, October 16, 1916, Binet test, Commitment papers, September 28, 1916, Jessie Mallory File, Colony Record no. 693.

60. G. W. Mallory to Priddy, November 23, 1916; Priddy to G. W. Mallory, November 27, 1916, Colony Record no. 691.

61. Deposition Jessie Mallory, 44–45, *Ex Parte Mallory;* and History and Medical Notes, March 1–19, 1917; and Infirmary Nurses Record, 3, Colony Record no. 691.

62. Furlough Agreement, Mrs. J. W. Murphy, May 26, 1917; and Priddy to Roller, October 16, 1917, Colony Record no. 691.

63. See deposition of Bowles, p. 62; Alonzo Bowles to Priddy, March 9, 1917, Priddy to Alonzo Bowles, March 13, 1917, *Ex Parte Mallory*.

64. Priddy to Howard Hall, March 28, 1917, Colony Record no. 691.

65. Deposition, Willie Mallory, 35, *Ex Parte Mallory*.

66. Irene Mallory to Priddy, April 22, 1917; Priddy to Irene Mallory, April 23, 1917; and Priddy to H. P. Wales [sic], May 11, 1917, Colony Record no. 691.

67. Priddy to G. W. Mallory, September 4, 1917, Colony Record no. 691.

68. Priddy to the Honorable J. Hoge Ricks, October 30, 1917. Colony Record no. 691.

69. Priddy to Sarah B. Roller, October 30, 1917, Colony Record no. 691.

70. See Act of March 20, 1916, ch. 388, *Virginia Acts of Assembly,* 1916, 662–68; see section 5, 664.

71. Priddy to H. C. Dawson, November 15, 1917, Colony Record no. 691.

72. H. D. Coghill, Board of Charities and Corrections, to Priddy, October 31, 1917; James Hoge Ricks, Juvenile and Domestic Relations Court, to Priddy, November 3, 1917; and Priddy to H. D. Coghill, November 3, 1917, Colony Record no. 691.

73. Summons and Notice of Delivery, November 6, 1917, Sheriff A. D. Watts. Record, *Mallory v. Priddy.* Virginia law at the time permitted delivery of legal papers by the sheriff to absent landlords by posting them on the front door of a residence.

74. Letter from George Mallory to Albert Priddy, November 5, 1917, reproduced unedited. Record, *Mallory v. Priddy.*

75. *Minutes of the Meeting of the Special Board of Directors of the State Colony for Epileptics and Feebleminded,* November 8, 1917.

76. Priddy to J. Hoge Ricks, November 10, 1917, Colony Record no. 691.

77. Priddy to George Mallory, November 13, 1917, Record, *Mallory v. Priddy.*

78. Plea in Abatement, *Willie Mallory v. A. S. Priddy,* affidavit endorsed November 13, 1917, Record, *Mallory v. Priddy;* and Priddy to George E. Caskie, Esq., November 13, 1917, Colony Record no. 691.

79. Complaint of Trespass on the Case, Record, *Mallory v. Priddy.*

80. "Mallory Obtains Writ of Habeas Corpus: Supreme Court of Appeals Orders His Five Children Produced at January Term," *Richmond (VA) Times-Dispatch,* November 17, 1917, 10.

81. R. W. Ivey to Caskie and Caskie, November 19, 1917, Exhibit D to Petition for Writ of Habeas Corpus, Record, *Mallory v. Colony.*

82. G. E. Caskie to R. W. Ivey, Esq., November 21, 1917, Exhibit C to Petition for Writ of Habeas Corpus; and petition filed November 24, 1917, Record, *Mallory v. Colony.*

83. *Virginia Acts of Assembly* 1916, ch. 312, sec. 9, p. 666.

84. Grounds of Defence [sic], Record, *Mallory v. Colony.*

85. Willie Mallory to Priddy, November 17, 1917; and Albert Priddy to Willie Mallory, November 26, 1917, Colony Record no. 691.

86. Priddy to J. Hoge Ricks, November 17, 1917; Priddy to Mastin, November 17, 1917; and Priddy to H. D. Coghill, November 24, 1917, Colony Record no. 691.

87. *Minutes of the Meeting of the Special Board of Directors of the State Colony for Epileptics and Feebleminded,* August 12, 1912, June 9, 1916, March 14, April 6, 1917.

88. I. P. Whitehead to Priddy, 1917, Colony Record no. 691.

89. *Schloendorff v. New York Hospital,* 211 N.Y. 125 (1914): "Every human being of adult years and sound mind has a right to determine what shall be done with his own body." See also *Mohr v. Williams,* 95 Minn. 261 (1905); and *Rolater v. Strain,* 137 Pac. 96 (Okl. 1913).

90. *Rishworth v. Moss,* 159 S. W. 122 (Tex. 1913).

91. *Pratt v. Davis,* 224 Ill. 300 (1906). For the law of consent in that era, see Paul A.

Lombardo, "Phantom Tumors & Hysterical Women: Revising Our View of the *Schloendorff* Case," *Journal of Law Medicine and Ethics*, vol. 33, Winter 2005, 791–801.

92. "Dr. Hathaway Sued for Atrocious Treatment of Inmate of Asylum," and "Butchery of Helpless is Officially Reported," *Helena (MT) Daily Independent*, October 5, 1924, 1, 16.

93. H. D. Coghill to Priddy, November 26, 1917; and Priddy to George E. Caskie, February 4, 1918, Colony Record no. 691.

94. Priddy to Sarah B. Roller, February 15, 1918; Priddy to Mrs. K. R. Matthews, February 8, 1918, and February 18, 1918 (including a ten-dollar advance); and W. R. Putney [MD] to Priddy, February 20, 1918, Colony Record no. 691.

95. Arguments in *Ex Parte Mallory* filed January 17, 1918, at 5–6. Record, *Ex Parte Mallory*.

96. Verdict, Record, *Mallory v. Priddy*.

97. Aubrey Strode to Don Preston Peters, July 19, 1939, AES Papers.

98. Leila Zenderland, *Measuring Minds: Henry Herbert Goddard and the Origins of American Intelligence Testing* (Cambridge, U.K.: Cambridge University Press, 1998) 289.

99. W. I. Prichard, "History: Lynchburg Training School and Hospital," *Mental Health in Virginia*, vol. 11, Autumn, 1960, 40–46, esp. 46.

100. *Minutes of the Meeting of the General Board of State Hospitals (VA)*, October 12, 1926.

101. *Ex Parte Mallory*, 122 Va. 298 (1918); and *Mallory v. Virginia Colony for the Feebleminded*, 123 Va. 205 (1918).

102. Priddy to Parke L. Poindexter, Justice of the Peace, December 22, 1917, Colony Record no. 561.

103. *Minutes of the Meeting of the Special Board of Directors of the State Colony for Epileptics and Feebleminded*, April 9, 1918, September 19, 1918.

CHAPTER 6. Laughlin's Book

1. Charles B. Davenport to Dr. H. H. Laughlin, December 21, 1920, Harry H. Laughlin Papers, Special Collections Department, Truman State University, Kirksville, Missouri, hereafter, HHL Papers.

2. "Biological Aspects of Immigration," statement of Harry H. Laughlin, *Hearings before the Committee on Immigration and Naturalization, House of Representatives, Sixty-Eighth Congress*, April 16, 1920, Washington, D.C.: Government Printing Office, 1921.

3. *Statistical Directory of State Institutions for the Defective, Dependent, and Delinquent Classes* (Washington, D.C.: Government Printing Office, 1919).

4. Harry H. Laughlin, "Population Schedule for the Census of 1920," *Journal of Heredity*, vol. 10, May 1919, 208–10.

5. Harry H. Laughlin, "The Socially Inadequate: How Shall We Designate and Sort Them?" *American Journal of Sociology*, vol. 27, July 1921, 54, 57, 68.

6. Thomas W. Salmon to Katherine B. Davis, May 28, 1920. Bureau of Social Hygiene Papers, Rockefeller Archive Center, Tarrytown, New York, hereafter BSH Papers.

7. J. Norris Meyers to Katherine B. Davis, May 29, 1920; J. Norris Meyers to Katherine B. Davis, June 21, 1920; Katherine B. Davis to Harry H. Laughlin, July 3, 1920; Laughlin to Davis, October 4, 1920, BSH Papers.

8. Davenport to Laughlin, September 1, 1920, HHL Papers.

9. Raymond B. Fosdick to Katherine B. Davis, January 31, 1921; and Davis to Fosdick, February 2, 1921, BSH Papers.

10. Editor's note to "Eugenical Sterilization in the United States," *Social Hygiene*, vol. 6, October 1920, 499–533, quotation at 499.

11. Laughlin to George Streeter, February 8, 1921, Carnegie Institute files, Alan Mason Chesney Medical Archives, Johns Hopkins University.

12. Laughlin to Davenport, April 13, 1921, HHL Papers.

13. Harry H. Laughlin, "The Present Status of Eugenical Sterilization in the United States," in *Eugenics in Race and State: Volume II, Scientific Papers of the Second International Congress of Eugenics* (Baltimore: Williams and Wilkins, 1923) 286–91, quotation at 291.

14. H. Laughlin, *The Second International Exhibition of Eugenics* (Baltimore: Williams and Wilkins, 1923), 67 (fig. 4, "Eugenical Classification of the Human Stock"), 153 (fig. 46, "Eugenical Sterilization in the United States").

15. A. H. Estabrook, *The Second International Exhibition of Eugenics* (Baltimore: Williams and Wilkins, 1923), 146 (fig. 43, "The Old Americans and the Tribe of Ishmael"), 148 (fig. 44, "The Nams"), 150 (fig. 45, "The Jukes").

16. John Henry Wigmore, "The Most Famous City Court in the World," *Illinois Law Review*, vol. 6, 1911–1912, 591–92.

17. Harry H. Laughlin, *Eugenical Sterilization in the United States* (Chicago: Psychopathic Laboratory, Municipal Court of Chicago, 1922).

18. "Harry Olson, Advocate, Judge and Scientist," *Journal of the American Judicature Society*, vol. 19, June 1935, 47–49, quotation at 47, 48.

19. For Olson's election to the Chicago Court, see Michal R. Belknap, *To Improve the Administration of Justice: A History of the American Judicature Society* (Chicago: American Judicature Society, 1992) 14, 15.

20. Dr. John Harvey Kellogg of the Battle Creek Sanitorium was among the eugenical leaders who also led the temperance movement. See David J. Pivar, *Purity Crusade: Sexual Morality and Social Control, 1868–1900* (Westport, CT: Greenwood Press, 1973) 117.

21. Vice Commission of Chicago, *The Social Evil in Chicago* (Chicago: Gunthorp-Warren, 1911) 291.

22. Michael Willrich, "The Two Percent Solution: Eugenic Jurisprudence and the Socialization of American Law, 1900–1930," *Law and History Review*, vol. 16, 1997, 63–111, esp. 80.

23. *Third Annual Report* (Chicago: Municipal Court, 1909) 48.

24. Harry Olson, "The Chicago Fight against Prostitution," *Social Hygiene,* vol. 6, January 1920, 279–80.

25. "Understanding Criminals," *Journal of the American Judicature Society,* vol. 2, October 1918, 83.

26. Olson, "Chicago Fight," 280.

27. "The Psychopathic Laboratory—Chief Justice Olson Reveals Secret of Criminal Defectiveness," *Journal of the American Judicature Society,* vol. 4, June 1920, 27.

28. "Carnegie Board to Aid City in Fight on Crime," *Chicago Daily Tribune,* March 2, 1922, 5.

29. "Would Sterilize Socially Inadequate," *New York Times,* January 4, 1923, 2.

30. Laughlin, *Eugenical Sterilization,* v–vi.

31. Laughlin, *Eugenical Sterilization,* 322–23, 336.

32. Laughlin, *Eugenical Sterilization,* 337–39.

33. Laughlin, *Eugenical Sterilization,* 339–40.

34. Laughlin, *Eugenical Sterilization,* 342–50.

35. Laughlin, *Eugenical Sterilization,* 362, 371.

36. Olson to Laughlin, January 18, 1923, Harry Olson Papers, Northwestern University Archives, Evanston, Illinois, hereafter, HO Papers.

37. "Statue for Chicago Courtroom," *Chicago Daily News,* January 27, 1923, Charles and Sophia Haag Collection, Augustana College Library, Rock Island, Illinois. The sculpture now resides in the Gallery of Augustana College.

38. "Law to Sterilize All Unfit to be Sought This Year," *Chicago Daily Tribune,* January 4, 1923, 7.

39. Dr. W. A. Evans, "A Study of Heredity," *Chicago Daily Tribune,* August 7, 1923, 6.

40. Scrutator, "Breeding Better Folks Held Way to Lower Taxes," *Chicago Daily Tribune,* December 30, 1923, A10.

41. "The Week," *New Republic,* vol. 33, January 17, 1923, 185.

42. Robert H. Gault, "Review and Criticisms," *Journal of Criminal Law and Criminology,* vol. 13, 1922–1923, 157–58.

43. Paul Popenoe, "Eugenical Sterilization," *Journal of Heredity,* vol. 14, October 1923, 310.

44. C. C. Spray, "Review: *Eugenical Sterilization in the United States* by Harry Hamilton Laughlin," *Illinois Law Review,* vol. 18, 1923–1924, 69, 72.

45. Harry Olson to Charles Evans Hughes, January 3, 1923, HO Papers. Hughes, then secretary of state, promised Olson to read the book "at the first opportunity." Charles E. Hughes to Harry Olson, January 9, 1923, HO Papers.

46. H. H. Laughlin to William Howard Taft, February 15, 1907, January 28, 1909, April 1, 1914, February 19, 1921, William Howard Taft Collection, Library of Congress, hereafter WHT Collection.

47. Harry Olson to William Howard Taft, January 3, 1923, HO Papers; and William Howard Taft to Harry Olson, January 5, 1923, WHT Collection.

48. E. G. Conklin to Harry Olson, January 27, 1923, HO Papers.

49. G. W. Brown (Superintendent, Eastern State Hospital) to Harry Olson, December 3, 1923; and H. H. Laughlin to Harry Olson, March 28, 1923, attached "Sterilization Book List, March 19, 1923," HO Papers.

CHAPTER 7. A Virginia Sterilization Law

1. For a full list of states with such laws as of 1922, see Harry H. Laughlin, "Report of the Committee to Study and to Report on the Best Practical Means of Cutting Off the Defective Germ-Plasm in the American Population: Legal, Legislative, and Administrative Aspects of Sterilization," *Eugenics Record Office Bulletin,* no. 10b (Cold Spring Harbor, NY: Eugenics Record Office, 1914) 98a (table: "Organization and Procedure Provided by Existing Sterilization Laws").

2. The four states that saw a governor's veto were Pennsylvania (1905), Oregon (1909), Vermont (1913), and Nebraska (1913). Oregon later passed a bill authorizing sterilization in 1913, but it was revoked by referendum before it went into effect. New Jersey (1913) and Iowa (1914) were two states where courts invalidated the measure. Iowa revised and enacted a new law in 1915. The five states that passed legislation successfully were South Dakota (1917), North Carolina (1919), Alabama (1919), Montana (1923), and Delaware (1923). New York's 1912 law was repealed in 1920. The three states that saw court reversals were Oregon (1917), Michigan (1918), and Indiana (1921). An Idaho bill was vetoed in 1919 and another Pennsylvania bill was vetoed in 1921. See Harry H. Laughlin, *Eugenical Sterilization in the United States* (Chicago: Psychopathic Laboratory, Municipal Court of Chicago, 1922) 26, 35, 38, 40, 45, 47, 50, 142, 143, 145, 146.

3. Jonas Robitscher, ed., *Eugenic Sterilization* (Springfield, IL: Charles C. Thomas, 1973) 123.

4. Laughlin, *Eugenical Sterilization,* 90–91.

5. Transcript, *Celebration of Dr. J. S. DeJarnette's Fiftieth Anniversary of Continuous Service at the Western State Hospital,* July 21, 1939, 30. Joseph DeJarnette Papers, Western State Hospital, Staunton, Virginia, hereafter, JDJ Papers.

6. *Minutes of the Meeting of the Special Board of Directors of the State Colony for Epileptics and Feebleminded,* October 12, 1920.

7. Paul A. Lombardo, "Eugenical Sterilization in Virginia: Aubrey Strode and the Case of *Buck v. Bell,*" Ph.D. dissertation, University of Virginia, 1982, 139–42.

8. *Virginia Acts of Assembly,* 1920, chs. 164, 262.

9. Strode to Dr. Don Preston Peters, July 19, 1939, AES Papers.

10. Colony Report, 1922–23, 27–28.

11. See Strode to Priddy, January 5, 1922; and Whitehead to Strode, January 31, 1922, AES Papers.

12. School notes from Louisa B. Hubbard Papers, University of Virginia Library, Special Collections.

13. For details of the Hubbard/Strode courtship, see Lombardo, "Eugenical Sterilization in Virginia," 149–53.

14. See Strode to Don Preston Peters, July 19, 1939, AES Papers.

15. Wm. Maybee to Priddy, February 15, 1915, Colony Record no. 457.

16. *Minutes of the Meeting of the General Board of State Hospitals,* May 12, August 9, 1922; May 10, August 10, September 17, 1923.

17. See, for example, *Virginia Acts of Assembly,* 1919, ch. 33, which authorized a special budget appropriation to meet deficits at hospitals for the insane.

18. *Minutes of the Meeting of the General Board of State Hospitals,* September 17, 1923.

19. "Governmental Economy Bills Are Introduced" and "Governor Urges Saving Rather than Increasing Taxes," *Richmond (VA) Times-Dispatch,* January 10, 1924, 1, 8.

20. "Slow, Steady Growth, Particularly in the South," *Richmond (VA) Times-Dispatch,* October 28, 1923, sec. 2, p. 2; "Record Crops Grown Despite Poor Start," *Richmond (VA) Times-Dispatch,* November 17, 1923, 15; "Labor Conditions in This State Are Good," *Richmond (VA) Times-Dispatch,* November 17, 1923, 18.

21. Aubrey Strode to Don Preston Peters, July 19, 1939, AES Papers.

22. Laughlin, *Eugenical Sterilization,* 59.

23. *Virginia Acts of Assembly,* 1924, ch. 394, 569–71, "An Act to provide for the sexual sterilization of inmates of State institutions in certain cases."

24. See, for example, the law to expand the Colony to accept feebleminded "women of child bearing age." *Virginia Acts of Assembly,* 1912, ch. 196. For Strode's legislation, see *Virginia Acts of Assembly,* 1916, chs. 104, 106, 388.

25. Laughlin, *Eugenical Sterilization,* 446–51. The last section of Strode's law allowing therapeutic latitude to physicians also repeated, almost verbatim, a section from Laughlin's Model Law.

26. Aubrey E. Strode, "Sterilization of Defectives," *Virginia Law Review,* vol. 11, February 1925, 296–301. Unable to locate some of the sources he had quoted, Strode clarified that the citations were taken directly from Laughlin's book. Strode to Charles P. Nash, January 23, 1925, AES Papers.

27. "Committee Favors Sterilization Bill," *Richmond (VA) News Leader,* February 13, 1924, 1.

28. "Sterilization Bill Reported in House," *Richmond (VA) Times-Dispatch,* February 28, 1924, 14.

29. "Helen Keller's Address Thrills House Members," *Richmond (VA) Times-Dispatch,* February 15, 1924, 1.

30. "To Further Aims of Racial Law," *Richmond (VA) Times-Dispatch,* March 19, 1924, 2.

31. See, for example, "Conference Body Is Appointed to Thresh Out the Differences," *Roanoke (VA) World-News,* March 8, 1924, 1–2.

32. "Delegates Pass 104 House Bills in 1 Roll Call," *Richmond (VA) Times-Dispatch,*

January 23, 1924, 1; and "General Assembly Adjourns, Sine Die," *Richmond (VA) Times-Dispatch,* March 19, 1924, 12.

33. *Virginia Acts of Assembly,* 1924, ch. 394.

34. "America's New Hall of Fame for Science," *Chicago Tribune,* May 18, 1924, J3.

35. *Minutes of the Meeting of the Special Board of Directors of the State Colony for Epileptics and Feebleminded,* August 6, 1924.

36. The General Board approved the test of the law and agreed to prorate the costs of the litigation among all five state institutions; Governor Trinkle was present at the meeting. *Minutes of the Meeting of the General Board of Hospitals,* August 13, 1924.

37. *Minutes of the Meeting of the Special Board of Directors of the State Colony for Epileptics and Feebleminded,* September 9–10, 1924.

CHAPTER 8. Choosing Carrie Buck

1. "Deposition of Mrs. Alice and J. T. Dobbs," in "Official Interrogatories and Papers of Commitment" for Carrie E. Buck, January 23, 1924, 21–26. These papers are summarized in the official *Briefs and Records* of *Buck v. Bell,* submitted to the Virginia Supreme Court of Appeals, hereafter *Buck Record.*

2. "Deposition of Mrs. Alice and J. T. Dobbs," *Buck Record,* 22, 24, 25.

3. "Deposition of Mrs. Alice and J. T. Dobbs," *Buck Record,* 22, 26, 30–31.

4. "Certificate of the Mental Examiner," *Buck Record,* 26.

5. "Order of Commitment to State Colony for Epileptics and Feeble-minded," *Buck Record,* 28.

6. Homer Ritchie to A. S. Priddy, March 10, 1924, and Priddy to Ritchie, March 13, 1924, Colony Record no. 1692.

7. Caroline Wilhelm to Priddy, March 11, 1924, Colony Record no. 1692.

8. Priddy to Wilhelm, March 14, 1924, Colony Record no. 1692.

9. Reference to J. F. Sables, Green Street, Charlottesville, as most recent local address for Carrie Buck, *Buck Record,* 18.

10. Priddy to Wilhelm, May 7, 1924; and Wilhelm to Priddy, May 5, 1924, Colony Record no. 1692.

11. Wilhelm to Priddy, May 27, 1924; and "Physical Examination of a Patient on Admission to the State Colony," June 5, 1924, Colony Record no. 1692.

12. "History and Clinical Notes, Carrie E. Buck," July 20, 1924, Colony Record no. 1692.

13. The Standard Revision of the Binet-Simon test was administered to Carrie, with the first two years of the Kuhlman Revision and the Supplementary Form Board and Construction tests. See June 5, 1924, Colony Record no. 1692.

14. "Register of Students," McGuffey School, Charlottesville, Virginia, 1916–1917; see also Records of the Midway School, 1913–1915, and McGuffey School, 1915–1918. During her school years, Carrie was registered under the name of her foster parents, Dobbs.

15. University of Virginia Hospital Ledgers, 1908, 1914; Claude Moore Health Science Library, Department of Special Collections. Doris Buck's later records incorrectly list her birth date as 1911.

16. "History and Clinical Notes," April 6, 1920, Colony Record no. 1141.

17. *Buck Record*, 17–20.

18. *Minutes of the Meeting of the Special Board of Directors of the State Colony for Epileptics and Feebleminded*, September 9–10, 1924.

19. Priddy made the following entry the week after the board meeting: "This girl being desirous of taking advantage of the sterilization law and duly notified in accordance with the law, was present at the meeting of the Board of the Colony on September 10, 1924 when the petition of the Superintendent for her sterilization was heard and granted and sterilization ordered. The Board and its Attorney, desiring to make a test case of the constitutionality of the law, her guardian, Mr. R. G. Shelton, who was present, appealed from the order of the Board to the Circuit Court of Amherst County." "History and Clinical Notes," September 18, 1924, Colony Record no. 1692.

20. *Buck Record*, 33–35.

21. *Buck Record*, 36.

22. Priddy to Wilhelm, September 18, 1924; and Mary Duke to Priddy, September 22, 1924, Colony Record no. 1692.

23. Priddy to Edith Furbush, September 18, 1924, Colony Record no. 1692. Before long, Furbush would criticize the Laughlin's sterilization scheme, questioning professional ability to determine who was "socially inadequate" or "cacogenic." See Edith M. Furbush, *Mental Hygiene*, vol. 10, 1926, 866–67, review of Harry H. Laughlin, *Eugenical Sterilization: 1926* (New Haven: American Eugenics Society, 1926).

24. Strode to Laughlin, September 30, 1924, AHE Papers.

25. Strode to Priddy, October 7, 1924, Colony Record no. 1692.

26. Laughlin to Strode, October 3, 1924, Colony Record no. 1692. Laughlin included a "Family Tree Folder" and "Single Trait Sheets" for Strode's use.

27. Priddy to Strode, October 14, 1924; and Priddy to Laughlin, October 14, 1924, Colony Record no. 1692.

28. Priddy to Laughlin, October 14, 1924, Colony Record no. 1692.

29. Caroline Wilhelm to Priddy, October 15, 1924, Colony Record no. 1692.

30. Strode to Estabrook, October 10, 1924, AHE Papers.

31. Strode to Priddy, October 29, 1924, Colony Record no. 1692.

32. Priddy to DeJarnette, November 1, 1924, Colony Record no. 1692.

33. Priddy to Strode, November 1, 1924, Colony Record no. 1692.

34. Jessie Estabrook to A. H. Estabrook, October 23, 1924; and Strode to Laughlin, November 5, 1924, AHE Papers.

35. *Buck Record*, 39.

36. Aubrey Strode to Arthur Estabrook, November 6, 1924, AHE Papers.

37. Arthur Estabrook to Aubrey Strode, November 8, 1924; and Strode to Estabrook, November 15, 1924, AHE Papers.

CHAPTER 9. Carrie Buck versus Dr. Priddy

1. Testimony of Anne Harris, *Buck Record*, 51–56.

2. *Virginia Acts of Assembly*, 1918, ch. 233, 411, "An Act to provide for public health nursing and medical inspection and health inspection of school children"; comments on the benefits of the course made by E. G. Williams, M.D., in Paul G. Pope, "Factors Involved in Medical Inspection of School Children," *Southern Medical Journal*, vol. 14, March 1921, 205–10, quotation at 209.

3. Testimony of Miss Eula Wood, Miss Virginia Beard, and Miss Virginia Landis, *Buck Record*, 57–60.

4. Testimony of John W. Hopkins, *Buck Record*, 60–63.

5. Testimony of Samuel Dudley, *Buck Record*, 63–65.

6. Testimony of Miss Caroline E. Wilhelm, *Buck Record*, 65–69.

7. Testimony of Miss Mary Duke, *Buck Record*, 69–71.

8. *Buck Record*, 51.

9. *Buck Record*, 52, 57.

10. *Report of the Western State Hospital* (Richmond, VA: Superintendent of Public Printing, 1908) 10, hereafter, *Report, WSH*.

11. Paul A. Lombardo, "Involuntary Sterilization in Virginia: From *Buck v. Bell* to *Poe v. Lynchburg*," *Developments in Mental Health Law*, vol. 3, 1983, 13–21.

12. *Report, WSH*, 1909, 17; 1911, 11; 1913, 11–12.

13. *Report, WSH*, 1915, 9–10; 1919, 10.

14. *Report, WSH*, 1920, 13. The poem was reproduced numerous times, including one printing in the state medical society journal as part of a DeJarnette retrospective on sterilization. J. S. DeJarnette, "Eugenical Sterilization in Virginia," *Virginia Medical Monthly*, vol. 57, January 1931, 678–80, quotation at 680. Doctors in other states also reprinted the poem. See Edward J. Larson, *Sex, Race, and Science: Eugenics in the Deep South* (Baltimore: Johns Hopkins University Press, 1996) 44–45.

15. Testimony of Dr. J. S. DeJarnette, *Buck Record*, 71–82.

16. Henry Herbert Goddard, *The Kallikak Family: A Study of the Heredity of Feeble-Mindedness* (New York: MacMillan, 1912) 96.

17. Arthur H. Estabrook and Ivan E. McDougle, *Mongrel Virginians: The WIN Tribe* (Baltimore: Williams and Wilkins, 1926).

18. Louisa Hubbard to "Sol" Estabrook, February 1, 1923, AHE Papers; Louisa Hubbard to Aubrey Strode, February 13, 1923, Estabrook to Hubbard, February 10, 1923, AES Papers.

19. Testimony of A. H. Estabrook, *Buck Record*, 82–93.

20. Testimony of Dr. A. S. Priddy, *Buck Record*, 94–102.

21. *Report, State Epileptic Colony, 1924–25, 8.*

22. "Deposition of Harry Laughlin," *Buck Record,* 37–50.

23. *Buck Record,* 41.

CHAPTER 10. Defenseless

1. Henry Herbert Goddard, *Feeble-Mindedness: Its Causes and Consequences* (New York: Macmillan, 1914) 499.

2. Goddard, *Feeble-Mindedness,* 29.

3. See, for example, Charles B. Davenport, *The Feebly Inhibited: Nomadism, or the Wandering Impulse, with Special Reference to Heredity* (Washington, D.C.: Carnegie Institution, 1915), 126 (fig. 2, "Pedigree Charts"), 136 (fig. 28), 141 (fig. 45), and 142 (fig. 46).

4. "Posterity of Ada Jukes: Analysis of Pedigree for Inheritance for Shiftlessness," [1914?], Eugenics Record Office files, American Philosophical Society.

5. Charles B. Davenport et al., "The Study of Human Heredity," *Eugenics Record Office Bulletin,* no. 2 (Cold Spring Harbor, NY: Eugenics Record Office, 1911) 5.

6. Davenport, *Feebly Inhibited,* 141 (fig. 43, "Pedigree Charts").

7. Albert Wiggam, "New Styles in Ancestors," *World's Work,* vol. 55, December 1927, 142–50, quotation at 148.

8. Statement of Harry H. Laughlin, "Biological Aspects of Immigration," *Hearings before the Committee on Immigration and Naturalization, House of Representatives,* April 16–17, 1920 (Washington, D.C.: Government Printing Office, 1920) 4–5, 9–10.

9. Theodore Roosevelt, *The Foes of Our Own Household* (New York: George H. Doran Co., 1917) 259.

10. See *Marriage Register, City of Charlottesville, Va.* ("Frank W. Buck to Emma Adeline Harlow, September 23, 1896").

11. See, for example, *Scott v. Hillenberg,* 85 Va. 245 (1888) (law presumes that child born in wedlock is legitimate, even though the father was absent for an extended period); *Reynolds v. Adams,* 125 Va. 295 (1919) (law presumes legitimacy).

12. "Register of Students," McGuffey School, Charlottesville, Virginia, 1916–1917.

13. Author interview with Carrie Buck, December 27, 1982.

14. See "Register of Students," Midway School, Charlottesville, Virginia, 1915.

15. Charles Giametta, "They Told Me I Had to Have An Operation," *Charlottesville (VA) Daily Progress,* February 26, 1980, A1; and Wendy Blair, narrator, "To Raise the Intelligence of the State," *Horizons,* National Public Radio, 1980.

16. Paul A. Lombardo, "Eugenical Sterilization in Virginia: Aubrey Strode and the Case of *Buck v. Bell,*" Ph.D. dissertation, University of Virginia, 1982, 57–61.

17. "Unwritten Law for Loving: One of His Attorneys Appeals to Virginia Chivalry," *New York Times,* June 29, 1907, 2.

18. M. Wills to Aubrey Strode, April 10, 1899, January 1, 1900, March 4, 1900, AES Papers.

19. *Code of Virginia*, 1923, §4414. "Offenses against the person: Rape."

20. *Code of Virginia*, 1923, §4410. "Offenses against the person: Seduction of female of previous chaste character."

21. *Warren Wallace Smith vs. Charles F. Williams*, no. 12,018, *Transcript on Appeal from the Clark Circuit Court, to the Supreme Court of the State of Indiana*, January 13, 1920, 4.

22. *Williams v. Smith*, 190 Ind. 526 (1921).

23. A. H. Estabrook, "The Indiana Survey," *Journal of Heredity*, vol. 8, April 1917, 156–59; *Mental Defectives in Indiana: Second Report of the Indiana Committee on Mental Defectives* (Indianapolis: Wm. B. Burford, 1919) 6; and Arthur H. Estabrook, "The Tribe of Ishmael," in *Eugenics, Genetics and the Family: Scientific Papers of the Second International Congress of Eugenics* (Baltimore: Williams and Wilkins, 1923) 398–404.

24. Harry H. Laughlin, *Eugenical Sterilization in the United States* (Chicago: Psychopathic Laboratory, Municipal Court of Chicago, 1922) 255, 256.

25. "Pedigree Chart of Warren Wallace Smith," Laughlin, *Eugenical Sterilization*, 320a.

26. *Laws of New York*, 1912, ch. 445, "An Act to amend the public health law, in relation to operations for the prevention of procreation," April 16, 1912.

27. Record on Appeal, *Frank Osborn v. Lemon Thomson et al. Board of Examiners of Feeble-Minded, Criminals and other Defectives*, Supreme Court of the State of New York, Appellate Division, Third Department, "Minutes of Trial," September 17, 1915, 117, hereafter *Osborn Record*.

28. Charles Davenport, "Animal Morphology in its Relation to other Sciences," *Science*, vol. 20, November 25, 1904, 697–706, quotation at 698.

29. *Osborn Record*, 125, 132, 139.

30. *Osborn Record*, 194.

31. *Osborn Record*, 26–32.

32. Laughlin, *Eugenical Sterilization*, 216–41.

33. *Osborn v. Thomson*, 169 N.Y.S. 638 (1918).

34. J. E. Wallace Wallin, "Who is Feebleminded? A Reply to Mr. Kohs," *Journal of Criminal Law and Criminology* vol. 7, 1916–1917, 56–78; and Samuel C. Kohs, "Who is Feeble-Minded? A Rejoinder and a Rebuttal," *Journal of Criminal Law and Criminology*, vol. 7, 1916–1917, 219–26.

35. Henry H. Goddard, "The Binet Test in Relation to Immigration," *Journal of Psycho-Aesthenics*, vol. 22, 1913, 105–7, quotation at 106.

36. Record Sheet for the Stanford Revision of the Binet-Simon Tests, Colony File no. 1692.

37. The Buck photos are preserved in the Arthur Estabrook Papers and analyzed in Paul A. Lombardo, "Facing Carrie Buck," *Hastings Center Report*, vol. 33, 2003, 14–17.

38. Testimony of A. H. Estabrook, *Buck Record*, 89.

39. Deposition of Harry Laughlin, *Buck Record*, 40.

40. Testimony of Dr. A. S. Priddy, *Buck Record*, 41.

41. Colony Report, 1913, 18.

42. Joel D. Hunter, "Sterilization of Criminals: Report of Committee H of the Institute," *Journal of Criminal Law and Criminology*, vol. 5, 1914–1915, 514–39, quotation at 524.

43. H. E. Jordan, "Surgical Sex-Sterilization: Its Value as a Eugenic Measure," *American Journal of Clinical Medicine*, vol. 20, 1913, 983–87, esp. 986.

44. Daniel J. Kevles, *In the Name of Eugenics: Genetics and the Uses of Human Heredity* (New York: Alfred A. Knopf, 1985) 148.

45. Allan Chase, *The Legacy of Malthus: The Social Costs of the New Scientific Racism* (New York: Alfred Knopf, 1977) 310–13.

46. H. S. Jennings, "Heredity and Environment," *Scientific Monthly*, vol. 19, September 1924, 225–38, quotation at 228.

47. See Elof Axel Carlson, *The Unfit: A History of a Bad Idea* (Cold Spring Harbor, NY: Cold Spring Harbor Laboratory Press, 2001) 342–43.

48. R. C. Punnett, "Eliminating Feeblemindedness," *Journal of Heredity*, vol. 8, October 1917, 464–65.

49. Raymond Pearl, "Sterilization of Degenerates and Criminals Considered from the Standpoint of Genetics," *Eugenics Review*, vol. 11, April 1919, 1–6, quotation at 5.

50. C. C. Little, "The Relation between Research in Human Heredity and Experimental Genetics," *Scientific Monthly*, vol. 14, May 1922, 401–14, quotation at 404.

51. *New Republic*, vol. 45, December 30, 1925, 148.

52. Samuel J. Holmes, *A Bibliography of Eugenics* (Berkeley: University of California Press, 1924) 2.

53. J. B. S. Haldane, "Eugenics and Social Reform," *New Republic*, vol. 39, June 25, 1924, 118–19, quotation at 118.

54. The University of Virginia Law library had the complete set of the *Journal of Criminal Law and Criminology* in 1924, as did the Johns Hopkins University Library. That journal, along with periodicals such as the *American Bar Association Journal*, covered the debate over sterilization from 1910 through the time of the *Buck* trial in exhaustive detail.

55. Colony Report, 1911, 10; 1913, 18; 1915, 17; 1916, 12; 1922–23, 28.

CHAPTER 11. On Appeal

1. Albert Priddy to Caroline Wilhelm, November 28, 1924, Colony File no. 1692.

2. Estabrook to Strode, December 11, 1924, AHE Papers.

3. Estabrook to Strode, December 27, 1924, AES Papers.

4. Priddy to Strode, December 12, 1924, Colony File no. 1692.

5. Priddy to Whitehead, December 12, 1924; Priddy to Caroline Wilhelm, November 28, 1924; and Priddy to Strode, December 12, 1924, Colony File no. 1692.

6. Strode to Charles Nash, July 24, 1924, AES Papers.

7. Aubrey E. Strode, "Sterilization of Defectives," *Virginia Law Review*, vol. 11,

February 1925, 296–301; and Aubrey Strode to Chas. P. Nash, January 23, 1925, AES Papers.

8. Strode to S. L. Ferguson, January 19, 1925, AES Papers. The formal decision from Judge Gordon was issued on April 13, 1925, some two months after he announced the result from the bench.

9. *Minutes of the Meeting of the General Board of State Hospitals*, February 11, 1925.

10. "John Hendren Bell," *The National Cyclopaedia of American Biography*, vol. 25 (New York: James T. White and Co., 1936) 421–22. Bell died of heart disease at the age of fifty-one, after eighteen years of service at the Colony. See *Virginia Medical Monthly*, vol. 61, January 1935, 624.

11. Strode to Dr. J. H. Bell, February 12, 1925; and J. H. Bell to Strode, February 13, 1924, Colony File no. 1692.

12. J. S. DeJarnette, "Eugenics in Relation to the Insane, the Epileptic, the Feeble-minded, and Race Blending," *Virginia Medical Monthly*, vol. 52, August 1925, 290–92, quotations at 291, 292.

13. W. A. Plecker, "Racial Improvement," *Virginia Medical Monthly*, vol. 52, November 1925, 486–90.

14. "Heredity Is Subject of Rev. Frank Pratt," *Richmond (VA) Times-Dispatch*, April 20, 1925, 9; and "Rise of the Unfit Threatens Civilization: Dr. Pratt Delivers Second Discourse on Problem of Heredity," *Richmond (VA) Times-Dispatch*, April 27, 1925, 8.

15. Supreme Court of Appeals of Virginia, September Term, 1925, *Carrie Buck v. Dr. J. H. Bell*, Brief for Appellee, 13.

16. *Buck v. Bell*, Brief for Appellee, 21–25.

17. *Buck v. Bell*, Brief for Appellee, 27–28.

18. *Buck v. Bell*, Brief for Appellee, 35–36.

19. *Buck v. Bell*, Brief for Appellee, 43.

20. Supreme Court of Appeals of Virginia, September Term, 1925, *Carrie Buck v. Dr. J. H. Bell*, Reply Brief for Appellant.

21. *Smith v. Board of Examiners of Feebleminded*, 85 N.J. L. 46 (1913); *Davis v. Berry*, 216 Fed. 413 (S.D. Iowa, 1914) *rev'd for mootness; Berry v. Davis*, 242 U.S. 468 (1916); *Haynes v. Lapeer*, 166 N.W. 938 (Mich. 1918); *Osborn v. Thompson*, 169 N.Y. Supp. 638 (1918); and *Mickle v. Hendrichs*, 262 F. 687 (Nev. 1918).

22. "Constitutional Law—Due Process," *Michigan Law Review*, vol. 20, 1921–1922, 101–2.

23. *Buck v. Bell*, 143 Va. 310 (1925) at 317, 319, 323.

24. "Sterilization Act Constitutional," *Virginia Law Register*, vol. 11, March 1926, 691–93, quotations at 691, 693.

25. *Minutes of the Meeting of the Special Board of Directors of the State Colony for Epileptics and Feebleminded*, December 7, 1925.

26. See *Minutes of the Meeting of the Special Board of Directors of the State Colony for*

Epileptics and Feebleminded. Estabrook payments are noted on July 11, 1925 ($298.80), October 12, 1925 ($276.47), April 12, 1926 ($295), and November 9, 1926 ($42 and $257).

27. Aubrey Strode to Senator Carter Glass, November 12, 1924, AES Papers.

28. See, for example, *Thomas v. Turner's Administrator,* 87 Va. 1 (1890).

29. Clarence Darrow, "The Edwardses and the Jukeses," *American Mercury,* vol. 6, October 1925, 147–57, quotations at 153, 155–57.

30. Clarence Darrow, "The Eugenics Cult," *American Mercury,* vol. 8, June 1926, 129–37, quotations at 134, 135, 137.

31. Harvey Wickham, *The Misbehaviorists: Pseudo-Science and the Modern Temper* (New York: Dial Press, 1928) 204, 210.

32. Henry Goddard, "Who Is a Moron?" *Scientific Monthly,* vol. 24, January 1927, 41–46, quotations at 42, 43, 44; and Henry H. Goddard, "Feeblemindedness: A Question of Definition," *Journal of Psycho-Aesthenics,* vol. 33, 219–27, quotations at 223, 225.

33. George Ordahl, "Heredity in the Feebleminded," *Transactions of the Commonwealth Club of California,* vol. 21, June 22, 1926, 168–74, quotation at 174.

34. Margaret Sanger, "The Function of Sterilization," *Birth Control Review,* October 1926, 299. From an address given at Vassar College, August 5, 1926.

35. Harper Leech, "Sees in Eugenics Way to Cut Cost of Government," *Chicago Daily Tribune,* September 14, 1926, 24.

CHAPTER 12. In the Supreme Court

1. *Jacobson v. Massachusetts,* 197 U.S. 11 (1905).

2. Wendy Parmet et al., "Individual Rights versus the Public's Health: 100 Years after *Jacobson v. Massachusetts,*" *New England Journal of Medicine,* vol. 352, February 17, 2005, 652–54.

3. U.S. Supreme Court, October Term, 1926, *Carrie Buck v. J. H. Bell,* Brief for Plaintiff in Error, 6, 17, 18.

4. *Union Pacific Railway v. Botsford,* 141 U.S. 250 (1891) at 251.

5. "Bill Requires Kansans to Have $1000 to Wed," *Washington Post,* February 10, 1927, 3.

6. "Sterilization Law in Indiana," *Washington Post,* March 12, 1927, 9; and "Governor Vetoes Sterilization Bill," *Journal of the American Medical Association,* vol. 88, April 30, 1927, p. 1423.

7. "German Renaissance," *Time,* October 4, 1926, 20–21.

8. William Howard Taft to Helen Taft Manning, March 15, 1926, cited in Alpheus Thomas, *William Howard Taft: Chief Justice* (New York: Simon and Schuster, 1965) 40, 275.

9. Judith Icke Anderson, *William Howard Taft: An Intimate History* (New York: W.W. Norton, 1981) 197–98.

10. "Yale as Battle-Ground," *New York Times,* April 4, 1880, 1.

11. Herbert Spencer, *The Study of Sociology* (New York: Appleton, 1874) 343.

12. Annie L. Cot, "'Breed Out the Unfit and Breed in the Fit': Irving Fisher, Economics and the Science of Heredity," *American Journal of Economics and Sociology,* vol. 64, July 2005, 793–826, quotation at 795. Sumner recommended Fisher for a position during Taft's presidency; see Sumner to Taft, January 13, 1909, and Taft to Sumner, January 18, 1909, WHT Collection.

13. See, for example, "Irving Fisher's Weekly Index," *Wall Street Journal,* November 15, 1926, 2.

14. See "Remington Rand Meeting Routine," *Wall Street Journal,* October 19, 1927, 16.

15. Fisher to Laughlin, October 30, 1934, HHL Papers.

16. Mark H. Haller, *Eugenics: Hereditarian Attitudes in American Thought* (New Brunswick, NJ: Rutgers University Press, 1963) 173.

17. Irving Fisher, "Impending Problems of Eugenics," presidential address of the Eugenics Research Association, *Scientific Monthly,* vol. 13, September 1921, 214–31 at 216, 231.

18. Jonathan Peter Spiro, "Patrician Racist: The Evolution of Madison Grant," Ph.D. dissertation, University of California, Berkeley, 2000, 15.

19. Taft to Fisher, May 18, 1908, WHT Collection.

20. Henry Gannett, ed., *Report of the National Conservation Commission* (Washington: Government Printing Office, 1909); Senate Document no. 676, 60th Congress; and "National Vitality," 620–751, chs. 6, 10.

21. *Report of the National Conservation Commission,* 672, 673, 674, 719, 723.

22. Christine R. Whittaker, "Chasing the Cure: Irving Fisher's Experience as a Tuberculosis Patient," *Bulletin of the History of Medicine,* vol. 48, 1974, 398–415.

23. *Report of the National Conservation Commission,* 748–51.

24. Irving Fisher, *National Vitality: Its Conservation and Preservation* (Washington, D.C.: Government Printing Office, 1910).

25. Cot, "'Breed out the Unfit,'" 817n14.

26. Taft to Fisher, July 8, 1913, WHT Collection.

27. Eugene Lyman Fisk to R. L. Noak, April 23, 1934, ERO records, Cold Spring Harbor Laboratory Archives, hereafter CSHL.

28. Fisher to Taft, May 4, 1914, attachment listing board members, WHT Collection.

29. Irving Fisher and Eugene Lyman Fisk, *How to Live: Rules for Healthful Living Based on Modern Science* (New York: Funk and Wagnalls, 1915).

30. Laura Davidow Hirshbein, "Irving Fisher and the Life Extension Institute, 1914–1931," *Canadian Bulletin of Medical History,* vol. 16, 1999, 89–124, esp. 95.

31. *Eugenical News,* vol. 1, January 1916, 4.

32. Fisher and Fisk, *How to Live,* 167, 323.

33. "How to Live," *New York Times,* January 2, 1916, enclosed in letter, Fisher to Taft,

January 14, 1916, WHT Collection. See also "Scientists Agree with Dr. Depew that Men Ought to Live to Be 100 by Observing Rules of Health," *New York Times,* November 26, 1916, p. ES3.

34. Fisher to Taft, October 16, 1926, and Taft to Fisher, October 19, 1926, WHT Collection.

35. "Now for Eugenic Wooing," *New York Tribune,* January 13, 1914.

36. Charles B. Davenport, *Heredity in Relation to Eugenics* (New York: Henry Holt, 1913) 47.

37. "Mr. Justice Holmes," *New Republic,* vol. 46, March 17, 1926, 83–84; and Elizabeth Shepley Sargeant, "Oliver Wendell Holmes," *New Republic,* vol. 49, December 8, 1926, 59–64, quotation at 59.

38. Justice Oliver Wendell Holmes, "The Path of the Law," *Harvard Law Review,* vol. 10, 1897, 457–78, quotation at 470.

39. Oliver Wendell Holmes, "The Soldier's Faith," quoted in Sheldon M. Novick, "Justice Holmes Philosophy," *Washington University Law Quarterly,* vol. 70, 1992, 703–53, quotation at 728.

40. Novick, "Justice Holmes Philosophy," 709–12.

41. "The Gas-Stoker's Strike," *American Law Review,* vol. 7, 1873, 582–84, quotation at 583. Sheldon Novick identifies Holmes as the anonymous author; see Novick, *Honorable Justice: The Life of Oliver Wendell Holmes* (Boston: Little Brown, 1989) 431n22–23. Holmes clearly rejects Spencerian utilitarianism while endorsing the social Darwinist aspects of Spencer's thought.

42. Oliver Wendell Holmes to Frederick Pollock, April 23, 1910, *Holmes-Pollock Letters: The Correspondence of Mr. Justice Holmes and Sir Frederick Pollock 1874–1932,* vol. 1, ed. Mark De Wolfe Howe (Cambridge, MA: Harvard University Press, 1942) 169–71.

43. Herbert Spencer, *Social Statics* (London: John Chapman, 1851) 325, 228; 93: "there exists in man what may be termed an *instinct of personal rights*—a feeling that leads him to repel anything like an encroachment upon what he thinks his sphere of original freedom."

44. *Lochner v. New York,* 198 U.S. 45 (1905) at 76.

45. Holmes to Laski, May 24, 1919, in *Holmes-Laski Letters: The Correspondence of Mr. Justice Holmes and Harold J. Laski, 1916–1935,* ed. Mark DeWolfe Howe, vol. 1 (Cambridge, MA: Harvard University Press, 1953), 207.

46. Holmes to Wigmore, November 19, 1915, quoted in *Justice Oliver Wendell Holmes: The Shaping Years, 1841–1870,* ed. Mark DeWolfe Howe (Cambridge: Harvard University Press, 1957) 25.

47. Harold Laski, "The Scope of Eugenics," *Westminster Review,* vol. 174, 1910, 25–34 at 25, 26, 27, 28, 32, 34.

48. See, for example, Holmes to Laski, December 9, 1921: "In short I believe in Malthus—in the broad—not bothering about details"; June 14, 1922: "Is not the present

time an illustration of Malthus?"; and September 16, 1924: "I am a devout Malthusian, as you know"; June 14, 1922; July 17, 1925, in *Holmes-Laski Letters,* ed. Howe, vol. 1, 385, 431, 658–59, 761.

49. Holmes to Clare Fitzpatrick, Lady Castletown, August 19, 1897, quoted in Novick, "Justice Holmes Philosophy," 729.

50. Holmes to Frankfurter, September 3, 1921, in *Holmes and Frankfurter: Their Correspondence, 1912–1934,* ed. Robert M. Mennel and Christine L. Compston (Hanover, NH: University Press of New England, 1996) 124–25.

51. Strode argued *U.S. v. Leary* 245 U.S. 1 (1917) and *Leary v. Leary* 253 U.S. 94 (1920). Strode sent several letters and telegrams to the Clerk of the Supreme Court to determine exactly when oral arguments would be scheduled for *Buck v. Bell.* See Strode to Clerk of the Supreme Court, January 15, 1927, Strode to William R. Stansbury, February 25, March 28, April 11, 16, and 18, 1927; and Whitehead to William R. Stansbury, March 30, 1927, U.S. Archives, case file of *Buck v. Bell,* no. 31681.

52. Taft to Holmes, April 23, 1927, quoted in Novick, *Honorable Justice,* 351–52.

53. Holmes to Laski, April 29, 1927, in *Holmes-Laski Letters,* ed. Howe, vol. 2, 938–39.

54. Holmes to Felix Frankfurter, October 24, 1920, Robert Mennel and Christine L. Compston, eds., *Holmes and Frankfurter: Their Correspondence, 1912–1934,* 95.

55. Laski to Holmes, May 7, 1927, in *Holmes-Laski Letters,* ed. Howe, vol. 2, 941.

56. All quotations from the Holmes opinion, *Buck v. Bell,* 274 U.S. 200 (1927).

57. "Memorial Day," address delivered May 30, 1884, in *The Essential Holmes: Selections from the Letters, Speeches, Judicial Opinions, and Other Writings of Oliver Wendell Holmes, Jr.,* ed. Richard A. Posner (University of Chicago Press, 1992) 80–87, quotation at 87.

58. Quoted in G. Edward White, "The Rise and Fall of Justice Holmes," *University of Chicago Law Review,* vol. 39, 1971–1972, 51–77, quotation at 53.

59. Joseph P. Pollard, "Justice Holmes, Champion of the Common Man," *New York Times,* December 1, 1929, 65 ("Justice Holmes is an aristocrat with a genuine interest in the welfare of the common man").

60. "Oliver Wendell Holmes," *New Republic,* vol. 49, December 8, 1926, 59.

61. Alexander Johnson, "Concerning a Form of Degeneracy. II. The Education and Care of the Feeble-Minded," *American Journal of Sociology,* vol. 4, January 1899, 463–73, quotation at 466: "In every institution there began to be an accumulation of inmates at or past the legal age limit, who yet were so manifestly unfit for self control that the managers felt it a wrong both to them and to the community to dismiss them."

62. Charles Richmond Henderson, "Ethical Problems of Prison Science," *International Journal of Ethics,* vol. 20, April 1910, 281–95, quotation at 285.

63. "Hope of Better Brains for All," *New York Times,* September 27, 1912, 9; and "The Feeble-Minded in Schools," *New York Times,* October 8, 1912, 12. See also "Fund to Forward Mental Hygiene," *New York Times,* November 10, 1912, 12.

64. Rev. Mabel Irwin, "Parenthood a Privilege, Not a Punishment, Says Woman Pastor," *Washington Post,* October 24, 1915, A8.

65. Davenport, *Heredity in Relation to Eugenics,* 256.

66. J. Ewing Mears, "Asexualization as a Remedial Measure in the Relief of Certain Forms of Mental, Moral, and Physical Degeneration," *Boston Medical and Surgical Journal,* vol. 161, October 21, 1909, 584–86, extract at 585.

67. Joseph Mayer, quoted in Charles P. Bruehl, *Birth Control and Eugenics in the Light of Fundamental Ethical Principles* (New York: Joseph F. Wagner, Inc., 1928) 128.

68. Robert Post, "The Supreme Court Opinion as Institutional Practice: Dissent, Legal Scholarship, and Decisionmaking in the Taft Court," *Minnesota Law Review,* vol. 85, 2001, 1267–1384, quotation at 1319.

69. Bleeker Van Wagenen [sic], "Preliminary Report of the Committee of the Eugenic Section of the American Breeders' Association to Study and to Report on the Best Practical Means for Cutting Off the Defective Germ-Plasm in the Human Population," *Problems in Eugenics: Papers Communicated to the First International Eugenics Congress* (London: Eugenics Education Society, 1912) 460–79, quotation at 478.

70. "The Nine Old Men," installment 11, *Sheboygan (WI) Press,* March 5, 1937, 7.

71. "Name Pierce Butler in $2,737,468 Suit," *Wisconsin Rapids (WI) Daily Tribune,* August 12, 1931, 1; "Nevada Woman Asks Part of Big Estate," *Bismarck (ND) Tribune,* August 12, 1931, 7; and "The Nine Old Men," installment 13, *Sheboygan (WI) Press,* March 8, 1937, 13

72. Conrad Black, *Franklin Delano Roosevelt: Champion of Freedom* (New York: Public Affairs, 2003) 405. For comments on anti-Semitism, see John M. Scheb II, "James Clark McReynolds," *The Oxford Companion to the Supreme Court of the United States,* ed. Kermit L. Hall (New York: Oxford University Press, 1992) 542.

73. See Sheldon M. Novick, *Honorable Justice: The Life of Oliver Wendell Holmes* (Boston: Little, Brown, 1989) 278–79.

74. Taft to Helen Heron Taft, May 6, 1927, WHT Collection.

75. *Olmstead v. U.S.,* 277 U.S. 438 (1928) at 472. Brandeis in dissent.

76. Holmes to Laski, May 12, 1927, Howe, *Holmes-Laski Letters,* vol. 2, 941–42.

77. Holmes to Lewis Einstein, May 19, 1927, Papers of Oliver Wendell Holmes Jr., Library of Congress.

78. Holmes to Laski, July 23, 1927, Howe, ed., *Holmes-Laski Letters,* vol. 2, 964.

CHAPTER 13. Reactions and Repercussions

1. "Safely through the Gamut," *Charlottesville (VA) Daily Progress,* May 3, 1927, 4.

2. "Upholds Operating on Feeble-Minded," *New York Times,* May 3, 1927, 19.

3. "Supreme Court Upholds Sterilization of Unfit," *Los Angeles Times,* May 3, 1927, 1; and "Sterilization of Defectives is Held Legal," *Chicago Daily Tribune,* May 3, 1927, 1.

4. "Sterilization Law of Virginia Upheld," *Baltimore Evening Sun,* May 2, 1927, 1; and "Va. Sterilization Act is Sustained," *Baltimore Sun,* May 3, 1927, 12.

5. "Supreme Court Upholds Sterilization Laws," *Boston Daily Globe*, May 2, 1927, 23.

6. "Sterilization," *Time*, May 16, 1927.

7. Russell Briney, "Sterilization of Defectives, Aim," *Louisville Courier-Journal*, May 27, 1927. Clipping, AHE Papers.

8. Raymond Clapper, "Sterilization Law Upheld by Court," *Atlanta Constitution*, May 3, 1927, 15.

9. "Virginia Sterilization Law Upheld by Court," *Washington Post*, May 3, 1927, 2.

10. "Virginia Act Upheld by Supreme Court," *Richmond (VA) Times-Dispatch*, May 3, 1927, 1.

11. J. C. Baskerville, "See Remedy for Imbecility in Recent Virginia Decision," *Gastonia (NC) Daily Gazette*, May 6, 1927, 11; and editorial, "To Protect the World against the Morons," *Davenport (IA) Democrat and Leader*, May 3, 1927, 6.

12. Editorial, "The Sterilization Law," *Kingsport (TN) Times*, May 31, 1927, 4.

13. Editorial, "Eugenics Wins in South," *Helena (MT) Daily Independent*, May 8, 1927, 4.

14. "Super-Race Seen following Court Sterilization OK," *Waterloo (IA) Evening Courier*, May 3, 1927, 17.

15. "A 'Sterilization' Ruling," *Montgomery (AL) Advertiser*, May 10, 1927.

16. "Supreme Court Upholds Sterilization Law," *Journal of the American Medical Association*, vol. 88, May 28, 1927, p. 1737.

17. James A. Tobey, "Law and Legislation: United States Supreme Court Upholds Sterilization," *American Journal of Public Health*, vol. 17, July 1927, 773–74, quotation at 774.

18. "To Halt the Imbecile's Perilous Line," *Literary Digest*, vol. 75, May 21, 1927, 11.

19. Victoria Claflin Woodhull Martin, *Stirpiculture, or the Scientific Propagation of the Human Race* (London, 1888) 8, 27; Victoria C. Woodhull Martin, *The Rapid Multiplication of the Unfit* (New York, 1891) 38; "Vote at Age of 25 for Women Early Enough, She Holds," *Washington Post*, May 8, 1927, 19; and "Says Voting at 25 is 'Young Enough,'" *New York Times*, May 8, 1927, E6.

20. Editorial, "Not a Panacea," *Decatur (IL) Herald*, June 12, 1927, 6.

21. Editorial, "Sterilization Upheld," *Kingston (NY) Daily Freeman*, May 11, 1927, 4.

22. Editorial, "Society's Self Protection," *Philadelphia Evening Bulletin*, reprinted in *Edwardsville (IL) Intelligencer*, May 11, 1927, 4.

23. "Sterilize Unfit, Say Physicians," *Lincoln (NE) State Journal*, Wednesday, May 25, 1927, 12.

24. "The Constitutionality of the Compulsory Asexualization of Criminals and Insane Persons," *Harvard Law Review*, vol. 26, 1912–1913, 163–65, quotation at 163.

25. "Harvard Declines a Legacy to Found Eugenics Course," *New York Times*, May 8, 1927, 1. See also "Harvard Declines Eugenic Request," *Journal of the American Medical Association*, vol. 88, June 18, 1927, p. 1973.

26. Unsigned note, Mears file, HHL Papers.

27. "Study Here Is Expected to Give Lead," *Pasadena (CA) Star-News,* May 3, 1927, reprint; E. S. Gosney to Laughlin, May 18, 1927, HHL Papers.

28. *Vancouver Sun,* May 25, 1927, 8, quoted in Angus McLaren, *Our Own Master Race: Eugenics in Canada, 1885–1945* (Toronto: McClelland and Stewart, 1990) 102.

29. Anon to J. H. Bell, May 4, 1927, Colony File no. 1962.

30. Colony Report, 1927–1928, 9.

31. Strode to W. Pendleton Sandridge Jr., June 24, 1927, AES Papers.

32. *Minutes of the Meeting of the Special Board of Directors of the State Colony for Epileptics and Feebleminded,* June 9, 1927.

33. Th. Laboure, "Is Vasectomy then Unlawful?" *Ecclesiastical Review,* vol. 44, January-June 1911, 574–83, quotation at 581.

34. "The Morality and Lawfulness of Vasectomy," *Ecclesiastical Review,* vol. 44, January-June 1911, 562–71, quotation at 570.

35. "The Morality and Lawfulness of Vasectomy," *Ecclesiastical Review,* vol. 45, July-December 1911, 71–77, quotation at 75.

36. Steven M. Donavan, "Summing up the Discussion on Vasectomy," *Ecclesiastical Review,* vol. 45, July-December 1911, 313–19, quotation at 316.

37. Charles Bruehl, "The State and Eugenical Sterilization," *Homiletic and Pastoral Review,* vol. 27, October 1926, 1–9, quotation at 6; Charles Bruehl, "The Morality of Sterilization," *Homiletic and Pastoral Review,* vol. 27, November 1926, 113–19, quotation at 113; and Charles Bruehl, "Sterilization and Heredity," *Homiletic and Pastoral Review,* vol. 27, December 1926, 225–31, quotation at 230.

38. Charles Bruehl, "Practical Objections against Legalized Sterilization," *Homiletic and Pastoral Review,* vol. 27, January 1927, 341–48, quotation at 342n1; Charles Bruehl, "Eugenical Education," *Homiletic and Pastoral Review,* vol. 27, March 1927, 569–76, quotation at 571; and Charles Bruehl, "Eugenics in the Christian Sense," *Homiletic and Pastoral Review,* vol. 27, May 1927, 801–9, quotation at 807.

39. Charles P. Bruehl, *Birth Control and Eugenics in the Light of Fundamental Ethical Principles* (New York: Joseph F. Wagner, Inc., 1928) 1, 70. Bruehl's book was a compilation of the eight articles he wrote in the *Homiletic and Pastoral Review* in 1926 and 1927.

40. William I. Lonergan, S.J., "The Morality of Sterilization Laws," *America,* vol. 34, March 13, 1926, 515–17.

41. "Unjustified Sterilization," *America,* vol. 37, May 14, 1927, 102.

42. John A. Ryan, "Unprotected Natural Rights," *Commonweal,* vol. 5, June 15, 1927, 151–52.

43. Holmes to Harold Laski, April 25, 1927, in *Holmes-Laski Letters,* vol. 2, ed. Mark DeWolfe Howe (Cambridge: Harvard University Press, 1953) 937–38.

44. See, for example, "Virginia Sterilization Act Held Constitutional," *Journal of the American Medical Association,* vol. 86, June 5, 1926, p. 1790.

45. Charles F. Dolle to Reverend John J. Burke, May 28, 1927, Papers of the National Conference on Catholic Men, Catholic University of America Archives, Washington, D.C., hereafter, NCCM Papers.

46. Dolle to Burke, June 3, 1927; and JJB, June 3, 1927 (Burke to file), NCCM Papers.

47. Memorandum, Dolle to Burke, June 8, 1927, "Petititon in Carrie Buck case," NCCM Papers.

48. I would like to thank Sharon Leon for pointing out the Burke/Dolle correspondence for me. She first identified the origins of the petition for rehearing in "Beyond Birth Control: Catholic Responses to the Eugenics Movement in the United States, 1900–1950," Ph.D. dissertation, University of Minnesota, 2004.

49. Irving Whitehead, "Petition for Rehearing and Argument," *Buck v. Bell*, 274 U.S. 200 (1927) 10.

50. See *The X-Ray*, vol. 3, September 1926, 1.

51. *Jacobson v. Massachusetts*, 197 U.S. 11 (1905); and Whitehead, "Petition for Rehearing and Argument," 2–4.

52. Whitehead, "Petition for Rehearing and Argument," 4.

53. See Lombardo, "Hysterical Women and Phantom Tumors," *Journal of Law Medicine and Ethics*, vol. 33, 2005, 791–801.

54. *Truax v. Corrigan*, 257 U.S. 312 (1921) at 313, cited in Whitehead, "Petition for Rehearing and Argument," 10, 11.

55. See generally, Record and Briefs, *Smith v. Command*, 231 Mich. 409 (1925).

56. *Smith v. Command*, at 144, 147.

57. See, for example, C. E. Shults Jr., "Constitutional Law: Sterilization of Mental Defectives," *Cornell Law Quarterly*, vol. 11, 1925–1926, 74–78.

58. Memorandum, Charles F. Dolle to John J. Burke, June 8, 1927, "Petition in Carrie Buck case," NCCM Papers.

59. Charles Elmore Cropley, "Report of Survey by the Clerk of Rules and Practice in Relation to Petitions for Rehearing," January 7, 1947 (unpublished), Library of the Supreme Court, Washington, D.C.

60. *Report WSH*, 1927, 9.

61. "Memorandum of Denial," October 10, 1927, U.S. Archives file of *Buck v. Bell;* and "Review of Buck Case is Denied," *Richmond (VA) Times-Dispatch,* October 11, 1927, 16.

62. Raymond Pearl, "The Biology of Superiority," *American Mercury*, vol. 12, November 1927, 257–66, quotations at 260, 266.

63. "Says Eugenists Do Cause Harm," *Richmond (VA) News Leader,* October 24, 1927, 1, 12.

64. Donald Lines Jacobus, "Haphazard Eugenics," *North American Review,* February 1928, vol. 325, 168–74, quotations at 169, 174.

65. "Dr. Charles Mayo, Famous Surgeon, Favors Sterilization," *Madison Wisconsin Journal,* October 3, 1928. See also Eugene Perry Link, *The Social Ideas of American Physicians, 1776–1976* (Selingsgrove, PA: Susquehanna University Press, 1992) 183.

66. "Eugenic Asexualization," *Journal of the American Medical Association,* vol. 90, May 5, 1928, p. 1462.

67. Davenport to Estabrook, November 21, 1924, CBD Papers. Davenport also mentioned the problem of "loneliness" that was endemic to fieldwork away from home.

68. Strode to Estabrook, December 11, 1924; Strode to Estabrook, November 6, 1924, AHE Papers; and Priddy to Strode, December 12, 1924, Colony Record no. 1692.

69. Estabrook to Strode, December 27, 1924, AES Papers.

70. Lousia Hubbard to Arthur H. Estabrook, February 1, 1923, AHE Papers.

71. Arthur H. Estabrook and Ivan E. McDougle, *Mongrel Virginians: The WIN Tribe* (Baltimore: Williams and Wilkins, 1926).

72. Ivan McDougle to Arthur Estabrook, January 27, 1924, AHE Papers.

73. Davenport to Estabrook, November 21, 1924; and Davenport to Mrs. Estabrook, November 28, 1928, CBD Papers.

74. Estabrook to Davenport, June 23, 1929; and Davenport to Estabrook, June 27, 1929, CBD Papers.

75. See, for example, Arthur H. Estabrook, "End-Result of Cancer Cases Treated in Philadelphia Hospitals in 1923, as Shown by Special 1930 Follow-Up Studies," *American Journal of Cancer,* vol. 16, September 1932, 1206–29.

76. "In the Wake of the News," *New York Evening Post,* October 16, 1933, clipping, AHE Papers.

CHAPTER 14. After the Supreme Court

1. Patient record, "Infirmary of State Colony" October 19–October 27, 1927, and "Continued Notes, Carrie Buck," October 19–November 12, 1927, Colony File no. 1692.

2. *Minutes of the Meeting of the Special Board of Directors of the State Colony for Epileptics and Feebleminded,* November 10, 1927.

3. Miss B. Stuart Meredith to J. H. Bell, April 12, 1926, Colony File no. 1968.

4. Bell to Meredith, April 13, 1926, Colony File no. 1968.

5. Sterilization Order, Doris Buck, December 10, 1927, Colony File no. 1968.

6. Mrs. Coleman to Dr. Bell, December 19, 1927; and Bell to Mrs. Coleman, December 20, 1927, Colony File no. 1692.

7. Mrs. Coleman to Mrs. Berry, January 5, 1928, Colony File no. 1692.

8. Operation Record, January 25, 1928, Colony File no. 1968.

9. Mrs. J. R. Gentry to Bell, November 18, 1926; and Bell to Mrs. J. R. Gentry, November 20, 1926, August 30, 1927, and January 25, 1928, Colony File no. 1968.

10. Mrs. J. R. Gentry to Bell, January 26, 1928; and Bell to Mrs. J. R. Gentry, January 27, 1928, Colony File no. 1968.

11. Bell to Mrs. J. T. Dobbs, January 12, 1928; Mrs. Dobbs to Bell, February 13, 1928; and Bell to Mrs. Dobbs, February 13, 1928, Colony File no. 1692.

12. Bell to A. T. Newberry, February 20, 1928; Bell to A. T. Newberry, February 22,

1928; Mrs. Newberry to Bell, May 25, 1928; and Bell to Buck, August 16, 1928, Colony File no. 1692.

13. Buck to Bell, December 8, 1928; and Mrs. Newberry to Bell, December 9, 1928, Colony File no. 1692.

14. Bell to Mrs. Newberry, December 11, 29, 1928, Colony File no. 1692.

15. Carrie Buck to Roxie Berry, August 2, 1928, Colony File no. 1968.

16. Mrs. J. R. Gentry to Bell, August 28 and September 6, 1928; Mrs. W. E. Hall to Bell, January 20, 1929; Doris Buck to Bell, January 20, 1929; Mrs. Mary V. Hudson to Doris Buck, May 31, 1929; Mrs. Mary V. Hudson to Bell, May 31, 1929; and Doris Buck to Bell, December 3, 1930, Colony File no. 1968.

17. Carrie [Buck] Eagle to Bell, May 17, 1932, Colony File no. 1692.

18. "Vivian Alice Dobbs," Record of Class Grades, Venable School, Charlottesville, Virginia, 1930–1931, 1931–1932. See also Stephen Jay Gould, "Carrie Buck's Daughter," *Natural History,* vol. 93, July 1984, 1–18.

19. *Certificate of Death, Commonwealth of Virginia,* "Vivian Alice Elaine Dobbs," July 3, 1932.

20. Paul Popenoe to Bell, March 16, 1933; Bell to Buck, March 21, 1933; Buck to Bell, March 27, 1933; and Bell to Popenoe, May 12, 1933, Colony File no. 1968.

21. Popenoe to Bell, May 23, 1933, Colony File no. 1968.

22. Bell to Red Cross, June 6, 1933; Bell to Dobbs, June 20, 1933; and Margaret Faris to Bell, June 21, 1933, Colony File no. 1968.

23. Bell to Faris, June 21, 1933; Faris to Bell, June 27, 1933; and Bell to Faris, June 29, 1933, Colony File no. 1968.

24. Paul Popenoe, "The Progress of Eugenic Sterilization," *Journal of Heredity,* vol. 25, January 1934, 19–26.

25. Laughlin to Charles T. Vorhies, May 10, 1927, HHL Papers.

26. Laughlin to Clerk of the Supreme Court, May 10, 1924; and Laughlin to Clerk of the Supreme Court, July 7, 1924, case file of *Buck v. Bell,* U.S. Archives.

27. Michael Willrich, "The Two Percent Solution: Eugenic Jurisprudence and the Socialization of American Law, 1900–1930," *Law and History Review,* vol. 16, 1998, 63–111, esp. 108. Olson had approached Laughlin with an offer of a position on the Psychopathic Laboratory staff; see Olson to Laughlin, February 28, 1924, HO Papers.

28. Laughlin to Olson, November 15, 1927; and Olson to Laughlin, November 17, 1927, HHL Papers.

29. J. A. Jaffray, Edmonton, Alberta, to Olson, June 8, 1927, HO Papers.

30. Harry Olson, "The Menace of the Half-Man," *Proceedings of the Third Race Betterment Conference,* January 2–6, 1928 (Battle Creek, MI: Race Betterment Foundation, 1928) 122–47, quotation at 146.

31. *Twenty-Third Annual Report of the Municipal Court of Chicago* (Chicago: Municipal Court, 1930) 14.

32. Harry H. Laughlin, *The Legal Status of Eugenical Sterilization* (Chicago: Municipal Court, 1929) 18, 53, 61, 78.

33. Harry H. Laughlin, "Legal Status of Eugenical Sterilization," *Birth Control Review,* vol. 9, March 1928, 78–80, quotation at 79.

34. Harry H. Laughlin, "The Progress of American Eugenics," *Eugenics,* vol. 2, February 1929, 3–16, quotations at 5, 6.

35. Harry H. Laughlin, "The Eugenical Aspects of Deportation," hearings before the Committee on Immigration and Naturalization House of Representatives, 70th Congress, 1st sess., February 21, 1928.

36. Allan Chase, *The Legacy of Malthus: The Social Costs of the New Scientific Racism* (New York: Alfred Knopf, 1977) 289–301; and Frances Hassencahl, "Harry H. Laughlin, Expert Eugenics Agent," Ph.D. dissertation, Case Western Reserve University, 1970.

37. Laughlin to E. S. Gosney, August 8, 1927, HHL Papers.

38. See Harry H. Laughlin, "Relation of a Proposed Bureau of Eugenics in a Proposed Recasting of the Administrative Structure of the United States Government—1929," 1929, HHL Papers.

39. *Twenty-Fourth and Twenty-Fifth Annual Reports* (Chicago: Municipal Court, 1931) 8.

40. E. R. K., "Sterilization of Habitual Criminals and Feeble-Minded Persons," *Illinois Law Review,* vol. 5, 1910–1911, 578; and Willrich, "Two Percent Solution," 106.

41. Laughlin to John J. Sonsteby, July 19, 1935, HHL Papers. See also Paul A. Lombardo, "The 'American Breed': Nazi Eugenics and the Origins of the Pioneer Fund," *Albany Law Review,* vol. 65, 2002, 743–830.

CHAPTER 15. Sterilizing Germans

1. Paul Weindling, *Health, Race, and German Politics between National Unification and Nazism, 1870–1945* (Cambridge: University Press, 1989) 128.

2. Sheila Faith Weiss, "The Race Hygiene Movement in Germany," *Osiris,* vol. 3, 2nd series, 1987, 193–236, quotations at 207, 208.

3. Leila Zenderland, *Measuring Minds: Henry Herbert Goddard and the Origins of American Intelligence Testing* (Cambridge, U.K.: Cambridge University Press, 1998) 432n88.

4. Marvin Miller, *Terminating the Socially Inadequate* (Commack, NY: Malmud-Rose, 1996) 61.

5. See Donald Pickens, *Eugenics and the Progressives* (Nashville: Vanderbilt University Press, 1968) 52–53. On the London encounter, see Allan Chase, *The Legacy of Malthus: The Social Costs of the New Scientific Racism* (New York: Alfred Knopf, 1977) 19, 136.

6. Davenport to Laughlin, December 21, 1920, HHL Papers.

7. Erwin Baur, Eugen Fischer, and Fritz Lenz, *Grundriss der menschlichen Erblich-*

keitslehre und Rassenhygiene (1921), vol. 1, *Menschliche Erblichkeitslehre*, 3rd ed. (Munich: Lehmann, 1927), translated as *Human Heredity* (London: George Allen and Unwin, 1931). See Robert Proctor, *Racial Hygiene: Medicine under the Nazis* (Cambridge: Harvard University, 1988) 50, 60.

8. Harry H. Laughlin, "National Eugenics in Germany: A Consideration of the Eugenical Aspects of the Constitution of the German Republic," *Eugenics Review,* January 1921, 1–3. Reprint, HHL Papers.

9. Laughlin to Davenport, April 13, 1921, HHL Papers.

10. H. H. Laughlin to Prof. Fritz Lenz, October 25, 1928, HHL Papers.

11. Harry Laughlin to Madison Grant, January 13, 1934, HHL Papers.

12. Adolf Hitler, *Mein Kampf,* trans. Ralph Manheim (1924; Boston: Houghton Mifflin, 1971) 249, 254, 255, 402, 403, 404, 438–41.

13. G. Von Hoffman, "Eugenics in Germany," *Journal of Heredity,* vol. 5, October 1914, 435–36.

14. Weiss, "Race Hygiene Movement," 224–25.

15. Weiss, "Race Hygiene Movement," 229.

16. See "Eugenical Sterilization in Germany," *Eugenical News,* vol. 18, September-October 1933, 89–93, quotation at 90.

17. Introduction by Dr. Falk Ruttke, "Address by Dr. Wilhelm Frick, Reichsminister for the Interior," *Pamphlet no. 1 of the Reichs Committee for Public Health Service,* June 28, 1933; and Laughlin to Grant, January 13, 1934, HHL Papers.

18. "German Population and Race Politics," *Eugenical News,* vol. 19, 1934, 33–38, quotation at 36–37.

19. "The New German Law against Dangerous Habitual Criminals," *Eugenical News,* vol. 19, May-June 1934, 79.

20. C. Thomalia [*sic*], "The Sterilization Law in Germany," *Eugenical News,* vol. 19, November-December 1934, 137–40, quotation at 137. Thomalla's role is described in Paul Weindling, *Health, Race, and German Politics between National Unification and Nazism: 1870–1945* (Cambridge, U.K.: Cambridge University Press, 1989) 412–13.

21. "Germany: Nearly Half Million 'Unfit' Will Be Sterilized," *Newsweek,* December 30, 1933, 11–12; "Reich Eugenics Court Orders Trio Sterilized," *Washington Post,* March 7, 1934, 4; and C. C. Hurst, "Germany's Sterilization Law: What It Might Accomplish," *New York Times,* August 5, 1934, 8.

22. See C. Thomalla, "The Sterilization Law in Germany," *Scientific American,* vol. 151, September 1934, 126–27, quotation at 127.

23. Paul Popenoe, "The German Sterilization Law," *Journal of Heredity,* vol. 25, July 1934, 257–60, quotation at 260; see also Stefan Kühl, *The Nazi Connection: Eugenics, American Racism, and German National Socialism* (New York: Oxford University Press, 1994), 44–45.

24. "Sterilization and its Possible Accomplishments," *New England Journal of Medicine,* vol. 211, 1933, 379–80.

25. J. H. Kempton, "Bricks without Straw," *Journal of Heredity,* vol. 24, December 1933, 463–66, quotation at 463.

26. Eliot Slater, "German Eugenics in Practice," *Eugenics Review,* vol. 27, 1936, 285–95, quotation at 295.

27. J. H. Landman, *Human Sterilization* (New York: Macmillan, 1932); and J. H. Landman, "Race Betterment by Human Sterilization," *Scientific American,* vol. 150, June 1934, 292–95.

28. "Nazi Decree Revives Sterilization Debate," *Literary Digest,* vol. 117, January 13, 1934, 17, 33.

29. W. W. Peter, *American Journal of Public Health and the Nation's Health,* vol. 24, March 1934, 187–91, quotations at 187, 189, 190.

30. Robert Cook, "A Year of German Sterilization," *Journal of Heredity,* vol. 26, December 1935, quotations at 485, 489.

31. H. S. Jennings, "Discussion and Correspondence: Proportions of Defectives from the Northwest and from the Southeast of Europe," *Science,* vol. 59, March 14, 1924, 256–57; see also Frances Janet Hassencahl, "Harry H. Laughlin, 'Expert Eugenics Agent' for the House Committee on Immigration and Naturalization, 1921 to 1931," 288–89.

32. Garland E. Allen, "The Eugenics Record Office at Cold Spring Harbor, 1910–1940: An Essay in Institutional History," *Osiris,* vol. 2, 2nd series, 1986, 225–64, esp. 250–51.

33. A. V. Kidder, "Report of the Advisory Committee on the Eugenics Record Office," June 28, 1935, 2–6, Carnegie Papers, Alan Mason Chesney Medical Archives, Johns Hopkins University, Baltimore, Maryland.

34. A. V. Kidder, to Dr. George L. Streeter, June 27, 1935, Papers of Earnest A. Hooton, Peabody Museum, Harvard University; Cambridge, Massachusetts; and John C. Merriam to Dr. A. V. Kidder, June 27, 1935, Carnegie Papers, Johns Hopkins.

35. L. C. Dunn to President Merriam, July 3, 1935, Carnegie Institute of Washington Papers, Cold Spring Harbor Laboratory Archives, Cold Spring Harbor, New York.

36. In an earlier letter, Laughlin arranged to have copies of *Eugenical Sterilization in the United States* sent to Fischer. See Laughlin to John J. Sonsteby, July 19, 1935, and Sonsteby response, July 26, 1935. The pedigree chart is chart 11, "Further Studies on the Historical and Legal Development of Eugenical Sterilization in the United States," International Congress for the Scientific Investigation of Population Problems, HHL Papers.

37. Proctor, *Racial Hygiene,* 112–14.

38. Laughlin to Fischer, July 31, 1935, HHL Papers.

39. A note on the published conference Proceedings, described as "a treasury of data and opinions on this great and growing field of study," appeared in *Eugenical News,* September-October 1936, 118–19.

40. Weindling, *Health, Race, and German Politics,* 524–25.

41. "Der Offentliche Gesundheitsdienst," December 5, 1935, draft copy, HHL Papers.

42. See *Eugenical News,* January-February 1936, inside cover. Copies of the Ruttke paper in typescript are available in the HHL Papers.

43. Weindling, *Health, Race, and German Politics,* 389, 524; and Weiss, *Race Hygiene,* 199n24.

44. "Sterilization Law Is Termed Humane: Author Says German Statute Extends 'Neighborly Love' to Future Generations," *New York Times,* January 22, 1934, 6.

45. "Praise for Nazis," *Time,* September 9, 1935, 20–21.

46. "U.S. Eugenist Hails Nazi Racial Policy," *New York Times,* August 29, 1935, 5; and "American Speakers Shun Berlin Meeting," *New York Times,* August 30, 1935, 8.

47. Laughlin to Frederick Osborn, September 17, 1930, HHL Papers; and "Praise for Nazis," *Time,* September 9, 1935, 20–21.

48. J. H. Bell, "Eugenical Sterilization," paper read at the meeting of the American Psychiatric Association, May 13, 1929, 1–11, quotations at 2, 11, JDJ Papers.

49. J. H. Bell, "The Protoplasmic Blight," read at the Medical Society of Virginia, October 22, 1929, 1–6, quotations at 4, 6, DeJarnette Papers; J. H. Bell, *The Biological Relationship of Eugenics of the Development of the Human Race* (Richmond, VA: Division of Purchase and Printing, 1930) 9; J. H. Bell, "Eugenic Control and Its Relationship to the Science of Life and Reproduction," read at the Virginia Medical Society, October 7, 1931, 1–9, quotations at 5, 8, University of Virginia Library, Department of Special Collections; J. H. Bell, "Status of the Feebleminded and Epileptic in Virginia," *Virginia Medical Monthly,* vol. 59, October 1932, 387–89; and J. H. Bell, "Sterilization as a Contraceptive," *Virginia Medical Monthly,* vol. 60, November 1933, 483–84.

50. "John Hendren Bell," *The National Cyclopaedia of American Biography,* vol. 25 (New York: James T. White and Co., 1936) 421–22; and *Virginia Medical Monthly,* vol. 61, January 1935, 624.

51. John Bell, *24th Annual Report of the Colony for Epileptics and Feeble-Minded,* 1933, 12–13.

52. E. R. Mickle and C. E. Holderby, "Eugenic Sterilization," *Virginia Medical Monthly,* vol. 57, September 1930, 387–89.

53. *Report WSH,* 1908, 10, 11.

54. J. S. DeJarnette, "Eugenic Sterilization in Virginia," *Virginia Medical Monthly,* vol. 57, January 1931, 678–80, quotations at 679, 680.

55. *Report WSH,* 1931, 10; 1933, 2.

56. *Report WSH,* 1935, 9.

57. "Delegates Urge Wider Practice of Sterilization," *Richmond (VA) Times-Dispatch,* January 16, 1934. Clipping, JDJ Papers.

58. Chas. W. Putney, "Eugenic Sterilization," *Virginia Medical Monthly,* vol. 62, March 1936, 705–10, esp. 708.

59. Marie E. Kopp, "Legal and Medical Aspects of Eugenic Sterilization in Germany," *American Sociological Review*, vol. 1, October 1936, 761–70, quotation at 770.

60. G. B. Arnold, "Eugenic Sterilization of the Epileptic and the Mentally Deficient," *Virginia Medical Monthly*, vol. 67, January 1940, 45–47.

61. "Virginia Reports," *Greenville (MS) Delta Democrat Times*, March 14, 1939, 2.

62. Paul A. Lombardo, "'The American Breed': Nazi Eugenics and the Origins of the Pioneer Fund," *Albany Law Review*, vol. 65, 2002, 743–830.

63. Charles Davenport to Sir Francis Galton, May 28, 1910, CBD Papers.

64. Jordan to Davenport, March 21, 1910, CBD Papers.

65. H. E. Jordan, "Heredity as a Factor in the Improvement of Social Conditions," *American Breeders Magazine*, vol. 2, 1911, 246–54, quotations at 248, 249, 251, 252; and H. E. Jordan, "Eugenical Aspects of Venereal Disease," *American Breeders Magazine*, vol. 3, 1912, 256–61, quotation at 257.

66. H. E. Jordan, "The Place of Eugenics in the Medical Curriculum," *Problems in Eugenics: First International Eugenics Congress* (London: Eugenics Education Society, 1912) 396–99, quotation at 398.

67. H. E. Jordan, "Eugenics: the Rearing of the Human Thoroughbred," *Cleveland Medical Journal*, vol. 11, 1912, 875–88. For details on Jordan's career in eugenics, see Gregory Michael Dorr, *Segregation's Science: Eugenics and Society in Virginia, 1785 to the Present* (Charlottesville: University of Virginia Press, 2008).

68. Jordan coauthored *War's Aftermath: A Preliminary Study of the Eugenics of War* (Boston: Houghton Mifflin Company, 1914) with David Starr Jordan, chancellor of Stanford University and another prominent eugenicist. For details of eugenics teaching at the University of Virginia, see Gregory Michael Dorr, "Assuring America's Place in the Sun: Ivey Foreman Lewis and the Teaching of Eugenics at the University of Virginia, 1915–1953," *Journal of Southern History*, vol. 66, Spring 2000, 257–96.

69. On Bean, see Stephen Jay Gould, *The Mismeasure of Man* (New York: Norton, 1981) 77–82.

70. See, for example, Jack Manne, *A Study of Feeblemindedness in a Closely Inbred Mountain Family*, Master of Science thesis, University of Virginia, 1934, 53.

71. Lombardo, "'American Breed,'" 775–780.

72. Schneider to Laughlin, May 16, 1936; and Laughlin to Schneider, May 28, 1936, HHL Papers.

73. "The German Universities," *New York Times*, April 12, 1936, sec. 4, p. 8.

74. Laughlin to Borchers, November 30, 1936, HHL Papers.

75. Lombardo, "'American Breed,'" 798.

76. H. H. Laughlin, "Eugenics in Germany: Motion Picture Showing How Germany Is Presenting and Attacking her Problems in Applied Eugenics," *Eugenical News*, vol. 22, July-August 1937, 65.

77. Laughlin notes, "Eugenics in Germany," March 12, 1937, HHL Papers.

78. Minutes, Pioneer Fund Board of Directors, March 22, 1937, HHL Papers.

79. Garland Allen, "The Eugenics Record Office at Cold Spring Harbor, 1910–1940: An Essay in Institutional History," *Osiris,* vol. 2, 2nd Series, 1986, 225–63, esp. 254.

80. V. Bush to Dr. A. F. Blakeslee, Director, Department of Genetics, April 28, 1939, Eugenics Record Office Papers, Cold Spring Harbor Laboratory.

81. V. Bush to Dr. A. F. Blakeslee, Director, Department of Genetics, June 8, 1939, Carnegie Institution of Washington Papers, Cold Spring Harbor Laboratory.

82. "Harry H. Laughlin, Geneticist and Author, 62, Once with Carnegie Institute, Dies," *New York Times,* January 28, 1943, 20.

83. Carleton E. MacDowell, "Charles Benedict Davenport, 1866–1944: A Study of Conflicting Influences," *Bios,* vol. 17, March 1946, 3–50.

84. Chas. B. Davenport to Soren Hanson, October 14, 1925, CBD Papers.

85. Charles Davenport to Henry Godard Leach, January 12, 1934 (?). Eugenics Record Office Archives, Cold Spring Harbor Laboratory.

86. MacDowell, "Charles Benedict Davenport, 1866–1944," at 34.

87. Carrie Eagle to John Bell, August 19, 1933, Colony File no. 1692.

88. John Bell to Carrie Eagle, August 21, 1933, Colony File no. 1692.

89. Carrie Eagle to G. B. Arnold, October 25, 1940; and G. B. Arnold to Carrie Eagle, October 29, 1940, Colony File no. 1692.

90. "History and Clinical Notes," April 28, 1944, Colony File no. 1141.

91. Aubrey Strode to Peyton Cochran, June 13, 1938, AES Papers.

92. For example, Strode to Bell, September 19, 1931, AES Papers: "I thank you for . . . your addresses on Eugenics and Mental Diseases." Fragments of correspondence and a bank draft from Louisa Strode to Arthur Estabrook dated September 30, 1934, survive in the AES Papers.

93. H. Minor Davis to Strode, January 10, 1936; and Ingolf E. Rasmus to Aubrey Strode, January 20, 1940, AES Papers.

94. Paul A. Lombardo, "Eugenical Sterilization in Virginia: Aubrey Strode and the Case of *Buck v. Bell,*" Ph.D. dissertation, University of Virginia, 1982, 249–50.

95. *Amherst (VA) New-Era Progress,* June 3, 1946, 2. Details of Strode's life were also summarized in a memorial in the *Reports of the Virginia State Bar Association,* vol. 62, 1946, 267–69; and *The National Cyclopedia of American Biography,* vol. 34, 1948, 98–99.

96. "Judge Aubrey E. Strode, Virginia Jurist, State Ex-Senator, Once on Pershing's Staff," *New York Times,* January 28, 1946, 19.

CHAPTER 16. *Skinner v. Oklahoma*

1. "Asylum Chief Is Opposed to Sterilization," *Oklahoma City Times,* March 9, 1929, 1.

2. Main testimony, *In the Matter of the Sterilization of Samuel W. Main,* May 27, 1932, 15, 32.

3. *In re Main*, 162 Okla. 65 (1933).

4. Inmate Ward 8, *Behind the Door of Illusion* (New York: MacMillan, 1932) 229, 230, 232, 234, 243.

5. "Award Is Due for Best Case of Reporting," *Daily Oklahoman*, May 29, 1933, 4A.

6. "Edith Johnson's Column," *Daily Oklahoman*, July 28, 1933.

7. Philip Jenkins, "Eugenics, Crime, and Ideology: The Case of Progressive Pennsylvania," *Pennsylvania History*, vol. 51, 1984, 64–78, esp. 72.

8. See *Annual Reports of the Chicago Municipal Courts*, 1921–1931.

9. "Calls Thin Woman an Imperfect Type," *New York Times*, January 9, 1914, 3.

10. "Sterilization of Habitual Criminals," *Oklahoma Session Laws*, 1933, ch. 46.

11. "Easing a Criminal's Last Hour," *Daily Oklahoman*, October 21, 8.

12. "Shall Poor Pay this Penalty?" *Daily Oklahoman*, December 8, 1934, 8.

13. "Insanity Curb Law Defended by Its Author," *Daily Oklahoman*, April 17, 1934, 12.

14. *Daily Oklahoman*, April 11, 1935, 10; and "Ritzhaupt Brings Speakers to Stock Poultry Clinic," *Guthrie (OK) Leader*, October 17, 1944.

15. "County Inmate of Prison to Test Sex Law," *Oklahoma City Times*, July 29, 1934.

16. "300 Convicts Are Subject to State Sterilization Law," *Oklahoma City Times*, July 28, 1934.

17. "Convicts Show Opposition as News Is Heard," *Oklahoma City Times*, March 21, 1934.

18. *Ada (OK) Evening News*, May 20, 1934, 5.

19. "Sterility Law Due For Test," *Daily Oklahoman*, July 30, 1934, 5.

20. Irvin Hurst, "Court to Get First Test of New Sex Law," *Daily Oklahoman*, July 31, 1934, 1.

21. "Convict Fights Sterilization," *Ada (OK) Evening News*, August 5, 1934, 6.

22. "State Solon Gets Threats," *Oklahoma City Times*, December 15, 1934; and *State of Oklahoma Session Laws of 1935*, ch. 26, "Sterilization of Habitual Criminals."

23. "Convicts Get Aid in Battle on New Law," *Daily Oklahoman*, November 7, 1934, 10.

24. Morris Fishbein, "Fishbein Cautious on Sterilization," *New York Times*, March 3, 1935, N1.

25. *Ada (OK) Evening News*, March 31, 1935, 1.

26. Sterilization of Habitual Criminals, 1935 *Okla. Sess. Laws*, ch. 26.

27. "Eugenics Law to be Tested at McAlester," *Daily Oklahoman*, April 14, 1936, 10.

28. "Sterilization Flayed," *Time*, November 16, 1936, 80–81.

29. "Off the Record," *Ada (OK) Evening News*, July 7, 1936, 8.

30. "Briggs Plans Fight on Law," *Daily Oklahoman*, May 20, 1936, 2; "Convicts Pay Briggs $1000," *Daily Oklahoman*, May 19, 1936, 1; and "State to Test Criminal Sex Law in Court," *Oklahoma City Times*, May 1, 1936.

31. "Behind Bars," *Daily Oklahoman*, September 17, 1936, 4.

32. "Sterilization Law Aid to Society, Convict Believes," *Oklahoma City Times*, June 17, 1936.

33. *State of Oklahoma v. Jack T. Skinner*, "Case Made," May 27, 1937, 102, 123, hereafter *Skinner Transcript*.

34. Plaintiff's Exhibit C, *Skinner Transcript*, 122, 124–25, 137.

35. *Skinner Transcript*, 70–88.

36. *Skinner Transcript*, 128–30.

37. *Skinner v. State*, 189 Okla. 235 (1941) at 241.

38. D. M. LeBourdais, "Purifying the Human Race," *North American Review*, vol. 238, November 1934, 431–37, esp. 433.

39. *Casti Canubii*, Encyclical of Pope Pius XI on Christian Marriage, December 31, 1930 ("And more, they wish to legislate to deprive these of that natural faculty by medical action despite their unwillingness").

40. Abraham Myerson et al., *Eugenical Sterilization: A Reorientation of the Problem* (New York: MacMillan Co., 1936) 88.

41. "Against Sterilization," *New York Times*, January 26, 1936, E8.

42. "$500,000 Operation," *Time*, January 20, 1936, 42–46.

43. "Ann Cooper Hewitt Sues Her Mother," *New York Times*, January 7, 1936, 3.

44. Wendy Kline, *Building a Better Race: Gender, Sexuality, and Eugenics from the Turn of the Century to the Baby Boom* (Berkeley: University of California Press, 2001).

45. "Finishing Schools," *Time*, November 8, 1937, 1, 16; and "180,000 in Germany Are Now Sterilized," *New York Times*, February 5, 1935, 7.

46. "Atlanta Doctors to Drive for Sterilization Bill," *Atlanta Constitution*, February 4, 1934, 1A.

47. "Winship Signs Sterilizing Bill," *New York Times*, May 15, 1337, 2.

48. George Gallup and Claude Robinson, "American Institute of Public Opinion—Surveys, 1935–1938," *Public Opinion Quarterly*, vol. 2, July 1938, 373–98, quotation at 390–91.

49. Theodore N. Kaufman, *Germany Must Perish* (Newark, NJ: Argyle Press, 1941).

50. "A Modest Proposal," *Time*, March 24, 1941, 96.

51. Donald F. Lach, "What They Would Do about Germany," *Journal of Modern History*, vol. 17, September 1945, 227–43, quotation at 228.

52. "Text of Goebbels Article on 'Hard and Relentless' War," *New York Times*, November 10, 1941, 3; and "Goebbels Tactics Hint at Nazi Woes," *New York Times*, September 27, 1942, 13.

53. "Text of Hitler's Heroes' Day Talk," *New York Times*, March 22, 1943, 8.

54. "Nazis Try Scaring Germans to Fight," *New York Times*, December 20, 1942, 39.

55. Frederick Osborn, "Eugenics and National Defense," *Journal of Heredity*, vol. 32, June 1941, 203–4.

56. "Action Taken on Cases," William O. Douglas Papers, Library of Congress Manuscript Division, hereafter WOD Papers.

57. "Court Curious about State's Sterilizing Law," *Daily Oklahoman*, May 7, 1942, 15.

58. *Skinner v. State*, 189 Okla. 235 (1941), *pet. for cert. filed*, 315 U.S. 789 (December 4, 1941) (No. 782); Record at 4–5.

59. *Skinner v. Oklahoma*, 316 U.S. 535 (1942) 538.

60. *Skinner v. Oklahoma*, at 541, 542.

61. Justice William O. Douglas, Supreme Court Conference Notes, April 11, 1942, WOD Papers.

62. *Skinner v. Oklahoma*, at 537, footnote 1.

63. Myerson et al., *Eugenical Sterilization*, 150–52.

64. Leon F. Whitney, "The Source of Crime," *Christian Work Magazine*, March 1926, 2, 13.

65. Leon F. Whitney, *The Case for Sterilization* (New York: Frederick Stokes Co., 1939) 165; unpublished autobiography of Leon F. Whitney, [1970?], Whitney Papers, 204–5, American Philosophical Society, Philadelphia, Pennsylvania; and J. H. Kempton, "Sterilization for Ten Million Americans," *Journal of Heredity*, vol. 25, October 1934, 415–18, quotation at 418.

66. Myerson et al., *Eugenical Sterilization*, 180.

67. *Skinner v. Oklahoma*, at 544–45.

68. *Skinner v. Oklahoma*, at 546.

69. "Transcriptions of Conversations between Justice William O. Douglas and Professor Walter F. Murphy" Princeton University Library, available at http://infos-hare1.princeton.edu/libraries/firestone/rbsc/finding_aids/douglas/douglas7b.html, accessed September 22, 2004.

70. "State Convicts Win Seven Year Fight on Sterilization Law: U.S. Tribunal Holds Law Is Discriminatory," *Daily Oklahoman*, June 2, 1942, 7.

71. "News Notes on Human Heredity and Eugenics," *Eugenical News*, vol. 27, September 1942, 17–18.

72. "Sterility Law Still in Force," *Daily Oklahoman*, April 10, 1948, 20.

73. James E. Hughes, "Eugenic Sterilization in the United States: A Comparative Summary of Statutes and Review of Court Decisions," Supplement No. 162, *Public Health Reports* (Washington, D.C.: Government Printing Office, 1940).

74. "Nearsightedness Could Be Wiped Out by Eugenics," *Science NewsLetter*, June 20, 1942, 387.

75. "Hooton Calls for Segregation of Unfit, 'Mercy Death' for Hopelessly Diseased," *Boston Herald*, July 26, 1940.

76. William L. Laurence, "Calls on Sciences to Save Humanity," *New York Times*, April 25, 1941, 8.

77. Henry E. Sigerist, *Civilization and Disease* (Chicago: University of Chicago Press, 1943) 85, 104–6.

78. See, for example, J. S. DeJarnette, "Eugenics in Relation to the Insane, the Epi-

leptic, the Feebleminded, and Race Blending," *Virginia Medical Monthly,* vol. 52, August 1925, 290–92; and J. S. DeJarnette, "Eugenic Sterilization in Virginia," *Virginia Medical Monthly,* vol. 57, January 1931, 678–80. On the clinics, see *Report WSH,* 1929, 9; 1930, 11.

79. *Celebration of Dr. J. S. DeJarnette's Fiftieth Anniversary of Continuous Service,* July 21, 1939, 29, JDJ Papers.

80. DeJarnette to E. S. Gosney, November 24, 1943; and DeJarnette to Lois G. Castle, December 7, 1943, E. S. Gosney Papers and Human Betterment Foundation Records, California Institute of Technology Archives, hereafter HBF Papers.

81. See Bob Burhans, "DeJarnette Presses Campaign for Sterilization of Unfit," *Richmond (VA) News Leader,* January 23, 1947, 14A.

82. See, for example, Ralph M. Brown, Virginia Tech, to E. S. Gosney, February 13, 1940; A. Clair Sager, Juvenile Court, Richmond, Virginia, to Paul Popenoe, May 12, 1941; and Orland E. White, University of Virginia, to Human Betterment Foundation, October 5, 1941, HBF Papers.

CHAPTER 17. *Buck,* at Nuremberg and After

1. "Streicher Calls Mass Slayings of Jews Hitler's Last Act," *Austin (TX) American,* April 30, 1946, 2.

2. Walter Cronkite, "French Woman Tells of Horrors of German Camp," *Dunkirk (NY) Evening Observer,* January 28, 1946, 9.

3. "War Crimes Trial of Nazi Medical Experts Begins," *Middletown (CT) Times Herald,* December 9, 1946, 3.

4. Ulf Schmidt, *Justice at Nuremberg: Leo Alexander and the Nazi Doctor's Trial* (London: Palgrave Macmillan, 2004) 31–33.

5. Abraham Myerson et al., *Eugenical Sterilization: A Reorientation of the Problem* (New York: Macmillan Co., 1936).

6. Michael R. Marrus, "The Nuremberg Doctors' Trial in Historical Context," *Bulletin of the History of Medicine,* vol. 73, 1999, 106–23.

7. Karl Loewenstein, "Law in the Third Reich," *Yale Law Journal,* vol. 45, 1936, 779–815, quotation at 797–98.

8. Matthias M. Weber, "Ernst Rudin, 1874–1952: A German Psychiatrist and Geneticist," *American Journal of Medical Genetics* (Neuropsychiatric Genetics) vol. 67, 1996, 330.

9. *Eugenical News,* vol. 21, January-February 1936, inside cover.

10. Paul Julian Weindling, *Nazi Medicine and the Nuremberg Trials: From Medical War Crimes to Informed Consent* (London: Palgrave Macmillan, 2004) 41, 71.

11. "State Sterilization," *Newsweek,* November 28, 1949, 46; see also Leo Alexander, "Public Mental Health Practices in Germany: Sterilization and Execution of Patients Suffering from Nervous or Mental Disease," National Archives Record Group 331, Records of Allied Operational and Occupational Headquarters, World War II, 3.

12. *Trials of War Criminals before the Nuremberg Military Tribunals under Control Council Law No. 10*, vol. 1 (Washington, D.C.: Government Printing Office, 1949) 37, 48–50.

13. Weindling, *Nazi Medicine*, 101–2, 159.

14. Joseph Debicki, "X Rays and the Problem of Races," *British Journal of Radiology*, vol. 30, 1925, 486–89, 488.

15. Karl Brandt defense documents no. 53, 121–24, esp. 122; no. 54, 125–30, esp. 127, Nuremberg Trials Project, Harvard University. Available at http://nuremberg.law. harvard.edu, accessed December 10, 2006.

16. *Trials of War Criminals before the Nuremberg Military Tribunals under Control Council Law No. 10*, vol. 1 (Washington, D.C.: Government Printing Office, 1949) 989.

17. "Hoffman Defense Exhibit 61," *Trials of War Criminals before the Nuremberg Military Tribunals under Control Council Law No. 10*, vol. 4 (Washington, D.C.: Government Printing Office, 1949) 1158–60.

18. Laughlin to Charles Davenport, April 13, 1921, HHL Papers.

19. Harry H. Laughlin, *Eugenical Sterilization in the United States* (Chicago: Psychopathic Laboratory, Municipal Court of Chicago, 1922) 112, 412–13, 422–23.

20. "Patient Notes, May 28, 1936," JDJ Papers.

21. Paul A. Lombardo, "'Of Utmost National Urgency': The Lynchburg Hepatitis Study, 1942," in *In the Wake of Terror: Medicine and Morality in a Time of Crisis*, ed. J. Moreno (Cambridge, MA: MIT Press, 2003).

22. E. E. Southard, "Social Research in Public Institutions," in *Proceedings, 43rd National Conference on Charities and Corrections* (Chicago: Hildmann, 1916) 376–87, quotations at 377, 386–87.

23. Paul A. Lombardo, "Tracking the 'Mongoloid' Chromosome: Theophilus Painter and Eugenic Research," abstract, *American Association for the History of Medicine Annual Meeting*, 2002.

24. Walter Berns, "*Buck v. Bell*: The Sterilization Decision and Its Effect on Public Policy," Master's thesis, University of Chicago, 1951, which appeared in an abbreviated form as "*Buck v. Bell*: Due Process of Law?" *Western Political Quarterly*, vol. 6, 1953, 762–75.

25. John B. Gest, "Eugenic Sterilization: Justice Holmes v. Natural Law," *Temple Law Quarterly*, vol. 23, 1950, 306–12, quotation at 312.

26. J. E. Coogan to Eugenics Record Office, November 12, 1940, J. E. Coogan Papers, Mercy Special Collections, University of Detroit, hereafter JEC Papers.

27. J. E. Coogan to Superintendent, State Colony, October 19, 1942; and L. Harrell to J. E. Coogan, October 26, 1942, JEC Papers.

28. John E. Coogan, "Eugenists in Virginia Betray Science and Democracy," *America*, vol. 68, January 16, 1943, 402–3.

29. J. E. Coogan, "Eugenic Sterilization Holds Jubilee," *Catholic World*, vol. 177, April 1953, 45–50; and "State Sterilization Law Great American Fraud," *Tucson Register*, May 1, 1953, clipping, JEC Papers.

30. Julius Paul, "Three Generations of Imbeciles Are Enough," unpublished manuscript, 1963.

31. "For Fewer Unfit," *Newsweek,* October 30, 1950, 50–51.

32. *Virginia House Bill,* no. 394, 1956, "A Bill to provide the sexual sterilization of females who give birth to certain illegitimate children."

33. *Report by the Commission to Study Problems Relating to Children Born out of Wedlock,* Senate Document 5, Commonwealth of Virginia, 1959.

34. *Report by the Commission to Study Problems Relating to Children Born out of Wedlock,* 23–24.

35. "Sterilization Laws of Virginia," *Report of the Virginia Legislative Council,* August 19, 1961 (Richmond, VA: Department of Purchases and Supply, 1961).

36. *Virginia Senate Bill,* no. 37, 1962, "A Bill to provide the sexual sterilization of females who give birth to certain illegitimate children under certain conditions."

37. *Virginia Acts of Assembly,* ch. 451, 1962, "An Act to authorize the performance by physicians and surgeons of certain operations upon the reproductive organs of certain persons," approved March 31, 1962.

38. "Reappraisal of Eugenic Sterilization Laws," *Journal of the American Medical Association,* vol. 173, July 16, 1960, 1245–50.

39. "91.5% of U.S. Physicians Say: Sterilization is Justified," *New Medical Materia,* vol. 4, November 1962, 19–20.

40. Gregory Michael Dorr, *Segregation's Science: Eugenics and Society in Virginia* (Charlottesville: University of Virginia Press, 2008).

41. Paul A. Lombardo, "Miscegenation, Eugenics, and Racism: Historical Footnotes to *Loving v. Virginia,*" *University of California Davis Law Review,* vol. 21, 1988, 421–52.

42. Lombardo, "Miscegenation, Eugenics, and Racism," 427–28.

43. *Virginia Acts of Assembly,* 1924, ch. 371, "An Act to Preserve Racial Integrity."

44. Gregory Michael Dorr, "Principled Expediency: Eugenics, *Naim v. Naim,* and the Supreme Court," *American Journal of Legal History,* vol. 42, 1998, 119–59.

45. *Loving v. Virginia,* 388 U.S. 1 (1966) at 12.

46. Paul Popenoe, "Sterilization in Practice," *The Survey,* vol. 74, June 1938, 202–4, quotation at 202.

47. Justin K. Fuller, "A Health Program for Prisons," *Proceedings of the Sixty-Eighth Annual Congress of the American Prison Association (1938)* (New York: American Prison Association, 1938) 311–22, quotation at 320.

48. Drummond Ayres Jr., "Sterilizing the Poor: Exploring Motives and Methods," *New York Times,* July 8, 1973, 154.

49. Judith Coburn, "Sterilization Regulations: Debate Not Quelled by HEW Document," *Science,* vol. 183, March 8, 1974, 935–39, quotations at 937, 938.

50. *Madrigal v. Quilligan,* No. CV–75–2057, *slip op.* (C.D. Cal. June 30, 1978).

51. Robert A. Williams Jr., "Encounters on the Frontiers of International Human Rights Law: Redefining the Terms of Indigenous Peoples' Survival in the World," *Duke*

Law Journal, vol. 1990, 660–704; and Robert A. Williams Jr., "American Indian Women Sterilized without Informed Consent," *Hastings Center Report,* vol. 7, February 1977, 2.

52. *Walker v. Pierce,* 560 F.2d 609 (1977) at 613.

53. Antonia Hernandez, "Chicanas and the Issue of Involuntary Sterilization: Reforms Needed to Protect Informed Consent," *Chicano Law Review,* vol. 3, 1976, 3–37, esp. 22.

54. *Stump v. Sparkman,* 435 U.S. 349 (1978).

55. "U.S. Announces New Rules on Medicaid Sterilizations," *New York Times,* November 8, 1978, A18.

56. *Wyatt v. Aderholt,* 368 F. Supp. 1382 (D.C.Ala., 1973).

57. Susan Okie, "Hospitals Ignore Order, Sterilizing Underaged," *Washington Post,* July 18, 1979.

CHAPTER 18. Rediscovering *Buck*

1. Virginia Acts, 1950 ch. 465; 1968, ch. 477; and 1974, ch. 296.

2. Leo Kirven to Julius Paul, July 15, 1980, Julius Paul Collection. In possession of the author.

3. George M. Stoddart, "Sterilization Scars Remain," *Winchester (VA) Evening Star,* February 14, 1980, 1; George M. Stoddart, "Sterilization Effect: A 51-Year-Mystery," *Winchester (VA) Evening Star,* February 15, 1980, 1; editorial, "Doris Figgins, et al.," *Winchester (VA) Evening Star,* February 26, 1980, 4.

4. Sandra G. Boodman and Glenn Frankel, "Over 7500 Sterilized by Virginia," *Washington Post,* February 23, 1980, A1; Ben A. Franklin, "Teen-Ager's Sterilization an Issue Decades Later," *New York Times,* March 7, 1980, A16; and "To Raise the Intelligence of the State," interview, Wendy Blair, National Public Radio, Washington, D.C., 1981.

5. *Report of the Virginia State Epileptic Colony* (Richmond, VA: Superintendent of Public Printing, 1930), 10.

6. Charles Giametta, "They Told Me I Had To Have an Operation," *Charlottesville (VA) Daily Progress,* February 26, 1980, A1.

7. "'I Wanted Babies Bad' Woman Told of Her Sterilization," *Charlottesville (VA) Daily Progress,* February 24, 1980, B1.

8. Bill McKelway, "Patient 'Assembly Line' Recalled by Sterilized Man," *Richmond (VA) Times-Dispatch,* February 24, 1980, B2.

9. Judy Goldberg to Jean L. Harris, April 1, 1980, case file, *Poe v. Lynchburg,* American Civil Liberties Union, Arthur Morris Law Library, University of Virginia.

10. Complaint in Civil Action no. 80-0172 in the U.S. District Court for the Western District of Virginia, Lynchburg Division, December 24, 1980.

11. John C. Keeney Jr. file memo, April 27, 1983, case file, *Poe v. Lynchburg,* American Civil Liberties Union, Arthur Morris Law Library, University of Virginia.

12. *Minutes of the Meeting of the General Board of State Hospitals,* July 13, 1944.

13. *Report WSH*, 1919, 10.

14. Joseph L. Kelly Jr., Assistant Attorney General, "Memorandum to the State Hospital Board, Forms and Procedures in Sterilization Cases," November 25, 1938, Julius Paul Collection.

15. *Minutes of the Meeting of the General Board of State Hospitals*, November 11, 1943, and April 13, 1944.

16. *Poe v. Lynchburg Training School and Hospital*, 518 F. Supp. 789 (1981) at 793.

17. Betty Booker and Bill McKelway, "Sterilizations Were Thought Beneficial, Nurse Recalls," *Richmond (VA) Times-Dispatch*, March 2, 1980, D1.

18. Geraldine Akard Andrews, "The Institutionalization of Public Welfare in Charlottesville, Virginia," M.A. Thesis, University of Virginia, 1947, 50–51.

19. Stephen Jay Gould, "Carrie Buck's Daughter," *Natural History*, vol. 93, July 1984, 14–18; Reprinted in *The Flamingo's Smile* (New York: Norton, 1985) 307–13.

20. Trombley's history of sterilization was published in the United Kingdom as *The Right to Reproduce: A History of Coercive Sterilization* (London: Weidenfeld and Nicolson, 1988). *The Lynchburg Story*, Bruce Eadie, producer; Stephen Trombley, director (Worldview Pictures, 1993).

21. Barbara R. Jasny and Donald Kennedy, "The Human Genome," *Science*, vol. 291, February 16, 2001, p. 1153.

22. International Human Genome Sequencing Consortium, "Initial Sequencing and Analysis of the Human Genome," *Nature*, vol. 409, February 15, 2001, 860–921, quotation at 860.

23. *House Joint Resolution* 607 (Va. 2001), "Expressing the General Assembly's regret for Virginia's experience with eugenics."

24. Peter Hardin, "Eugenics Edict Goes to Senate," *Richmond (VA) Times-Dispatch*, February 13, 2001, A1, A6.

25. Peter Hardin, "Documentary Genocide," *Richmond (VA) Times-Dispatch*, March 5, 2000, A1; Peter Hardin, "Seeking Sovereignty: Indians Face Barriers," *Richmond (VA) Times-Dispatch*, March 6, 2000, A1; Peter Hardin, "Virginia Indians Muster Support for Sovereignty," *Richmond (VA) Times-Dispatch*, March 19, 2000, A1; and Peter Hardin, "Segregation's Era of Science," *Richmond (VA) Times-Dispatch*, November 26, 2000, A1.

26. Bill Baskervill, Associated Press, "Grim Legacy," *Richmond (VA) Times-Dispatch*, February 6, 2001.

27. On the Charlottesville press, see editorial, "Sterilizations Forced by State Merit Apology," *Charlottesville (VA) Daily-Progress*, December 19, 2000, A10; Bob Gibson, "Van Yahres Calls for Apology," *Charlottesville (VA) Daily-Progress*, January 14, 2001, B1; and Bob Gibson, "Va. Senate to Mull Eugenics Bill," *Charlottesville (VA) Daily-Progress*, February 11, 2001, A1. On the state legislature, see Peter Hardin, "House 'Regrets' Eugenics," *Richmond (VA) Times-Dispatch*, February 3, 2001, A1; Peter Hardin, "Confronting an Ugly Legacy," *Richmond (VA) Times-Dispatch*, February 12, 2001, A1; Peter

Hardin, "Eugenics Effort Denounced," *Richmond (VA) Times-Dispatch*, February 15, 2001, A1.

28. Peter Hardin, "Eugenics Edict Goes to Senate," *Richmond (VA) Times-Dispatch*, February 13, 2001, A1.

29. Pamela Stallsmith, "House Regrets Eugenics," *Richmond (VA) Times-Dispatch*, February 3, 2001, A1, A6.

30. Peter Hardin, "Eugenics Edict Goes to Senate," *Richmond (VA) Times-Dispatch*, February 13, 2001, A1.

31. Craig Timberg, "Va. House Voices Regrets for Eugenics," *Washington Post*, February 3, 2001, A1.

32. David Usborne, "Virginia Now Regrets Sterilizing the 'Feebleminded,'" *The Independent* (London), February 16, 2001, 17.

33. A. J. Hostetler, "'Book of Life' a Real Stunner," *Richmond (VA) Times-Dispatch*, February 13, 2001, A1; and Peter Hardin, "Eugenics Edict Goes to Senate," *Richmond (VA) Times-Dispatch*, February 13, 2001, A1.

34. Editorial, "The Horror," *Richmond (VA) Times-Dispatch*, Feb. 20, 2001, A10.

35. Peter Hardin, "Rivals Support Apology by State," *Richmond (VA) Times-Dispatch*, December 13, 2000, A1, A6.

36. Peter Hardin, "Hospital's Name Changed," *Richmond (VA) Times-Dispatch*, March 23, 2001, A1; see also Peter Hardin, "'Lifelong service to . . . Va.,': DeJarnette Picture Distorted, Kin Says," *Richmond (VA) Times-Dispatch*, March 22, 2001, A1; and Peter Savodnik, "Mental Hospital Gets New Name," *Charlottesville (VA) Daily Progress*, March 23, 2001, A1.

37. Bonnie Rochman, "Sterilized by State Order," *Raleigh (NC) News and Observer*, April 15, 2001, 21A; and Michael Ollove, "The Legacy of Lynchburg," *Baltimore Sun*, May 6, 2001, 1F.

38. Tony Mauro, "In the Shadows of History," *Legal Times*, February 26, 2001, 12.

39. Stephen Buckley, "Human Weeds," *St. Petersburg (FL) Times*, November 11, 2001, A1.

40. *Virginia Senate Joint Resolution 79* (2002) "Commending Raymond W. Hudlow"; *Virginia House Joint Resolution 299* (2002) "Honoring the memory of Carrie Buck"; and *Virginia House Joint Resolution 148* (2002) "Establishing a joint subcommittee to study medical, ethical and scientific issues relating to stem cell research."

41. Carlos Santos, "Historic Test Case: Wrong Done to Carrie Buck Remembered," *Richmond (VA) Times-Dispatch*, February 17, 2002, B1; and Tony Mauro, "A Case to Remember," *Legal Times*, April 15, 2002, 20.

42. Peter Hardin, "Eugenics Gains Second Chance? New Age of Genetics Spurs Debate," *Richmond (VA) Times-Dispatch*, April 27, 2002, A1.

43. Eric Adler, "Shame Lingers Decades after Missourian Shaped Eugenics Movement," *Kansas City Star*, April 27, 2002, A1.

44. Carlos Santos, "Sterilized War Hero Honored," *Richmond (VA) Times-Dispatch*, May 2, 2002, B1.

45. *Buck v. Bell* Historic Marker, 800 Preston Avenue, Charlottesville, Virginia, 22903.

46. Peter Hardin, "Apology for Eugenics Set, Warner Action Makes Virginia First State to Denounce Movement," *Richmond (VA) Times-Dispatch*, May 2, 2002, A1.

47. Tony Mauro, "Did Eugenics Foreshadow Genetic Engineering?" *USA Today*, May 2, 2002, 11A; William Branigin, "Warner Apologizes to Victims of Eugenics, Woman Who Challenged Sterilizations Honored," *Washington Post*, May 3, 2002, B1; and Carlos Santos, "A Sad Reminder, Marker Honors State's 1st Eugenics Victim," *Richmond (VA) Times-Dispatch*, May 3, 2002, B1.

48. Bob Gibson, "A Shameful Effort, Governor Apologizes for Sterilization Law," *Charlottesville (VA) Daily Progress*, May 3, 2002, A1.

49. Editorial, "Safely Through the Gamut," *Charlottesville (VA) Daily Progress*, May 3, 1927.

50. Editorial, "Eugenics," *Richmond (VA) Times-Dispatch*, May 8, 2002, A12.

51. Leef Smith, "Robbed of the Promise of Life, Victim of Va.'s Old Sterilization Law Says Amends Can't End Pain over Loss," *Washington Post*, May 13, 2002, B1; Deborah Blum, "Reproductive Wrongs," *Los Angeles Times*, May 12, 2002, M6; and "Eugenics," interview of Stephen Selden of the University of Maryland concerning eugenics history, National Public Radio, May 3, 2002, available at www.npr.org/ram-files/atc/20020503.atc.07.ram, accessed January 22, 2008.

52. Laurence M. Cruz, "Governor Apologizes for Eugenics," *Salem (OR) Statesman Journal*, December 3, 2002, 1.

53. The news series is available at http://againsttheirwill.journalnow.com, accessed December 2, 2007. Kevin Begos explains the genesis of the series in "Obscure Footnote Uncovers State Eugenics Program," *Science Writers*, vol. 52, Spring 2003, 1–3.

54. Danielle Deaver, "WFU Medical School Apologizes again for Role," *Winston-Salem (NC) Journal*, November 4, 2003.

55. Kevin Begos, Danielle Deaver, and John Railey, "Easley Apologizes to Sterilization Victims," *Winston-Salem (NC) Journal*, December 13, 2002.

56. Dana Damico, "Easley Repeals Eugenics Statute: Two Women Sterilized under Old Law Attend Signing Ceremony," *Winston-Salem (NC) Journal*, April 18, 2003.

57. Deborah Blum, "Reproductive Wrongs," *Los Angeles Times*, May 12, 2002, M6. Another call for attention to California history was made by Peter Irons, "Forced Sterilization: A Stain on California," *Los Angeles Times*, February 16, 2003, M5.

58. Tom Abate, "California's Role in Nazis' Goal of 'Purification,'" *San Francisco Chronicle*, March 10, 2003, E1.

59. Carl Ingram, "State Issues Apology for Policy of Sterilization," *Los Angeles Times*, March 12, 2003.

60. *California Senate Concurrent Resolution* no. 47, September 12, 2003.

61. Jeremy Redmon, "Apology Asked for Sterilizations State Required," *Atlanta Journal-Constitution*, February 2, 2007, 1; and Gayle White, "Involuntary Sterilization in Georgia: Why Did It Happen?" *Atlanta Journal-Constitution*, February 4, 2007, 1.

62. Cynthia Tucker, "Apology for Sterilizations Is Necessary," *Atlanta Journal-Constitution*, February 7, 2007, A15.

63. *Georgia Senate Resolution* no. 247, "A Resolution expressing profound regret for Georgia's participation in the eugenics movement in the United States," adopted March 27, 2007.

64. Linda Stacy, "Eugenics Apology: Let the High-Risk Individuals Qualify," *Atlanta Journal-Constitution*, February 6, 2007, A10.

65. Ken Kusmer, "Indiana Apologizes for Role in Eugenics," *Fort Wayne (IN) News Sentinel*, April 13, 2007, 1.

66. Shari Rudavsky, "Looking at the History of Eugenics in Indiana," *Indianapolis Star*, April 13, 2007, 1.

67. "Sweden: Sterilization Payments Started," *New York Times*, June 29, 1999, A6.

68. Friedemann Pfafflin and Jan Gross, "Involuntary Sterilization in Germany from 1933 to 1945 and Some Consequences for Today," *International Journal of Law and Psychiatry*, vol. 5, 1982, 419–23, esp. 421.

69. *1928 Alberta Acts*, ch. 37, March 21, 1928.

70. *1937 Alberta Acts*, ch. 47, April 14, 1937.

71. "Nine Women Sterilized in B.C. Have Lawsuits Settled for $450,000," *Vancouver Sun*, December 21, 2005.

EPILOGUE. Reconsidering *Buck*

1. *Code of Virginia*, sec. 54.1-2975 to 2980.

2. *Connecticut General Statutes*, sec. 45a-690. Other states include Vermont, Idaho, and Delaware.

3. *Arkansas Code Annotated*, sec. 20-49-101 to 20-49-302.

4. *Griswold v. Connecticut*, 381 U.S. 479 (1965).

5. *Loving v. Virginia*, 381 U.S. 1 (1965).

6. *Roe v. Wade*, 410 U.S. 113 (1973).

7. *Roe*, at 153, 154.

8. Paul Lombardo, "Medicine, Eugenics, and the Supreme Court: From Coercive Sterilization to Reproductive Freedom," *Journal of Contemporary Health Law and Policy*, vol. 13, 1996, 1–25.

9. Washington State also retains an antique (and apparently defunct) 1909 statute that applies to those found "guilty of carnal abuse of a female person under the age of ten years, or of rape, or . . . adjudged to be an habitual criminal." It prescribes an

operation for "prevention of procreation" as additional punishment for such offenders. *West's Revised Code of Washington Annotated*, section 9.92.100.

10. *Mississippi Code*, section 41-45-1 *et seq.*

11. *Vaughn v. Ruoff*, 253 F.3d 1124 (8th Cir. 2001) at 1129.

12. *Hearings before the Committee on the Judiciary of the United States Senate, on the Nomination of Robert H. Bork to be Associate Justice of the Supreme Court of the United States,* September 15–30, 1987, 245, 118–20.

13. Robert H. Bork, *The Tempting of America; The Political Seduction of the Law* (New York: Touchstone Press, 1990) 62, 66.

14. *Gonzales v. Carhart*, 550 U.S. ___, 127 S.Ct. 1610 (2007).

15. The full quotation reads, "The life of the law has not been logic; it has been experience. The felt necessities of the time, the prevalent moral and political theories, institutions of public policy, avowed or unconscious, even the prejudices which judges share with their fellow men, have had a good deal more to do than the syllogism in determining the rules by which men should be governed. The law embodies the story of a nation's development through many centuries, and it cannot be dealt with as if it contained only the axioms and corollaries of a book of mathematics." Oliver Wendell Holmes Jr., *The Common Law* (Boston: Little Brown, 1923) 1.

16. *Planned Parenthood of Southeastern Pennsylvania v. Casey*, 505 U.S. 833 (1992) at 859.

17. *Armstrong v. Montana*, 296 Mont. 361 (1999) at 377.

18. Richard Lynn, *Eugenics: A Reassessment* (Westport, CT: Praeger, 2001) 129, 133.

19. Richard Lynn, *The Science of Human Diversity: A History of the Pioneer Fund* (Lanham, MD: University Press of America, 2001) 25; see also Lynn, *Eugenics*, 231–32.

20. *Colorado House Bill* no. 1252 (1994), "A bill for an Act concerning measures to assist in reducing the crime rate in the state."

21. John Sanko, "Legislature '94," *Rocky Mountain News*, February 13, 1994, 41A.

22. *Florida House Bill* no. 1451 (1994), "An act related to family planning and public assistance."

23. *Virginia Senate Bill* no. 679 (2006). Rosalind S. Helderman, "Surgical Option Urged for Sex Offenders," *Washington Post*, February 7, 2006, B5. The bill was later amended to direct state officials to study "physical or chemical castration" as features of a conditional release program for sex offenders. The Virginia legislature passed the amended bill, but Governor Tim Kaine vetoed the legislation. Tim Craig, "Kaine Vetoes Bill to Study Castration Study," *Washington Post*, April 11, 2007, B1; and Peter Hardin, "Eugenics Edict goes to Senate," *Richmond (VA) Times-Dispatch*, February 13, 2001, A1, A6.

24. *New Mexico Senate Bill* no. 1175 (2007).

25. At http://www.rodadair.com; March 14, 2007, accessed March 14th, 2007.

26. "Judge Gives Deadbeat Dads a Choice: Jail or a Vasectomy," May 6, 2004, *South Florida Sun-Sentinel*, 5.

27. The court order not to become pregnant was overturned on appeal. See *In the Matter of Bobbijean P.*, 6 Misc 3d 1012(A), 2005 NY Slip Op 50031(U), reversed, 46 A.D.3d 12, 2007 N.Y. Slip Op 07173.

28. Sharon Bernstein, "Mother in the Middle; By Offering $200 to Addicted Women Who Agree to Have Their Tubes Tied, Barbara Harris Has Gone from Waitress to Lightning Rod," *Los Angeles Times*, May 14, 2000, 11

29. H. L. Mencken, "On Eugenics," *Chicago Tribune*, May 15, 1927, H1.

30. H. L. Mencken, "Utopia by Sterilization," *American Mercury*, vol. 41, August 1937, 399–408, quotations at 399, 408.

31. Judge Ralph B. Robertson, "Letters," *Developments in Mental Health Law*, vol. 11, January-June 1991, 4.

32. "City Official Calls for Mass Sterilization," *Charleston (SC) Post and Courier*, October 1, 2006, 1.

33. L. C. Dunn, "Cross Currents in the History of Human Genetics," *American Journal of Human Genetics*, vol. 14, 1962, 1–13, quotation at 2.

A NOTE ON SOURCES

While *Buck v. Bell* is uniformly cited in histories of the U.S. eugenics movement, it has rarely been investigated in detail. Scholarly and popular histories of eugenics inevitably quote the Holmes opinion from the case; some even spend as much as a chapter discussing it. Stephen Trombley's *The Right to Reproduce: A History of Coercive Sterilization* (1988) provides one of the few thorough analyses of *Buck*, devoting two chapters to its origins and aftermath. The only full-length treatment of the case is *The Sterilization of Carrie Buck: Was She Feebleminded or Society's Pawn?* by J. David Smith and K. Ray Nelson (1989). Richard B. Sherman remarked on the absence of notes and lack of proper source documentation among other "serious limitations" of that book. Sherman, *Journal of Southern History*, August 1992, 565–66.

I relied on primary sources to examine *Buck*, in many instances returning to documents and records I had used earlier as foundation for my dissertation and the articles on eugenics I wrote in the twenty-five years following that study.* That preliminary work provided me opportunities to explore the history of eugenics in the United States and analyze several specific episodes within the larger *Buck* story. Revisiting those records forced me to question and revise many of my earlier conclusions about the case, and this study offers what I hope is an original perspective on it. I referred to Harry Laughlin's own *Eugenical Sterilization in the United States* (1922) for many details of early U.S. sterilization history. The pioneering work of Garland Allan, describing the founding of the Eugenics Record Office and the roles of Charles Davenport and Harry Laughlin within it, was critical to my account of the *Buck* case.

In the course of my research on *Buck*, I used several important archival collections including those of Truman State University (Harry H. Laughlin); the University of Virginia (Aubrey E. Strode, Louisa Hubbard Strode); Cold Spring Harbor Laboratory (Eugenics Record Office); the American Philosophical Society (Charles B. Davenport); the California Institute of Technology (Human Betterment Foundation); the State University of New York, Albany (Arthur H. Estabrook); the Rockefeller Archive Center (Bureau of Social Hygiene Pa-

355

pers); Northwestern University; the Chicago Historical Society (Harry Olson); the Library of Congress (William Howard Taft, Oliver Wendell Holmes Jr.); Western State Hospital (Joseph DeJarnette); Central Virginia Training Center (Mallory and Buck cases); Alan Mason Chesney Medical Archives, Johns Hopkins University (Carnegie Institute of Washington); and the Catholic University (National Council of Catholic Men).

Some figures on sterilization were taken from Julius Paul's unpublished manuscript, "Three Generations of Imbeciles Are Enough: State Eugenic Sterilization Laws in American Thought and Practice" (1965). Many of those figures were also printed in Dr. Paul's article, "State Eugenic Sterilization History," in Jonas Robitscher, ed., *Eugenic Sterilization* (Springfield, IL: Charles C Thomas, 1973).

---

*Paul A. Lombardo, "Eugenical Sterilization in Virginia: Aubrey Strode and the Case of *Buck v. Bell*," University of Virginia (University of Virginia, Ph.D. dissertation, 1982); "Involuntary Sterilization in Virginia: From *Buck v. Bell* to *Poe v. Lynchburg*," *Developments in Mental Health Law*, vol. 3 (July 1983), 13–21; "Settlement of *Poe v. Lynchburg* Ends Sterilization Era," *Developments in Mental Health Law*, vol. 5 (July 1984), 18; "Three Generations, No Imbeciles: New Light on *Buck v. Bell*," *New York University Law Review*, vol. 60 (April 1985), 30–62; "Miscegenation, Eugenics, and Racism: Historical Footnotes to *Loving v. Virginia*," *University of California at Davis Law Review*, vol. 21 (1988), 421–452; "Medicine, Eugenics, and the Supreme Court: From Coercive Sterilization to Reproductive Freedom," *Journal of Contemporary Health Law and Policy*, vol. 13 (Fall 1996), 1–25; "Pedigrees, Propaganda, Paranoia: Family Studies in Historical Context," *Journal of Continuing Education in the Health Professions*, vol. 21 (December 2001), 247–255; "Carrie Buck's Pedigree," *Journal of Laboratory and Clinical Medicine*, vol. 138 (October 2001), 278–282; "The American Breed: Nazi Eugenics and the Origin of the Pioneer Fund," *Albany Law Review*, vol. 65 (April 2002), 743–830; "Taking Eugenics Seriously: Three Generations of ??? Are Enough?" *Florida State University Law Review*, vol. 30 (Winter 2003), 191–218; "Facing Carrie Buck," *Hastings Center Report*, vol. 33 (March-April 2003), 14–17; and, with Gregory M. Dorr, "Eugenics, Medical Education, and the Public Health Service: Another Perspective on the Tuskegee Syphilis Experiment," *Bulletin of the History of Medicine*, vol. 80 (2006), 291–316.

INDEX

Abate, Tom, 264

abortion, right to, 247, 269. *See also* reproductive rights

Adair, Rod, 275–76

Addams, Jane, 83

Alabama, sterilization lawsuit in, 247–48

Alderman, Edwin, 210

Alexander, Leo, 236–37, 238

American Breeders Association Committee on Eugenics, 42, 54

American Civil Liberties Union (ACLU), 251–53

American Eugenics Society, 159, 211, 231

American Institute of Criminal Law and Criminology (AICLC), 54, 55

American Social Hygiene Association, 16

Anti-Sterilization League, 28

Arkansas, sterilization law in, 267

Arnold, G. B., 216

asexualization. *See* castration; eugenic sterilization; sterilization laws

Barker, Lewellys, 32, 42, 160, 170

Barr, Martin, 21

Bateson, William, 56

Baur, Erwin, 200

Bean, Robert Bennett, 211

Begos, Kevin, 263

Belfield, William, 55

Bell, Alexander Graham, 32, 56, 160

Bell, John, 105, 186, 190, 192, 217, 241; as advocate of eugenic sterilization, 208; and Carrie Buck's discharge from the Virginia Colony, 188–89; and Carrie Buck's sterilization, 185, 250; as physician at the Virginia Colony, 187–88; as named defendant in appeal of *Buck* case, 150, 176–77. *See also Buck v. Bell*

Berry, Roxie, 185, 187, 189

Binet, Alfred, 38

Binet-Simon intelligence test, 18, 74, 156

Blue, Rupert, 45

Blum, Deborah, 264

Bork, Robert, 271–72

bodily integrity, 167–68. *See also* reproductive rights

Boston, Charles A., 53–54

Brandeis, Louis, x, 172, 173

Brandt, Karl, 239

Briggs, Claude, 222, 223, 224, 225

Brooks, Rose, 263

Brown, Sam, 222

Brown v. Board of Education, 245

Bruehl, Charles, 177–78

Bryan, Mrs. William Jennings, 46

Buck, Carrie, 102, 113, 250–51; as candidate for sterilization, x–xi, 106–11, 132, 277–78; death of, 254–56; discharged from the Virginia Colony, 185–87; efforts of, to contact her mother, 214–16; marriage of, 189–90; sterilization of, ix–x, 185; trial of, xi, 1–6, 112–35; at the Virginia Colony, 103–4, 105, 106, 187. *See also Buck v. Bell; Buck v. Priddy*

Buck, Doris, ix, 3, 5, 106, 113, 115, 119, 250, 254; chosen for sterilization surgery,

as advocate of eugenic sterilization, 150–51, 209–10, 234–35; building named for, 258–59; as expert witness at Carrie Buck trial, 121–27, 136

Detamore, Charlie, 256

disabled people: laws protecting, 249

Dobbs, Alice, 103–4, 116, 119, 131, 133, 139–40, 144; as guardian of Vivian Buck, 186, 188, 190, 192

Dobbs, John, 103–4, 116, 133, 139–40, 190

Dolle, Charles F., 179

Donald, Mary, 257

Douglas, William O.: as author of opinion in *Skinner v. Oklahoma*, 229–32, 233, 246, 271

Draper, Wickliffe, 207, 210, 213

Dudley, Samuel, 115–16

due process, 157, 167, 231. *See also* Fourteenth Amendment

Dugdale, Richard, 4, 9–10, 36–37, 128

Duke, Mary, 119

Dunn, L. C., 206, 279

Easley, Mike, 263

Edison, Thomas, 26

Eighth Amendment, 151

epileptics: institutional care for, 13–15; sterilization as treatment for, 21, 26–27

Equal Protection clause, 27, 102, 169–70, 229

Erbkrank, 213

Estabrook, Arthur, 186, 217; as eugenics researcher, 36–38, 81; as expert witness at Carrie Buck trial, 1–6, 110, 111, 115, 116, 127, 128–30, 141–42, 144–45, 148, 149, 182–83; sexual indiscretions of, 183–84

Eugenical Sterilization in the United States (Laughlin), 78–90, 97, 134, 142, 194

eugenic marriage laws, 44, 45, 46, 59, 86, 158, 160, 175; racial aspect of, 24–46, 196; religious support for, 46, 59; in Virginia, 245–46

eugenics: and immigration policy, 78, 87, 193–94, 196, 200–201; negative connotations of, 205, 278; popular understandings of, 45–46, 56

eugenics movement, x–xii, xiii–xiv, 2–6; biblical support for, 40; disagreement within, 142–43; and the Eugenics Record Office (ERO), 30–38; objections to, 46, 155–56; in Germany, 199–207; Holmes as supporter of, 163–66; perceived as step toward social progress, 208, 277–79; scientists' concerns about, 44, 51–52, 53, 179–80, 181–82; as solution for dealing with crime and immorality, 7–11, 15–16, 36–38; Taft's involvement with, 158, 162–63

Eugenics Record Office (ERO), 54, 55, 80, 108, 136, 159, 160, 161, 182, 184; bulletin of, 49–50; establishment of, 30–31; philanthropic support for, 31–32; research sponsored by, 33–41

Eugenics Research Association, 37, 82, 159, 193–94, 211

eugenic sterilization: apologies for, 258–65, 274; compared with vaccination, 86, 180; experimental, 239–40; history of, 257–58, 261–66; justifications for, xiii–xiv, 1–6, 28–29, 31, 41, 53, 55, 61–64, 78–90, 96–97, 99–100, 136–38, 152, 156, 247, 274–77; Laughlin as advocate of, 42, 44–45, 47–51, 54, 55, 78–90, 99–100, 108, 142–43, 152; under the Nazis, xii–xiii; Olson as advocate of, 81–85; recent debate over, 258–61; as treatment for epilepsy, 21, 26–27; x-rays used for, 238–39; after World War II, 241–43. See also *Buck v. Bell*; *Buck v. Priddy*; eugenics movement; sterilization laws

Evans, W. A., 83, 87

feeblemindedness: definitions of, 5, 15, 19, 40, 156; diagnosis of, 144; hereditary

Hopkins, John W., 115
How to Live (Fisher and Fisk), 161–62
Hudlow, Raymond, 253, 259, 261
Hughes, Charles Evans, 89
Human Betterment Foundation, 176, 190
Human Genome Project, 257–58
Hurty, J. N., 24

idiots, 41
illegitimacy: associated with feeblemind-
    edness, 137; associated with race, 242
Illinois, attempts to enact sterilization
    law in, 84, 97, 198
imbeciles, 41
immigration policy, 78, 86, 137, 158,
    193–94, 196, 200–201
immorality, eugenics movement as solu-
    tion to, 7–11, 15–16, 36–38
Indiana: sterilization law in, 24–25, 43,
    53–54, 59, 60, 74, 98–99, 142, 158, 160;
    sterilization lawsuit in, 249
In re Main, 219
intelligence tests, 18, 38, 40, 144
interracial marriage, 196, 245–46
Iowa, sterilization law in, 28–29

Jackson, Robert, 232, 236, 237
Jacobson, Henning, 157, 169, 180
Jacobson v. Massachusetts, 86, 152, 157, 169,
    180, 269
Jennings, H. S., 146, 148, 204–5
Jeter, Mildred, 246
Johnson Immigration Restriction Act of
    1924, 196
Jordan, David Starr, 89–90
Jordan, Harvey Earnest, 145, 210–11
Journal of the American Medical Association,
    20, 52, 175, 182, 223
Jukes family, 4, 6, 9–10, 11, 36–38, 57, 59,
    128, 137

Kallikak family, 5–6, 38–40, 41, 123–24,
    129, 137, 199

Kansas: eugenic marriage law in, 158;
    sterilization law in, 21
Kantsaywhere (Galton), 7
Kaufman, Theodore, 228, 236
Keller, Helen, 100
Kellogg, John Harvey, 10, 20, 47, 48
Kerlin, Isaac, 21
Kidder, A. V., 205
Kitzhaber, John, 263
Kohs, Samuel, 53
Korematsu case, xiii

Landman, J. H., 204
Laski, Harold, 164–65, 166, 167, 178
Laughlin, Harry H., xii, 3, 32, 33, 37,
    137, 181, 239; as advocate of eugenic
    sterilization, 42, 44–45, 47–51, 54, 55,
    78–90, 99–100, 108, 142–43, 152, 161,
    193–98, 210, 220–21; as author of *Eu-
    genical Sterilization in the United States*,
    78–90, 97, 163; and the Carrie Buck
    trial, 109, 110, 111, 133–35, 145, 150; criti-
    cisms of, 204–6, 213; as influence in
    Germany, 200–203, 206, 211–13; plan
    of, for Bureau of Eugenics, 196–98. *See
    also* Model Sterilization Law
Lenz, Fritz, 200
Lester, E. F., 222
Lewis, Ivey, 211
Life Extension Institute, 160, 161
Little, Clarence Cook, 89–90, 147
Little, Lora C., 28
Lochner v. New York, 164
Lockyer, Bill, 264
Loewenstein, Karl, 237–38
London, Jack, 56
Loving, Richard, 246
Loving v. Virginia, 245–46, 269
Lydston, G. Frank, 47, 55–56
Lynchburg Story, The, 256–57
Lynchburg Training School (formerly the
    Virginia Colony), 250
Lynn, Richard, 274–75

Main, Samuel W., 219
Mallory, George, 65, 67, 68, 70, 71, 76
Mallory, Irene, 65, 68
Mallory, Jessie, 67, 71, 74, 76
Mallory, Nannie, 67, 68, 71–72, 76
Mallory, Willie, 65–68, 71, 72–73, 75–76
Mallory family, 65; commitment of, to
    insitutional care, 66–77
Mallory v. Priddy, 64–77, 92, 101, 104, 132
Malthusian theory, 165, 166
Mann, William Hodges, 59
marriage. *See* eugenic marriage laws;
    interracial marriage
Martin, Victoria Woodhull, 175
Mastin, Joseph, 17–19, 73–74
masturbation, 21, 24
Mayer, Joseph, 171
Mayo, Charles, 182
McCarley, T. H., 225
McCulloch, Oscar, 10–11, 24
McDougle, Ivan, 183
McReynolds, James, 172
Meadows, Jesse, 257, 263
Mears, J. Ewing, 22, 23, 171, 176
Mein Kampf (Hitler), 201, 203
menace to society. *See* criminal behavior;
    "socially inadequate"
Mencken, H. L., 276
Mendel, Gregor, 56, 258; work of, as
    applied to eugenic sterilization argu-
    ments, 30, 95, 122, 123, 125, 143, 146, 155
mentally ill people: changing attitudes
    toward, 278; institutional care for, 12.
    See also "socially inadequate"
Merriam, John C., 205–6
Michigan, sterilization law in, 21
Mickle, Perry, 28
Mickle v. Hendrichs, 28
Minnesota, sterilization law in, 97
miscegenation. *See* interracial marriage
Mississippi, sterilization law in, 270
Model Sterilization Law, xii, 3, 51–53, 54,

79, 80, 86–87, 89, 97, 99–100, 107, 108,
    142, 201
Montana, sterilization law in, 75
Moore, Hubert L., 224
Morgan, Thomas Hunt, 32, 56
morons, 38, 41, 156
Muir, Leilani, 265–66
Myerson, Abraham, 226, 231, 237

Nam family, 4, 36
National Academy of Sciences, 100–101
National Conservation Commission, 159
National Council of Catholic Men
    (NCCM), 179, 181
National Research Council, 100–101
National Society for the Promotion of
    Practical Eugenics, 46
Nazi Germany. *See* Germany
Nelson, K. Ray, 250
Nesbit, Evelyn, 82
Nevada, sterilization law in, 28
New Hampshire, sterilization law in, 97
New Jersey, sterilization law in, 26–28, 43,
    54, 98–99
Newberry, Mrs., 188–89
North Carolina, eugenic sterilization in,
    242
North Dakota, sterilization law in, 99
Nuremberg Trials, Nazi sterilization
    practices as issue at, 236–39

Ochsner, Albert, 23
Ohio, sterilization law in, 97
Oklahoma Habitual Criminal Steriliza-
    tion Act, 223
Oklahoma, sterilization law in, 219–20,
    221–33
Oliver, Mary Margaret, 264–65
Olson, Harry, 81–85, 87, 89, 90, 92, 134,
    151, 194, 198, 221
Ordahl, George, 156
Osborn, Frank, 143–44, 228

Skinner, Jack, 224–25, 230
Skinner v. Oklahoma, 224–33, 268, 269, 271–72
Smith, Alice, 26–27, 85
Smith, Warren Wallace, 141–42
Smith, Willie, 180–81
Smith v. Command, 180–81
Smith v. Williams, 141–42
social Darwinism, 158–59, 234
social hygiene movement, 16–17
"socially inadequate": as applied to Carrie Buck, ix, 107, 130, 135, 182; as term used to justify sterilization, 49–50, 53, 78, 87, 99–100, 113–14, 130, 137, 178, 197, 198, 250, 268, 270
South Carolina, sterilization lawsuit in, 248
South Dakota, sterilization law in, 91
Southard, Elmer E., 32, 160, 240
Spencer, Herbert, 84, 158–59, 163–64
Statistical Directory of State Institutions for the Defective, Dependent, and Delinquent Classes, 78
sterilization laws, ix–xiv, 91, 97, 217–18; advocates of, xiii–xiv, 14, 19, 21–29, 30–32, 42–45, 47–53, 54, 55–56, 80–81, 121–24; in Canada, 194, 265–66, 274–77; Catholic opposition to, 171–72, 177–79, 181, 226, 248; criticisms of, 53–57, 143–44, 146–47, 155–56, 158, 175–76, 203–4, 223–24, 233; current status of, 267–68; in Germany, xii–xiii, 193, 198, 202, 203–4, 208, 209–10, 236–39; in Indiana, 24–25, 43, 53–54, 59, 60, 74, 98–99, 142, 158, 160; international impact of, xii, 199–207, 217–18, 265–66; legal challenges to, 43, 92–93, 248–49, 251–54; in Oklahoma, 219–20, 221–33; opponents of, 28, 53–57, 58, 179–80, 223, 226, 230–31; race as factor in, 58, 247–48; in Virginia, x, xi, 14, 57, 58–60, 64–65, 76, 91–102, 106–11,

151–54, 157–58, 177, 209–10, 217, 243, 251–53. *See also* Buck, Carrie; *Buck v. Bell; Buck v. Priddy;* eugenic marriage laws; eugenic sterilization; *Mallory v. Priddy; Skinner v. Oklahoma; and names of individual states*
sterilization. *See* eugenic sterilization; salpingectomy; sterilization laws; vasectomy
Stone, Harlan, 173, 228–29, 231
Stone, William, 22
Streicher, Julius, 236
Strode, Aubrey, 1–6, 13, 19, 183, 234; as advocate of sterilization law in Virginia, 92–94, 97–98, 100, 101–2, 155; and appeal of *Buck* case, 149–50, 151–52, 154; and Carrie Buck's sterilization trial, 106, 108, 110, 111, 112–17, 119, 120, 123–25, 127–29, 132–33, 135, 136, 138, 140, 141, 147; and rehearing of *Buck* case, 177; Supreme Court arguments of, 157–58, 165–66; as Whitehead's friend, 216–17
Strode, Louisa Hubbard, 2, 94–95, 128, 183
Sumner, Walter Taylor, 46
Sumner, William Graham, 158
Sunday, Billy, 56–57
Supreme Court. *See* U.S. Supreme Court

Taft, William Howard: as Chief Justice of the Supreme Court, x, xii, 89, 158, 160, 165–66, 173, 180, 228
Thaw, Harry, 82–83
Thomalla, Kurt, 203
Thomas, Clarence, 273
Training School for the Feebleminded (Vineland, NJ). *See* Vineland Training School
Trinkle, E. Lee, 95–97, 100
Trombley, Steven, 256
Truax v. Corrigan, 180

CPSIA information can be obtained at www.ICGtesting.com
Printed in the USA
LVOW09s2204080415

433855LV00005B/288/P